International Library of Ethics, Law, and the New Medicine

Volume 63

Series editor

David N. Weisstub
University of Montreal Fac. Medicine, Montreal, Québec, Canada

The book series International Library of Ethics, Law and the New Medicine comprises volumes with an international and interdisciplinary focus. The aim of the Series is to publish books on foundational issues in (bio) ethics, law, international health care and medicine. The 28 volumes that have already appeared in this series address aspects of aging, mental health, AIDS, preventive medicine, bioethics and many other current topics. This Series was conceived against the background of increasing globalization and interdependency of the world's cultures and governments, with mutual influencing occurring throughout the world in all fields, most surely in health care and its delivery. By means of this Series we aim to contribute and cooperate to meet the challenge of our time: how to aim human technology to good human ends, how to deal with changed values in the areas of religion, society, culture and the self-definition of human persons, and how to formulate a new way of thinking, a new ethic. We welcome book proposals representing the broad interest of the interdisciplinary and international focus of the series. We especially welcome proposals that address aspects of 'new medicine', meaning advances in research and clinical health care, with an emphasis on those interventions and alterations that force us to re-examine foundational issues.

More information about this series at http://www.springer.com/series/6224

Monty Nelson · Marguerite Trussler
Editors

Fetal Alcohol Spectrum Disorders in Adults: Ethical and Legal Perspectives

An Overview on FASD for Professionals

Springer

Editors
Monty Nelson
Consulting Psychology Associates
Edmonton, AB
Canada

Marguerite Trussler
Office of the Ethics Commissioner
Edmonton, AB
Canada

ISSN 1567-8008 ISSN 2351-955X (electronic)
International Library of Ethics, Law, and the New Medicine
ISBN 978-3-319-20865-7 ISBN 978-3-319-20866-4 (eBook)
DOI 10.1007/978-3-319-20866-4

Library of Congress Control Number: 2015943042

Springer Cham Heidelberg New York Dordrecht London

Printed on acid-free paper

Springer International Publishing AG Switzerland is part of Springer Science+Business Media (www.springer.com)

Preface

The origins of this edited volume began in 2013. As part of a Canadian delegation of clinicians and judges spearheaded by Mr. Neil Skinner, executive legal counsel to the Provincial Court of Alberta, we attended the 33rd Congress of the International Academy of Law and Mental Health in Amsterdam in July 2013. There were two panels organized with speakers from Alberta and Saskatchewan on the topic of FASD. While there was some significant interest in the topic, we were left with the impression that much more needed to be done to educate health and legal practitioners around the world about the various legal and ethical challenges associated with those who have FASD. Dr. Floor Oosting and Mr. Chris Wilby from Springer Publishing approached Monty about considering assembling an edited volume with contributions from authors from around the world on the topic. With the encouragement of Dr. Egon Jonsson from the Alberta Institute of Health Economics, the journey to collect chapters from authors in the field was undertaken.

Marguerite agreed to assist Monty with the project; the combination of a retired judge and a psychologist as editors seemed very apt for such a volume. A call for papers went out in the fall of 2013. Most of the papers were received by the summer of 2014 and the editing process commenced. We are pleased that we have 28 contributors from seven countries, and want to graciously thank each of them for their contributions and patience with the editing process.

We sincerely thank series editor Dr. David Weisstub, as well as Dr. Oosting and Mr. Wilby for their efforts and support in publishing this work. Ms. Kathryn Kelly and Dr. Diane Black assisted in spreading the word about the book to potential authors. We are also thankful to our many colleagues (including Judge Peter Ayotte, Ms. Denis Milne, Ms. Audrey McFarlane, Ms. Teresa O'Riordan, Ms. Sharon Brintnell, Dr. Hasu Rajani, and Dr. Brent Symes) who over the years have influenced our understanding and involvement in the field of FASD. We have also learned extensively from the various individuals with FASD we have worked with. Two anonymous reviewers also spent extensive time reviewing an earlier draft of the volume, and providing thorough feedback and suggestions as well; their time and effort are greatly appreciated.

Finally, we thank our families for their patience while we brought together this work.

We dedicate this volume to those individuals with FASD, and those who work to assist them.

Contents

Part I Ethical Considerations and Implications in FASD

Fetal Alcohol Spectrum Disorder: Ethical, Legal and Moral Implications Associated with the Rights of the Unborn Child 3
Jenny Salmon

Why People with FASD Fall for Manipulative Ploys: Ethical Limits of Interrogators' Use of Lies 23
Stephen Greenspan and John H. Driscoll

Neurodevelopmental Disorder Associated with Prenatal Alcohol Exposure: Consumer Protection and the Industry's Duty to Warn 39
Susan D. Rich and Laura J. Riley

Experience in Involuntary Hospitalisation of Drug and Alcohol Addicted Mothers During Their Pregnancies at Borgestadklinikken, Norway 49
Egil Nordlie

Is It Ethical to Condone Low Levels of Drinking in Pregnancy? 65
Janni Niclasen

Ethical Challenges When Screening for and Diagnosing FASD in Adults 85
Ilona Autti-Rämö

Part II Clinical Perspectives on Criminal Behaviour in FASD

Neurocognitive Difficulties Underlying High Risk and Criminal Behaviour in FASD: Clinical Implications 101
Katherine Wyper and Jacqueline Pei

Medico-Legal Interventions in Management of Offenders with Fetal Alcohol Spectrum Disorders (FASD) 121
Mansfield Mela

Oral Language and Communication Factors to Consider When Supporting People with FASD Involved with the Legal System 139
Linda Hand, Megan Pickering, Sally Kedge and Clare McCann

Part III Judicial Perspectives

Evolution of the Diagnosis of Fetal Alcohol Spectrum Disorder from DSM-IV-TR to DSM-5: The Justice System in the United States—Time for a Paradigm Shift! 151
Anthony P. Wartnik, Jerrod Brown and Sarah Herrick

Smart Justice and FASD in Alaska: From Prevention to Sentence Mitigation 169
Teri Tibbett and Michael I. Jeffery

FASD & Justice: The Ethical Case for Effective Training and Knowledge Mobilization Practices for Frontline Justice Professionals in Canada 191
Michelle Stewart

Part IV Social and Policy Issues

Social Care and Fetal Alcohol Spectrum Disorders in the UK 209
Alison McCormick

FASD in a Canadian Aboriginal Community Context—An Exploration of Some Ethical Issues Involving the Access to FASD Service Delivery 223
Lori Vitale Cox

FASD and the Concept of "Intellectual Disability Equivalence" 241
Stephen Greenspan, Natalie Novick Brown and William Edwards

Part V Personal Perspectives and Stories

Minimizing Secondary Disabilities 269
Inger Thormann

The Challenges of Raising an FASD Child in Canada 285

I Did not Take on This Job of Parent in Order to Raise Criminals 293
Dianne O'Connor

Editors and Contributors

About the Editors

Monty Nelson Ph.D. has been a psychologist in Edmonton, Alberta, Canada since 1998. He is also an Adjunct Assistant Professor of Educational Psychology at the University of Alberta. He is in full-time private practice conducting neuropsychological assessments in Edmonton, rural Alberta, and the northern territories. He has been a consultant for various adult FASD clinics since 2008. In collaboration with the Lakeland Centre for FASD he also does training for other clinicians about the FASD diagnostic process. He has presented at provincial, national, and international conferences on assessments of adults with FASD.

Marguerite Trussler B.A., LL.B., LL.M., LL.D. (Hon.), Q.C. was a justice of the Court of Queen's Bench for 20 years. Following her retirement she became the chair of the Alberta Gaming and Liquor Commission where she developed an interest in the issues surrounding FASD. For a number of years, she chaired the Institute of Health Economics' steering committee for the Alberta initiatives on FASD and was the moderator of the Institute's 2013 consensus conference on Legal Issues of FASD. Since May of 2014, she has been the Ethics Commissioner for the Legislature of the Province of Alberta, Canada.

Contributors

Ilona Autti-Rämö is a pediatric neurologist with 30 years of clinical and research experience in FASD. She has been the secretary for the national screening committee and was involved in developing ethical evaluation of screening programs on national and EU level. She is currently the medical director at the Social Insurance Institute of Finland in Helsinki.

Jerrod Brown MA, MS, MS, MS is the Treatment Director for Pathways Counseling Center in St. Paul, Minnesota, USA. Jerrod is also the founder and CEO of the American Institute for the Advancement of Forensic Studies (AIAFS) and is the lead developer and program director of an online graduate degree program in Forensic Mental Health from Concordia University in St. Paul.

Natalie Novick Brown Ph.D. is a clinical and forensic psychologist in Seattle, Washington, USA. She is a founding partner of FASD Experts, through which she has evaluated numerous criminal defendants who have FASD. She has an adjunct appointment at the University of Washington, and is known for her scholarly work on suggestibility and executive functioning in people with FASD.

Lori Vitale Cox Ph.D. is from the Elsipogtog First Nation, NB, Canada. She is the Clinical Coordinator of the Eastern Door Diagnostic Team and the Director of the Nogemag Healing Lodge for Youth at Risk. She is also the author of "Medicine Wheel Tools for Screening, Diagnosis and Intervention of FASD" which was included in the Public Health Agency of Canada FASD Screening Toolbox.

John H. Driscoll Ph.D. is a school psychologist for the Vernon Connecticut (USA) Public Schools, and adjunct assistant professor of educational psychology at the University of Connecticut, where he teaches graduate courses in psychological assessment. He often testifies in civil legal proceedings, and has done research on the cognitive challenges facing youths involved in police interrogations.

William Edwards J.D. is a Deputy Public Defender in Los Angeles County, California, USA, where he spent several years representing people with FASD in the LA Mental Health Court. He served two consecutive terms on the US President's Committee on Intellectual and Developmental Disabilities, and has lectured on FASD to attorneys and judges around the world. He edited a two-issue special volume on FASD for the Journal of Psychiatry and Law.

Stephen Greenspan Ph.D. is emeritus professor of educational psychology at the University of Connecticut, and now resides in Colorado, USA. His writings and testimony on social vulnerabilities of people with intellectual disabilities involved in criminal justice proceedings won recognition from the American Association on Intellectual and Developmental Disabilities, which awarded him its Gunnar and Rosemary Dybwad Award for Humanitarianism.

Dr. Linda Hand is an experienced speech–language therapist in New Zealand. Sally, Linda and Clare have formed the Talking Trouble NZ Project which focuses on language and communication issues for children and young people involved (or at risk of becoming involved) in the legal system. Talking Trouble NZ is developing a range of projects involving research, professional development, speech–language therapy clinical services for children and young people and advocacy. Megan works clinically with children and young people with FASD.

Sarah Herrick MA, LP, LPCC, CCFC has worked with sexual abusers ranging in age from ten to the elderly since 1991, in residential, community mental health, and secured settings. Currently, she is working with civilly committed sexual abusers and is an adjunct professor with Concordia University's (St. Paul, MN, USA) Forensic Mental Health on-line graduate program.

Michael I. Jeffery J.D. was a Superior Court Judge who retired in 2014 after a 32-year judicial career at Barrow, Alaska, USA. He adapted his court activities for FASD-affected persons. He assisted with Alaska's FASD mitigating factor legislation and is a member of statewide committees dealing with FASD and juvenile justice. He graduated from Stanford University and Yale Law School.

Sally Kedge is an experienced speech–language therapist in New Zealand. Sally, Linda and Clare have formed the Talking Trouble NZ Project which focuses on language and communication issues for children and young people involved (or at risk of becoming involved) in the legal system. Talking Trouble NZ is developing a range of projects involving research, professional development, speech–language therapy clinical services for children and young people and advocacy. Megan works clinically with children and young people with FASD.

Dr. Clare McCann is an experienced speech–language therapist in New Zealand. Sally, Linda and Clare have formed the Talking Trouble NZ Project which focuses on language and communication issues for children and young people involved (or at risk of becoming involved) in the legal system. Talking Trouble NZ is developing a range of projects involving research, professional development, speech–language therapy clinical services for children and young people and advocacy. Megan works clinically with children and young people with FASD.

Alison McCormick (Ali) is a qualified GSCC Registered, Post Graduate Social Work Consultant living in Huntingdonshire, England. Her Social Work experience includes Management, Child Protection and Investigation, Care Management, Bereavement, Adoption and Fostering, Therapy and Behavior Management, as well as writing and training. She also speaks at Conferences and was previously a Children's Guardian advising and advocating for legal rights of children in Care Proceedings. She has also chaired the Birmingham Independent Review Mechanism Panel (IRM) for 8 years.

Dr. Mansfield Mela is an academic forensic psychiatrist in the University of Saskatchewan and a Founder of the forensic subspecialty in Canada. He consults at the Regional Psychiatric Center and the Saskatoon forensic psychiatric outpatient clinic in Saskatoon, Saskatchewan, Canada. He is a member of the Saskatchewan Review Board and is the head of the forensic interdisciplinary research team. His research focuses on the medicolegal implications of cognitive disorders, specifically fetal alcohol spectrum disorder and ADHD.

Janni Niclasen Ph.D. is a psychologist and an assistant professor at the Department of Psychology, University of Copenhagen, Denmark. Janni is a researcher with a particular interest in and knowledge about prenatal exposure to low doses of alcohol and child neurodevelopment.

Egil Nordlie became a Physician in 1972, in Oslo, Norway. Since 1978 he has been the chief physician at Borgestadklinikken, in Skien, Norway, one of the country's largest institutions for patients with alcohol and drug problems. Since the 1990s, he has been responsible for the units for pregnant women with addictions and for involuntarily hospitalized pregnant women. He has done research, written articles, and lectured nationally and internationally about these programs, as well as other topics related to alcohol and drug related problems.

Dianne O'Connor and her husband, Mickey, are the parents of nine children, six of whom have varying degrees of Fetal Alcohol Spectrum Disorders and are now young adults. Dianne is a board member of the NYS Citizen's Coalition for Children and is currently a kinship caregiver for two of her grandchildren.

Jacqueline Pei Ph.D. is an Associate Professor in the Department of Educational Psychology and Assistant Clinical Professor in the Department of Pediatrics at the University of Alberta in Edmonton, Canada. Also a practicing Registered Psychologist, Dr. Pei began her career as a criminologist and has continued in this vein, studying youth at risk, and interventions for those with FASD. Dr Pei currently leads the Intervention Network Action Team (iNAT) for the Canada FASD Research Network.

Megan Pickering is an experienced speech–language therapist in New Zealand. Sally, Linda and Clare have formed the Talking Trouble NZ Project which focuses on language and communication issues for children and young people involved (or at risk of becoming involved) in the legal system. Talking Trouble NZ is developing an a range of projects involving research, professional development, speech–language therapy clinical services for children and young people and advocacy. Megan works clinically with children and young people with FASD.

Susan D. Rich MD, MPH is a board certified psychiatrist specializing in diagnosis and treatment of Neurodevelopmental Disorder associated with Prenatal Alcohol Exposure (ND-PAE) in her clinical and forensic practice in Potomac, Maryland, USA. She has promoted the need for recognition of ND-PAE through psychiatric publications, national conference lectures, and grand rounds at universities.

Laura J. Riley is an attorney working on gender discrimination issues at the California Women's Law Center with a background in disability rights and employment law. She graduated from Columbia University before attending Law School at the University of Southern California where she developed an interest in studying the intersection of Fetal Alcohol Syndrome and the law through a class on Mental Health Law taught by Elyn Saks. She is from Los Angeles, California, USA.

Dr. Jenny Salmon is a social scientist and reseacher in the FASD field in Dunedin, New Zealand. She is also a Member of the University of Otago Centre for Research on Children and Families—research cluster.

Michelle Stewart Ph.D. is an Assistant Professor in Justice Studies at the University of Regina, Saskatchewan, Canada. Her research interests center on policing practices with a focus on political anthropology, social justice and science and technology studies. Her current research investigates risk and prevention with attention to FASD, including how risk and diagnostic tools can influence the ways in which FASD is understood in particular communities of practices.

Inger Thormann MA is a psychologist from Copenhagen, Denmark. Since 1973 Inger has worked with neglected and abused children, including the past 30 years at the observation and treatment centre Skodsborg Observations- og Behandlingshjem for infants. Inger has authored books for professionals and children and is active as a film consultant, lecturer and teacher. Inger co-founded the Dansk Institut for Spædbarnsterapi (Danish Institute for Infant Therapy) together with psychotherapist Inger Poulsen; they also recently co-authored the book "Therapy with Infants".

Teri Tibbett BLA, MFA is a writer living in Juneau, Alaska, USA. She was a legislative aide with the Alaska State Legislature and is the advocacy coordinator for the Alaska Mental Health Board and Advisory Board on Alcoholism and Drug Abuse. She works closely with the Alaska Mental Health Trust Authority and other stakeholders, and serves as the coordinator of the Alaska Fetal Alcohol Spectrum Disorders (FASD) Partnership.

Anthony P. (Tony) Wartnik was a trial judge in the United States for 34 years. He chaired the Washington State Governor's Advisory Panel on FAS/FAE and conducted training for judges on FASD. Tony is in Seattle, Washington, USA, and is recognized nationally and internationally as an expert on issues involving FASD and the Justice System, and the Law and related issues.

Katherine Wyper M.Ed. is a doctoral student in the Department of Educational Psychology at the University of Alberta, in Edmonton, Canada. Her research and clinical interests focus on the neurocognitive factors that underlie high risk and criminal behaviors in youth and adults, and she has a particular interest in Fetal Alcohol Spectrum Disorder.

Michelle Stewart PhD, is an Assistant Professor in Justice Studies at the University of Regina, Saskatchewan, Canada. Her research interests center on policing practices with a focus on political anthropology, social justice and science and technology studies. Her current research investigates risk and prevention with attention to FASD, including how risk and diagnosis [illegible] influence the ways in which [illegible]

[illegible]

[illegible]

[illegible]

Introduction

To work for the common good is the greatest creed.

Woodrow Wilson

This volume is an effort to combine the viewpoints of many toward a mutual goal of helping those with Fetal Alcohol Spectrum Disorder.

We have assembled some new authors and some of the best known experts in the field of FASD from around the world; our hope is to be of service to you, the reader. We have provided a collection of 18 chapters written by more than two dozen authors from Europe, North America, and New Zealand to discuss ethical and legal issues associated with FASD. The authors are clinicians (such as psychologists, psychiatrists, physicians, social workers, and speech–language pathologists), legal professionals (including judges and lawyers), and caregivers.

Various regions of the world are at different stages of learning about people with FASD. As well, different professions are progressing in their learning about this population; each region and discipline therefore offers its own unique perspective. We hope this collection of chapters provides an overview of the development of each discipline's efforts with this population, and by implication, where we as a society may need to head to more effectively assist.

The topic of FASD is replete with ethical issues; those born with FASD have higher rates of early learning and behavioral issues. As adults, they are at elevated risk to have legal issues emerge in their life. Given the high costs associated with diagnosis, management, and support of those with FASD, who should shoulder those costs? Where should we cast our gaze, and by implication, place responsibility for these issues? The volume plunders the depths of many facets of this population and we hope it will encourage discussion.

We have divided the book chapters into five broad areas: the first is consideration of the ethical issues in prevention of FASD. Who is responsible? Is it the brewing industry, and do they have a duty to warn? What about the mother who gave birth to the individual with FASD? Where does her responsibility lie? Does the unborn child have rights? Furthermore, does a country possess the right to involuntarily hospitalize mothers who are addicted?

The second part allows clinicians to provide their perspectives on criminal behavior in FASD. What are the cognitive and linguistic elements to consider in the legal problems that concern those with FASD? What interventions work for offenders?

In the third part, judicial and justice perspectives follow. How does the new DSM-5 change judicial perspectives? Can FASD be used as a mitigating factor in sentencing? What about the training that police have?

A discussion of social and policy issues is fourth. The limited access to FASD diagnostic services is discussed, as is the critical issue of difficulty in gaining access to limited social supports. Could the notion of an Intellectual Disability "Equivalence" transform the way we help those with FASD gain access to services?

The fifth and last part of the book provides more personal perspectives, and a view from the front lines; nonprofessional caregivers provide candid glimpses into the struggles they face in raising those with FASD.

It is truly an honor to be asked to assemble such a volume. We hope that it will be of assistance to health, social work and legal professionals in their work and will invite conversation and debate on the many issues pertaining to FASD. We also hope that it will educate those who know little about the subject with respect to its complexity and importance. However, we acknowledge that there are many accomplished authors, as well as many ethical and legal issues that are not represented in this volume. This volume is therefore a starting point for the various issues that emerge for those with FASD. It is not meant to be exhaustive; however, it is hoped that it will provide a useful contribution to help professionals and caregivers alike as we work toward our common goal of helping this population.

Edmonton, AB, Canada
Summer 2015

Monty Nelson
Marguerite Trussler

Part I
Ethical Considerations and Implications in FASD

Fetal Alcohol Spectrum Disorder: Ethical, Legal and Moral Implications Associated with the Rights of the Unborn Child

Jenny Salmon

Introduction

Fetal alcohol spectrum disorder (FASD) is an umbrella term used to describe a profound constellation of disabilities that may result when a pregnant woman consumes teratogenic ethanol (alcohol), causing the developing fetus to be neurodevelopmentally impaired (Douds et al. 2012). The use of the term 'fetus' (Latin) generally implies that a mammalian embryo[1] has developed to the point of being recognizable as belonging to its own species and this is usually taken to be the ninth week after fertilization. A fetus is also characterized by the presence of all the major body organs though they will not yet be fully developed and functional and may not all be situated in their final anatomical location.

The developing human organism is particularly sensitive to the conditions in the uterine environment. Environmental influences begin to shape children from the moment of conception. As the fertilized egg divides, creating zygote, embryo and fetus, the infant-to-be derives far more from its mother than simple nutrition. It also receives a veritable cocktail of hormones and other chemicals that can have a profound, lifelong impact on its mental and physical growth. As with any drug, there is a risk that alcohol ingested by a pregnant woman will cross the placenta into

[1]For most of recorded history, people have fundamentally disagreed about the moral status of the human embryo. In early times this was because people knew very little about what actually went on in the womb and so had very little idea what an embryo was. Later, the problem was that a pregnancy could not be recognized until it was well established and the embryo made its presence felt by causing unmistakable symptoms in the mother or by starting to move in the womb. At this stage of the pregnancy, it was natural to think of the embryo as a being that was able to do things and they assumed that this was also true of the very earliest (and unknown) stages of pregnancy (BBC Ethics. www.bbc.co.uk/ethics/abortion/child/historical.shtml. Accessed 21/6/14).

J. Salmon (✉)
Centre for Research on Children & Families—Research Cluster,
University of Otago, Dunedin, New Zealand
e-mail: jvsalmon@ihug.co.nz

M. Nelson and M. Trussler (eds.), *Fetal Alcohol Spectrum Disorders in Adults: Ethical and Legal Perspectives*, International Library of Ethics, Law, and the New Medicine 63, DOI 10.1007/978-3-319-20866-4_1

the bloodstream of her unborn child. The effects of such transmission on the fetus can be profound given alcohol's teratogenic effects. As various researchers have detailed, prenatal exposure to alcohol affects many crucial biochemical and cellular components of fetal development (Michaelis and Michaelis 1994). Cellular events that may be disrupted by alcohol include cell division and proliferation, cell growth and differentiation and the migration of maturing cells within the embryo. These changes can all have tragic results (Jacobson and Jacobson 1994).

However, it has not been established what level of prenatal alcohol exposure can cause disabilities (Abel 2006) although it is clear that both chronic alcohol use and intermittent binge drinking during pregnancy can cause FASD (Burd et al. 2010). Moreover, there is no consensus within the clinical community on the precise diagnostic criteria for FASD. The more narrowly defined and full-blown disorder termed fetal alcohol syndrome (FAS) is recognized in the Diagnostics and Statistical Manual of Mental Health Disorders (*DSMIV*[2]; Astley et al. 2004), but the broader spectrum of related disorders, encompassed by the term FASD are subject to interpretation, depending on which diagnostic criteria clinicians use (Chudley et al. 2005). Because this spectrum often includes conditions that can be diagnosed independently of FASD, the term eludes clinical diagnostic definition (Boland et al. 2002; Stratton et al. 1996).

The disorder causes a range of physical and mental health problems including facial abnormalities, behavioural problems, hearing and ear problems, learning disabilities and growth issues. Exposure can cause permanent brain alterations leading to lifetime disability neurologically and psychologically. The disorder is irreversible, but in certain cases its effects can be reduced (Ministry of Justice, New Zealand 2011). Mukerjee estimates that between 1 and 3 per cent of the United Kingdom (UK) population is affected by FASD (Daily Mail, UK, 2014). It is also stated that up to 7,000 children a year in Britain have been revealed to be affected by alcohol exposure in the womb (Drinkaware 2014). According to the Centres for Disease Control and Prevention, FASD is the leading cause of preventable birth defects and developmental disabilities worldwide (CDC 2012).

Furthermore, FASD is significantly correlated with criminal justice involvement (Koren et al. 2004; Streissguth et al. 2004, Fast and Conry 2004). The effects of FASD plague people throughout their lives, causing them to be at increased risk for delinquent and criminal behaviour and for contact with criminal justice systems (Sayal et al. 2009; Boland et al. 2002). Persons with FASD are at increased risk for antisocial behaviour from birth by virtue of their in utero exposure to alcohol (Bailey et al. 2004; Roebuck et al. 1999) and they frequently suffer from a broad constellation of secondary disabilities (Howell et al. 2006;

[2]In the new DSMV (2013) neurodevelopmental disorder associated with prenatal alcohol exposure (ND-PAE) is included in the appendix as conditions needing further study (p. 798) in order to empower judges and other legal professionals for decision-making involving afflicted individuals. ND-PAE is used as an example for "Other Specified Neurodevelopmental Disorder" code 315.8 (p. 86). http://www.casaforchildren.org/site/c.mtJSJ7MPIsE/b.8968413/k.9568/JP3_Rich_Brown.htm.

Carmichael Olson et al. 1997). In severe cases, sufferers of FASD exhibit profound antisocial, delinquent or aggressive behaviours and they lack the ability to foresee the consequences of their actions (Streissguth et al. 2004).

Nonetheless, FASD is a recognized medical disability which implicates sufferers' legal rights (Douglas 2010; Moore and Greene 2004) ranging from their rights against self-incrimination and their right to effective assistance of counsel to their right to introduce mental health in mitigation at sentencing. Persons suffering from FASD and related secondary disabilities often do not have the cognitive ability to form criminal intent, to confess or testify meaningfully or to benefit from traditional sentencing structures (SAMHSA 2006). Research demonstrates that many individuals with FASD are highly suggestible and easily led (Salmon 2007).

Fast and Conry (2009) provide a useful mnemonic device "ALARM" to summarize the myriad disorders frequently arising from FASD. Poor *Adaptive* functioning; *Language* deficits; *Attention* problems; deficiencies in *Reasoning* skills and poor *Memory*. Each of these five disabilities has profound implications for criminal justice practitioners and criminal defendants. However, in light of the confusion within the clinical community (which is still in evidence today), it is not surprising that there is no consensus among federal courts on the meaning or significance of an FASD diagnosis (SAMHSA 2006).

Background

FASD is not a new phenomenon. Concern about gestational harm from alcohol arose in North America some 40 years ago when researchers found a group of children with an apparent unique phenotype, coupled with a high dose of prenatal exposure to ethanol. Over this period of time much has been learned about the complexity, seriousness and frequency of alcohol on brain development of the fetus (Salmon 2007).

Whilst some researchers have estimated thresholds such as seven standard drinks per week for neurobehavioural deficits (Jacobson and Jacobson 1994), the point has been well made by Hanson (1981) that no safe level of alcohol intake during pregnancy has been established. This comment holds good today. For instance, a single episode of binge drinking by an expectant mother can result in her child being born with alcohol-related neurodevelopmental disorder (ARND) (Salmon 2007). Schmidt et al's (1990) research indicated that between one-third and one-half of women drink during pregnancy with the typical profile of the pregnant drinker being single, older and of a higher socio-economic status.

Current guidelines in many countries now state that expectant mothers should avoid alcohol, but that if they do choose to drink, then they should limit their consumption to one or two units a week. Moreover, in New Zealand (NZ), the government advises zero intake of alcohol during pregnancy and when planning to become pregnant (Salmon 2007).

A British study in claiming that pregnant women can drink up to one glass of wine a day without harming their child's neurodevelopment, is sending out mixed messages to women in the community. The study by University of Bristol researchers, which recruited 7,000 10 year-olds, argues that higher total alcohol consumption before and after pregnancy was associated with better performance by the children. The children underwent a 20 minute assessment, testing dynamic balance by walking on a beam and static balance, including standing on one leg with their eyes both open and closed. Seventy per cent of the mothers—those whose alcohol consumption had been monitored at 18 weeks and again at birth—had drunk no alcohol whilst pregnant. One in four mothers had consumed small amounts of alcohol: about one glass a week or moderate amounts anywhere from three to seven glasses per week. About one in 20 mothers drank more and one in seven drank four or more glasses at one point (FASD-Canadian-Link 2013).

Another British study by Yvonne Kelly et al. at University College, London in 2008 argued that light drinking in pregnancy may be beneficial for an unborn child. She further stated that the research had found light drinking by pregnant mothers did not increase the risk of behavioural difficulties or cognitive deficits (Belfast Telegraph 2008). The study focused on 12,495 children up to three years of age and 'light drinking' was defined as anyone who had 'no more than one or two drinks a week or no more than two drinks on an occasion.' This broad categorization includes patterns of drinking that have been shown to affect fetal brain development in laboratory research studies (Zhou 2008).

Emily Oster (2013) at the University of Chicago advocated drinking whilst pregnant e.g. a glass of wine per day, stating that it was safe. She also suggested that the available evidence does not support the recommendation to abstain from alcohol use during pregnancy. Consequently, Bristol University researchers, Kelly and Oster will likely add more confusion surrounding drinking alcohol when pregnant and contradict official recommendations from many government agencies to avoid alcohol altogether when expecting a child (FASD-Canadian-Link 2013).

Added to this, legal and medical professionals worldwide have become increasingly concerned about the number of children born with alcohol-related brain disorders (Ministry of Justice, NZ 2011). Advocates and researchers maintain that FASD diagnoses are critical at every stage of the criminal justice process (Streissguth et al. 2004; Steinhausen et al. 1993). Furthermore, it has been documented that mothers-to-be who harm their unborn child by drinking alcohol could be committing a criminal act, lawyers argue (Daily Telegraph, UK 2014).

Certainly there are internationally accepted ethical norms about human life. It is an acknowledged and fundamental principle that everyone has the right to life, freedom and personal security and that the right to life is protected by the law. This is what the United Nations (UN) Declaration of Human Rights of 1948 and the European Convention about Human Rights 1950 declare. Another ethical norm is the 1959 UN Declaration of the Child's Rights, where it is stated that "both the child and its mother shall be given special protection and special care…before and after birth" [including protection against harmful influence e.g. alcohol, narcotics] (Per Westman 1993, p. 2).

Moreover, for more than a decade, law enforcement personnel, judges and elected officials in the United States (US) have sought to punish women for their actions during pregnancy that may affect the fetuses they are carrying (Gallagher 1987). Women who are having children despite substance abuse problems have been a particular target, finding themselves prosecuted for such non-existent crimes as 'fetal abuse' and delivery of drugs (including alcohol) through the umbilical cord. In addition, pregnant women are being civilly committed or jailed and new mothers are losing custody of their children even when they would be capable parents. Certainly research by Salmon (2007) argues that some alcohol-abusing but capable biological mothers do have their children with FASD taken from them and as a result, are distraught.

Some proponents of these efforts are motivated by the misguided belief that they are "promoting fetal health and protecting children" (Pollitt 1990 cited Hoffman 1990, p. 11). Others hope to gain legal recognition of fetal rights—the premise that a fetus has separate interests that are equal to or greater than those of a pregnant woman (Hoffman 1995). Recognition of such rights would require women to subordinate their lives and health—including decisions about reproduction, medical care and employment—to the fetus. Some advocates of fetal rights have argued that children should be able to sue their mothers for prenatal injuries. Others would argue that the pregnant addict and the fetus seem to have competing interests, although this is not the case (Pollitt 1990). Because a pregnant woman abuses herself as well as her fetus when she consumes alcohol, the woman and the fetus have a joint interest—ending the abuse—rather than competing interests (Schroedel 2000).

It is the conflicts between human fetuses and their mothers that soon became and remain today the focus of the most heated debates in legal periodicals, popular media, state legislatures and courts.[3] On this note, it has recently been argued in the UK that women who harm their unborn babies by drinking alcohol during pregnancy could now be found guilty of a criminal offence. Indeed, as far as I am aware, there has never been a prosecution of a mother in England for poisoning her fetus (unless it led to the fetus's death, in which case, this would constitute an illegal abortion). Nonetheless, a new legal test case claims that a six-year old girl who suffered brain damage owing to in utero alcohol exposure is the victim of a crime. It believes that her mother 'poisoned' her by continuing to drink whilst pregnant—despite being warned of the risks including damaging the child who was diagnosed with FASD at birth when she displayed signs of intrauterine growth retardation, small stature and particular facial features. There was a failure to thrive and there was a risk of neurodevelopmental, intellectual and behavioural difficulties, organ dysfunction and other problems (CICA 2013).

The case was brought before the Court of Appeal by a council in North-West England and the girl awarded 500,000 GBP in criminal injuries compensation for

[3]But the terminology of 'maternal-fetal conflict' probably oversimplifies the situation in most cases. See Cheryl M Plambeck, Divided Loyalties: Legal and Bioethical Considerations of Physician-Pregnant Patient Confidentiality and Prenatal Drug Abuse. *Journal of Legal Medicine* (2002), 4-5.

severe brain damage suffered in the womb as a result of her mother drinking (SAF Ocean Indien 2014). However, the Criminal Injuries Compensation Authority, which provides financial compensation to the victims of crime has appealed against this decision claiming that the child was a fetus and therefore, 'not a person' at the time. The only relevant offence that it has been suggested was committed by the mother and of which the child could have been a victim for the purposes of this present case is that of maliciously administering poison so as to endanger life or inflict grievous bodily harm contrary to Sect. 23 of the Offences Against the Person Act 1861 (UK) which states as follows:

Whosoever shall unlawfully and maliciously administer to or cause to be administered to or taken by any other person any poison or other destructive or noxious thing, so as thereby to endanger the life of such person, shall be guilty of felony and being convicted thereof shall be liable…to be kept in penal servitude for any term not exceeding ten years (Social Security and Child Support Commissioners 2013).

Nonetheless, the upper tribunal found that the child did not constitute a victim of violence because at the relevant time the child was 'not a person' in legal terms, but a fetus. As such, no crime had actually been committed to underpin a compensation claim (Priaulx 2015).

In the interim, media interest focused on concerns that any successful appeal could lead to criminalisation of mothers for fetal injuries or impairments resulting from maternal behaviours during pregnancy (Larcher and Brierley 2014). It was thought that the case could potentially have far-reaching implications.

Similarly, the Northern Territory (NT) Government of Australia in an attempt to reduce the number of children born with FASD, is also considering legislation that could see pregnant women prosecuted or restrained if they drink dangerously because they are infringing on the rights of the unborn child. Many children in Australia suffer from severe disabilities caused by things done before they were born, but most are not entitled to compensation for the harm they suffered in the womb and there is no law to prevent prenatal injuries. However, the NT Government is currently exploring the antenatal rights of the unborn child and it could mean the introduction of laws to prevent women from drinking through pregnancy (ABC Net 2014). This brings into question a number of human rights. Furthermore, in New South Wales, law-makers are considering giving legal status to the fetus. In fact, 'Zoe's Law' has passed the lower house and is about to be debated in the upper house. The proposed legislation is an example of a fetal personhood law, whereby the fetus is considered to be a person and harming it leads to a criminal sentence (Savulescu and de Crespigny 2014).

Certainly, in NZ [whose law is inherited from the English legal system] (Ministry of Justice, NZ 2015), the tragic legacy of fetal exposure to alcohol threatens the core notion of maternal autonomy and raises the issue of to what extent (if any) the state may seek to control maternal behaviour in order to protect the unborn child (Webb 2000).

This paper now discusses the implications thus raised and who is responsible for same.

Moral Personhood

What do we mean when we say that a fetus is a 'human being'? When people talk about 'human life' they may mean:

- A member of the biological human species—having the human genetic code

However, they may mean something very different:

- A being that possesses certain human characteristics in addition to the human genetic code
  - (a) characteristics often suggested might be the ability to think, to imagine, to communicate
  - (b) but the lists of characteristics put forward may be designed to limit the definition of human in the way the speaker wants
- A being that is a 'moral person', i.e. one that has rights and probably duties too

Conversely, the time at which a fetus gets the right to life because it has achieved the relevant list of characteristics can vary from the moment of conception to the time of birth. In fact, for some philosophers, very young babies do not really qualify as having earned the right to life by possessing the right characteristics. Fortunately, for young children these philosophers concede that young babies do have the right to life as a result of tradition and law instead. So, it may not matter in terms of pure moral philosophy, since there are cases when individuals would agree that even if the fetus is a person, the consumption of alcohol by a pregnant woman may be morally justified (BBC, UK 2014). This idea is indeed contrary to the view of the Criminal Injuries Compensation Authority in North-West England.

Other people refer to the fetus as having the right to life because it is a 'potential human being'. The 'potential human being' argument gives the right to life to the unborn from the very earliest stage of development—the moment when the egg is fertilised. This argument renders irrelevant any concerns about what sort of being the fetus is at any particular stage of its development. However, there is no doubt that a fertilised egg is a potential human being since, if the pregnancy runs its full course, it will be born as a human baby; the question is whether being a potential human being gives the fetus any rights? Furthermore and as a general rule, it is not felt that potential properties are the same as actual properties or that potential rights are the same as actual rights. Children are potential adults, but that does not give them the same rights or obligations as adults (BBC, UK 2014).

Generally, society does differentiate between actual and potential beings—distinguishing between their rights also. Many of the rights of a potential person are only potential rights—they only become actual rights when the person becomes an actual person. For example, a 10 year old in the UK is a potential voter and has the potential right to vote, but he/she does not get the actual right to vote until he/she reaches his/her 18th birthday and becomes eligible to vote. One of the strongest arguments for giving the full rights of person to the fetus because it is a potential person flows from the status of a new born baby. At birth a new born baby possesses so few of the characteristics required for *moral personhood* that its right to life cannot be based on it being a 'moral person'. Nonetheless, people generally do accept that it has a right to life—even those who follow the 'moral person' line of thought. There are others however, who do not accept that the newborn has a right to life: some ethicists argue that abortion and infanticide are both morally permissible because 'moral personhood' and the right to life are acquired only after birth, but this is an uncommon position. This right to life seems to flow from the potential that the newborn has to become a 'moral person' and this in turn seems to support the argument that a potential 'moral person' has the right to life (BBC, UK 2014).

Interestingly, *Dietrich v. Northampton* established an important precedent that defined the parameters of fetal personhood tort cases for more than sixty years. In *Dietrich* a woman tried to hold the city of Northampton, Massachusetts, liable for the miscarriage of her fetus after she fell as a result of a defect in a public highway. The court ruled that the unborn child could not sustain life outside the mother and therefore, did not have the legal right to action (Schroedel 2000).

Of note, is that the heightened awareness in medicine and ethics has led doctors to perceive the fetus as an individual patient with needs distinct from those of its mother. The implications of these developments for both the legal and medical professions are profound… (Bennett 1991).

Establishing an Adversarial Relationship Between Mother and Fetus

The relationship between a pregnant woman and her fetus is unlike any other in law, medicine or ethics. Within the same body, there exists one person and one potential person with both similar and separate interests and for the fetus, developing rights. This set of circumstances gives mother and fetus a biological, psychological, moral and legal connection that is unique (Post 1996).

From a theoretical standpoint, Faludi (1991) argues that by heavily idealising the mother's role in its backwards-looking fantasy of the nuclear family, this backlash has often served to pit mother against fetus in an adversarial relationship. When measured against the ideologically constructed image of the perfect mother, any woman giving birth to an alcohol-affected baby is thus vulnerable to the Western tendency of mother-blaming (Caplan 1989).

In Wisconsin, law enforcement officials are directed to take a pregnant woman into custody whenever they believe her consumption of alcohol may affect the physical health of her unborn child. The state can even decide to involuntarily commit the woman until her child is born. Whilst Wisconsin's regime seems to be an effective solution to prenatal alcohol exposure, the use of coercive legislative tactics is inherently flawed. Not only does a punitive approach assume that a pregnant woman and her fetus occupy adversarial roles, but it also fails to address alcohol addiction as the root of the problem (Linder 2005 cited Palmer 1999). Even more troubling is the notion that states can intrude into the lives of pregnant women when the conduct at issue is a legal activity. Accordingly, as medical technology improves and more behaviour is deemed harmful to fetal development, there may be no logical end to the diminution of a pregnant woman's rights (Schroedel 2000).

Heriot (1996) in addressing bills introduced in Mississippi state legislature in 1990 and 1991 states that fetus is defined as "unborn child" (p. 178). Furthermore, the bill's authors assume that a fetus is an *unborn child* and want to persuade women that this is the case. Although the language of the proposed bills may have seemed to be neutral, House Bill 624, introduced in 1990, actually suggested an adversarial relationship between the 'mother' or 'pregnant woman' and the 'child' or 'fetus'. Though neither the words *parent* nor *mother* appears in the text cited below, medical references to prenatal exposure (e.g. alcohol) imply that the mother is a threat to her fetus:

It is the intent of the Legislature that the following shall constitute prima facie evidence of abuse or neglect of a child as the case may be:…(a) proof that a child has a medical diagnosis of failure to thrive syndrome is prima facie evidence of abuse; (b) proof that a child has a medical diagnosis of *FAS* is prima facie evidence of neglect; (c) proof that a child has a medical diagnosis at birth of withdrawal symptoms from any controlled substance is prima facie evidence of neglect (House Bill 624, 1990).

Even though the Bill does not distinguish between prenatal and postnatal exposures to the fetus/child, it is hard to escape the conclusion that the charges of neglect or abuse[4] brought at birth would be directed at the mother. Had it become law, this legislation would have paved the way for increased monitoring of pregnant women and offered the implicit potential (already being applied in some legal/legislative cases) for control over women to ensure the birth of 'healthy' children. In essence, it posed a fundamental dichotomy between the rights of the woman as a person and the rights of the fetus (Heriot 1996 cited Johnsen 1986, 1987; Paltrow 1990). Had the more restrictive of the Mississippi laws been passed as originally authored, for example, the rights of a fetus would have outweighed those of the woman who carried that fetus (Heriot 1996).

Warren (1989) offers a different rationale for why the fetus should not be placed in opposition to the woman who carries it. She concluded that definitions stating that personhood (see p. 10) begins at conception do not take into sufficient account the legal and moral status of women as individuals—individuals with

[4]See also GEN. ASSEMBLY OF PA., SENATE BILL NO. 575, S3 (1989) (proposing an expansion of Pennsylvania's definition of child abuse to include a "substance-abused child", defined as "a child who is born with fetal alcohol syndrome…The Controlled Substance, Drug, Device and Cosmetic Act"). Cited McGinnis, DM (1990) Prosecution of drug-exposed babies: Constitutional and criminal theory. *Univ PA Law Review, 139*(2), 505-39.

rights that should outrank those of the fetus. Warren, a feminist moral philosopher, points out the problems with the assumption that a fetus has the same rights as an infant. She states while "most contemporary philosophers believe that birth cannot make a difference to moral rights", it in fact does make a difference and has done so in the past (p. 46). In fact, Warren examines two common assumptions that underlie moral rights: (1) the intrinsic properties assumption, "the view that the only facts that can justify the ascription of basic moral rights or the moral standing to individuals are facts about the intrinsic properties of those individuals" and (2) the single-criterion assumption, which claims that "there is some single property, the presence or absence of which divides the world into those things which have moral rights or moral standing and those which do not" (p. 47). She believes that these two assumptions ignore the *social* relationships that could be taken into account in moral decision making. Because the courts rely on these assumptions in addressing issues of personhood, they do not address moral questions about personhood that are related to larger social issues (e.g. poverty). As a consequence, decisions about when life begins are framed as all-or-nothing decisions.

Significantly, these sanctions apparently did not pertain to fathers[5] as long as the neglect or abuse could be attributed to actions taken by the woman during a pregnancy. Such reasoning is based upon a rationale that makes the outcome of a pregnancy *uniquely the responsibility of the individual woman* (Heriot 1996).

Drinking During Pregnancy/Pre-conception—Who Is Responsible?

A knee-jerk response to the sad picture of a child suffering from FASD is often to punish the mother because she is deemed to be responsible. Most commentators agree, however, that subjecting women to criminal sanctions will not address the problem of children being exposed to alcohol in utero (Moss 1990; Merrick 1993). Some scholars question whether invoking the law to punish this kind of 'irresponsible' maternal behaviour would place society on the slippery slope to even burdensome restrictions on a pregnant woman's actions for the sake of her unborn child (Paltrow 1990) or indeed, that constraints would extend to include all fertile women of childbearing age, since it is evident that alcohol consumption can harm a fetus even before a woman knows that she is pregnant (Chernoff 1980; Field 1989).

For some feminist writers, any such talk of prosecuting pregnant women who drink is suspect, in that "punishing women for harming their fetuses—intentionally or unintentionally—resurrects patriarchal stereotypes of women whose value

[5]A father's age has been stated to influence whether a baby might be born alive or dead (Goodwin 2008 cited Wyrobeck et al. 2006). Thus, research has demonstrated that increasing paternal age is significantly associated with spontaneous abortion independent of maternal age and multiple other factors (Kleinhaus, Perrin et al. 2006).

is defined solely in terms of their reproductive capacity" (McNulty 1987, p. 319). Such understandings of responsibility reinforce traditional gender roles as defined by our society and place the blame squarely on women. Pollitt (1990) notes that in considering the outcome of pregnancies, sexist bias implies that if the mother isn't to blame, no one is to blame. Other feminist writers in recent years have identified that the preventable tragedy of FASD may arguably be seen as part of the more general 'backlash' (see p. 13) against women (Webb 2000).

Until now fathers' lifestyle choices have not had a causal link to FASD or seen any repercussion on their unborn children. Also, no FASD cases have been documented without maternal drinking or resulting from paternal drinking alone (Gearing et al. 2005). However, ground-breaking research by Lee et al. (2014) provided the first definitive evidence that fathers' drinking habits pre-conception could cause significant fetal abnormalities. These researchers studied male mice exposed to varying concentrations of alcohol and one control group exposed only to saline. Following exposure, the mice were mated and resulting fetuses examined. The findings revealed previously unknown and riveting evidence that paternal alcohol consumption can directly affect fetal development i.e. abnormal organ development and/or brain development. The authors believe that alcohol consumption affects genes in sperm which are responsible for normal fetal development.

Nonetheless, blaming mothers or fathers is an inaccurate, incomplete and unproductive approach that reduces responsibility to the individual level and ignores the larger social determinants of health, such as poverty or disadvantaged minority status (Gearing et al. 2005).

Rights of the Mother

The literature reveals that discussions of maternal substance abuse are typically infused by a rhetoric of rights (Webb 2000). On the one hand are those who argue that a mother has the right to pursue any lifestyle she chooses and that only the pregnant woman herself can make the intimate choices that affect her fetus. This is an argument based on a woman's right to privacy, bodily integrity and self-determination. It is bolstered by the fact that expectant mothers are naturally more concerned than anyone about the health and well-being of their unborn child. This reasoning is relied upon by those who contend that the state should never usurp the decision-making autonomy of the mother during pregnancy, for to do so would be to unfairly limit her freedom to pursue her own life-choices. Field (1989) sums up this approach when she writes: Pregnancy is a person event. Pregnancy should not become a legal event as well—an occasion for governmental intrusion.

Leonard (1998 cited Linder 2005) states that any regulation which punishes or restricts a pregnant woman's behaviour in order to protect an unborn fetus essentially subordinates the civil liberties of the woman to the rights of her fetus. This maternal-fetal conflict is especially severe when states attempt to regulate prenatal behaviour that is otherwise legal, such as alcohol consumption. Civil commitment schemes which regulate prenatal alcohol consumption, such as Wisconsin's protective custody statute, pose a number of constitutional and practical problems.

Accordingly, Dineen (1994) argues that prenatal alcohol abuse protections in America may violate both the Due Process and Equal Protection Clauses of the Constitution. Even if these protections pass constitutional scrutiny, they may still be rendered unworkable by the range of practical problems they face. She argues that a purely punitive approach to preventing FASD, which forces pregnant women into involuntary treatment, does not address the addiction at hand but instead, may deter women from seeking proper prenatal care. In addition, proscribing otherwise legal behaviour is the first step towards regulating any maternal conduct that may adversely affect the health of a fetus. Further, statutes that simply react to the problem of prenatal alcohol abuse by taking already pregnant women into custody are unlikely to prevent FASD since the fetus has most likely already suffered irreversible damage.

Significantly the 1980s witnessed the increasing intensity of anti-abortion activism in reaction to the 1973 *Roe v. Wade* US Supreme Court decision that legalized abortion as a woman's constitutional right (Oaks 2001). Research by Salmon (2007) purports that some women who fear their unborn child may have FASD, do attempt to abort the fetus.

Rights of the Father

There is scarcity of research and therefore, paucity of information in the literature on this issue [described as a conspicuously absent grey area] (Gearing et al. 2005) probably because there may be many ramifications—political, moral, social, legal and historical.[6] Furthermore, when it comes to the rights of the unborn child's supposed father, the issue becomes very sticky. However, English law commonly states that because the unborn child has no rights of its own at least until it is born, the father has no grounds or right to act on its behalf. Neither does it give the father custody rights of the unborn child (Per Westman 1993). Also it is documented that in Quebec legislation or case law, there is nothing to support the argument that the father's interest in a fetus he helped create gives him the right to veto a woman's decision in respect of the fetus that she is carrying (Hrcr.org. 1989).

In 1989, Jean-Guy Tremblay of Quebec filed an injunction against his girlfriend, Chantal Daigle, in an attempt to prevent her from obtaining an induced abortion in the 1989 Canadian case *Tremblay v. Daigle.* The Supreme Court of Canada ultimately ruled that there was no precedent for a man's right to protect his 'potential property', as a fetus was not found to have a right to life under the Quebec Charter of Human Rights and Freedoms (Greschner 1990).

---

[6]See Case centers on rights of unborn child's supposed father. *Daily Record* [St. Louis, MO/St. Louis Countian] 5 Feb 2003. *General One File.* Accessed 24 March 2015.

Furthermore, father's rights activists have long ago argued that men should have veto power over their partners' decisions to abort[7] (Drakich 1989); the philosopher George W Harris (1986), postulating that there were circumstances under which a woman's decision to have an abortion would be morally wrong because it could harm the father. In this regard, a new Chinese law in 2002 put a man's right to have a child on an equal footing with the right of his wife and a man sued his wife for infringing that right by undergoing an abortion (BBC, UK 2014).

Political Dimension and Rights of the Child

A fundamental issue is whether the rights of the unborn child are protected by the Charter of the UN, including:

The Universal Declaration of Human Rights 1948; the International Covenant on Civil and Political Rights 1966 and the Convention on the Rights of the Child 1989. The NZ Human Rights Commission on their website address this question under the 'right to life' (Jayawickrama 2002). The Commission states:

There have been differing interpretations internationally of whether the right to life applies to the unborn child. The ICCPR declares that 'every human being' has the inherent right to life, while in respect of other rights. The expressions used are 'everyone', 'every person', 'every child'—or even 'every citizen'. This use of different terminology has raised the question whether 'every human being' has a more expansive meaning than usually attributed to 'every person'; in particular, whether it also includes the 'unborn child'. At the national level this is determined generally by policy rather than by law. Of significance is the fact that there have been differing views internationally on whether the right to life includes the right to die (Jayawickrama 2002).

The text of the 1948 Universal Declaration of Human Rights clearly states that everyone has the right to life and that what is meant by everyone is 'every member of the human family'—that is, all human beings. Here is the nub of the matter. The Royal Commission regarded the unborn child as one of the weakest, the most vulnerable and most defenceless forms of humanity and should receive protection (The Life Resources Charitable Trust 2011).

On this note, the majority of fetal rights legislation in Wisconsin has focused on the consumption of alcohol as prenatal drug abuse which can negatively affect fetal health and the well-being of a fetus (Linder 2005 cited Dineen 2001). In fact, maternal alcohol consumption is considered even more dangerous than prenatal drug abuse because it is more prevalent in society and the leading cause of non-hereditary mental retardation. Moreover, sudden withdrawal from alcohol can harm or kill a fetus.

[7]See Unmarried biological fathers must be given notice and chance to comply with statutory requirements before terminating their parental rights: Heart of Adoptions, Inc. v J.A., No. SC07-738 (Fla. 2007). *Child Law Practice Newsletter* (2007) *26*(6), 82.

Several US states, including South Dakota and Wisconsin have passed criminal legislation directed at pregnant women who consume alcoholic beverages. Wisconsin's civil commitment laws, in particular, have sparked a great deal of controversy. These laws allow the state to take an expectant mother into custody if a public officer finds that she exhibits an habitual lack of self-control in the use of alcoholic beverages (Linder 2005).

Furthermore, since the state may intervene on behalf of an unborn child beginning at the time of fertilization, a pregnant woman can be taken into custody before she even knows that she is pregnant. Wisconsin's statutory scheme is a far more draconian approach to protecting fetal rights than previous legislation targeting the use of illegal substances. However, this approach to deterring harmful maternal behaviour threatens to cross constitutional boundaries (Linder 2005).

Nonetheless, there are those within society who point to the basic right of the child to begin life with a sound mind and body. This idea is often championed by geneticists (Glass 1975; Milunsky 1977 cited Webb 2000). It is also the right spoken of in the UN Declaration of the Rights of the Child adopted in 1959, to the effect that: The child…shall be entitled to grow and develop in health. This basic right is taken by some to mean that maternal action during pregnancy should be prescribed so as to provide for the well-being of the child.

Robertson (1983 cited Webb 2000) purports that the mother has, if she conceives and chooses not to abort, a legal and moral duty to bring the child into the world as healthy as is reasonably possible. She has a duty to avoid actions or omissions that will damage the fetus and child… In terms of fetal rights, a fetus has no right to be conceived, to be carried to viability. But once the mother decides not to terminate the pregnancy, the viable fetus acquires rights to have the mother conduct her life in ways that will not injure it. However, a number of scholars indicate that this idea is virtually non-sensical in the context of state intervention in pregnancy e.g. Hanigsberg (1991). Notwithstanding, a NZ Family Court decision recognised a fetus a few weeks before birth as having some rights. The mother was 15 and pregnant. There was evidence that her boyfriend had been violent towards her (The Life Resources Charitable Trust 2011).

Fletcher's situational ethics takes this point to its logical conclusion in proposing that regardless of what it means for the rights of those involved, one must calculate the gains and losses of various courses of action or inaction and then select that course which offers the greatest good for society. Ideally it is better to do the moral thing freely Fletcher contends, but sometimes it is more compassionate to force it to be done than to sacrifice the well-being of the many to egocentric rights of a few (Webb 2000 cited Fletcher 1974). (*Admittedly though, this strict brand of utilitarianism is likely to encounter as much, if not more resistance as any absolute-rights argument in the FAS and ARND context)*—[comments in italics Webb's].

Conclusion

Whilst I have attempted to unpack the ethical, legal and moral implications associated with the rights of the unborn child in relation to FASD, it must be firmly stated that the dangers of drinking alcohol during pregnancy were well known in Biblical and Greco-Roman times (Abel 1997). This implies that if society had paid attention to the bitter lessons of the past, the modern 'tragedy' might have been avoided. Physicians, writers and theologians had written about the effects of alcohol on the fetus "since Biblical times" intoned Hill and Tennyson (1980) who then chastised modern society with the inclusive moral lapse of failing "to heed the wisdom of our forefathers…" (p. 177). The authors concluded with a statement from a temperance tract dating to the 1860s that said it "still holds true today that parents are *responsible* for their children's infirmities, blindness and idiocy" (p. 198). Notably, the word 'parents' is used rather than reference being made to the biological mother, whom some writers and legal professionals still argue is in fact responsible for the child's disability and should therefore, be punished by law, as described by Gallagher (1987) and various earlier legal Acts which differ by country, as previously discussed.

Because there are many mixed messages within society being given to women of child-bearing age regarding pregnancy and the consumption of alcohol, it is no wonder that the disability is still prevalent worldwide. It is noteworthy that in 1981, the Surgeon General of the US advised 'women who are pregnant (or considering pregnancy) not to drink alcoholic beverages and to be aware of the alcoholic content of foods and drugs' (Food and Drug Administration 1981). The Surgeon General's warning stands in stark contrast to the official advice offered in other countries. For example, the British Royal College of Obstetricians and Gynaecologists issued guidelines in 1996 stating that 'no adverse effects on pregnancy outcome have been proven with a consumption of less than 120 grams of alcohol per week' and recommending that 'women should be careful about alcohol consumption in pregnancy and limit this to no more than one standard drink per day'. Although these recommendations have been challenged by some (Guerri et al. 1999), they are typical of the European stance on prenatal drinking (EUROMAC 1992). Moreover, it is not clear how punishing a woman for giving birth to a child with FASD is going to reduce the incidence of this birth defect. In fact, it would seem far more logical as a moral imperative, to find and help those women most at risk of adverse outcomes.

It is evident from what is documented above that there is no clear cut testimony in legal circles regarding the rights of the unborn child in relation to FASD with various countries abiding by different legislation or laws—this over many years in some cases. Ironically though, some individuals would argue, why should what amounts to a major public health issue be deemed a matter for the criminal justice system? How has society come to see women as the major threat to the health of their newborns and the womb as the most dangerous place a child will ever inhabit?

Any proposal to bring prohibition to the womb exists within a political matrix. Indeed, it would be almost impossible to enter a discourse on 'policing pregnancy' without buying into a political argument about the rights of those involved and the concepts that underlie those rights. Surely with new research evidence that men's drinking may also cause organ/brain dysfunction of the fetus, should they not be partly held accountable?

Nonetheless, the great debate presently is between the legal rights of the woman against the legal rights of the unborn child and there is a lot of resistance to monitoring the lifestyle of women who may or may not be pregnant. Criminalizing women who drink would set a profoundly dangerous legal precedent. Whether states have a right to protect the fetus from the actions of the mother is a question that requires answering. States should also give serious thought to the rights of the father since one-half of every child is the biological product of that individual and he deserves a voice. Thus, fathers and their involvement is a distinct and significant factor that unmistakably exerts positive and/or negative influence on FASD across a number of domains, specifically genetics, partnering relationships, family life and larger environmental and structural conditions.

FASD poses serious and clearly avoidable injury to the developing fetus and consequently, to society. Tackling the problem of fetal alcoholism through legislation however, presents several inherent difficulties. Prevention can quickly take on a very intrusive intervention into a woman's private life, threatening the right to make autonomous decisions about their lives and bodies.

Clearly, the only way to protect the rights of the expectant mother and unborn child is to follow a strategy of education and treatment aimed at stopping prenatal alcohol addiction before it starts. Unfortunately, this is easier said than done, especially since drinking alcohol is not a crime and in many countries (e.g. NZ; Australia), it is indeed a culture. Furthermore, threats of jail or the loss of their children may well drive women away from prenatal care, hospital deliveries and put health professionals under pressure to report women suspected of drinking to the police. This is not what is required—rather society and the legal profession should concentrate on a concerted effort to minimise the risk of harm to the unborn child.

References

ABC Net (News). 2014. Drinking alcohol during pregnancy targeted as NT Government considers rights of unborn child. Accessed 14 June 2014.

Abel, E.L. 1997. Was the fetal alcohol syndrome recognized in the ancient near East? *Alcohol and Alcoholism* 33: 3–7.

Abel, E.L. 2006. Fetal alcohol syndrome: A cautionary note. *Current Pharmaceutical Design* 12: 1521–1529.

Astley, S.J. 2004. *Diagnostic guide for fetal alcohol spectrum disorders: The 4-digit diagnostic code*. Seattle: University of Washington.

Bailey, B., V. Delany-Black, C. Covington, J. Ager, J. James, J. Hannigan, and R. Sokol. 2004. Prenatal exposure to binge drinking and cognitive and behavioral outcomes at age 7 years. *American Journal of Obstetrics and Gynecology* 191: 1037–1043.

BBC, UK. 2014. http://www.bbc.co.uk/ethics/abortion/philosophical/moralperson.shtml. Accessed 13 June 2014.

Belfast Telegraph. 2008. BelfastTelegraph.co.uk. Accessed 7 July 2014.

Bennett, B. 1991. Pregnant women and the duty to rescue: A feminist response to the fetal rights debate. *Law in Context* 1(1): 70–91.

Boland, F.J., A.E. Chudley, and B.A. Grant. 2002. The challenge of fetal alcohol syndrome in adult offender populations. *Forum on Corrections Research* 14(3). Accessed 17 July 2014.

Burd, L., D.K. Fast, J. Conry, and A.D. Williams. 2010. Fetal alcohol spectrum disorder as a marker for increased risk involvement with corrections. *The Journal of Psychiatry and Law* 38: 559.

Caplan, P.J. 1989. *Don't blame the mother: Mending the mother-daughter relationship*. New York: Harper & Row.

Carmichael Olson, H., A.P. Streissguth, P.D. Sampson, H.M. Barr, F.L. Bookstein, and K. Theide. 1997. Association of prenatal alcohol exposure with behavioral and learning problems in early adolescence. *Journal of the American Academy of Child and Adolescent Psychiatry* 36: 1187–1194.

Centres for Disease Control and Prevention. 2012. Fetal alcohol spectrum disorders (FASDs). Data and statistics. http://www.edc.gov/mcbddd/fasd/data.html. Accessed 16 July 2014.

Chernoff, G. 1980. The fetal alcohol syndrome: Clinical studies and strategies of prevention. In *Prevention of mental retardation and other developmental disabilities*, ed. M.K. McCormack. New York: Marcel Dekker.

Chudley, A.E., Conry, J., Cook, J.L., Loock, C., Rosales, T., and LeBlanc. 2005. Fetal alcohol spectrum disorder: Canadian guidelines for diagnosis. *CMAJ* 172, 5: suppl S1–S21.

CICA v First-tier Tribunal and CP (CIC). 2013. UKUT 0638 (AAC).

Daily Mail, UK. 2014. www.dailymail.co.uk/news/article-2550391/women-drink-alcohol-pregnancy-guilt. Accessed 18 June 2014. Also see www.drinkaware.co.uk.

Daily Telegraph, UK. 2014. http://www.telegraph.co.uk/health/healthnews/10612818/Drinking-alcohol-during-pregnancy. Accessed 20 June 14.

Dineen, C.E. 1994. Fetal alcohol syndrome: The legal and social responses to its impact on Native Americans. *Heinonline* 70 N.D.L. Rev. 1.

Douds, A.S., H. Stevens, and W.E. Sumner. 2012. Sword or shield? A systematic review of the roles FASD evidence plays in judicial proceedings. *Criminal Justice Policy Review* XX(X): 1–18.

Douglas, H. 2010. The sentencing response to defendants with foetal alcohol spectrum disorders. *Criminal Law Journal* 34: 221–239.

Drakich, J. 1989. In search of the better parent: The social construction of ideologies of fatherhood. *Canadian Journal of Women and the Law* 3(1): 69–87.

Drinkaware, UK. 2014. www.drinkaware.co.uk.

EUROMAC. 1992. A European concerted action: maternal alcohol consumption and its relation to the outcome of pregnancy and child development at 18 months. *International Journal of Epidemiology* 21(Suppl. 1): S1–S87.

Faludi, S. 1991. *Backlash: The undeclared war against american women*. NY: Crown Publishing Group.

FASD-Canadian-Link. 2013. FASD-Canadian-News@yahoogroups.com. Accessed 26 June 2014.

Fast, D.K., and J.L. Conry. 2004. The challenge of fetal alcohol syndrome in the criminal legal system. *Addiction Biology* 9: 161–666.

Fast, D.K., and J.L. Conry. 2009. Fetal alcohol spectrum disorders and the criminal justice system. *Developmental Disabilities Research Reviews* 15: 250–257.

Field, M. 1989. Controlling the woman to protect the fetus. *Law, Medicine & Health Care* 17: 114.

Fletcher, J.F. 1974. *The ethics of genetic control: Ending reproductive roulette*. Garden City: Doubleday.

Food and Drug Administration (FDA). 1981. Surgeon General's advisory on alcohol and pregnancy. *FDA Drug Bulletin* 11: 9–10.
Gallagher, J. 1987. Prenatal invasions and interventions: What's wrong with fetal rights? *Harvard Women's Law Journal* 10: 9–58.
Gearing, R.E., T. McNeill, and F.A. Lozier. 2005. Father involvement and fetal alcohol spectrum disorder: Developing best practices. *JFAS International* 3: e14.
Greschner, D. 1990. Abortion and democracy for women: A critique of *Tremblay v. Daigle. McGill Law Journal* 35: 633–669.
Guerri, C., E. Riley, and K. Stromland. 1999. Commentary on the recommendations of the Royal College of Obstetricians and Gynaecologists concerning alcohol consumption in pregnancy. *Alcohol and Alcoholism* 34: 497–501.
Goodwin, M. 2008. Prosecuting the womb. *The George Washington Law Review* 76: 1657.
Hanigsberg, J.E. 1991. Power and procreation: State interference in pregnancy. *Revue de droit dOttawa* 23(1): 36–70.
Hanson, J.W. 1981. Counselling for Fetal Alcohol Syndrome. In *Genetic screening and counselling: A multidisciplinary perspective*, ed. Applewhite et al. Illinois: Charles C Thomas, Springfield.
Harris, G.W. 1986. Fathers and fetuses. *Ethics* 96(3): 594–603.
Heriot, M.J. 1996. Fetal rights versus the female body: Contested domains. *Medical Anthropology Quarterly* 10(2): 176–194.
Hill, R.M., and L.M. Tennyson. 1980. An historical review and longitudinal study of an infant with fetal alcohol syndrome. In *Alcoholism: A perspective*, ed. F.S. Messiha, and G.S. Tyner, 177–201. Westbury: PJD Publications.
Hoffman, J. 1995. Pregnant addicted—and guilty. *New York Times Magazine,* August 19.
Howell, K.K., M.E. Lynch, K.A. Platzman, G.H. Smith, and C.D. Coles. 2006. Prenatal alcohol exposure and ability, academic achievement and school functioning in adolescence: A longitudinal follow-up. *Journal of Pediatric Psychology* 31(1): 116–126.
Hrcr.org/org/safrica/life/tremblay_daigle.html. 1989. *Tremblay v. Daigle*. Accessed 20 March 2015.
Jacobson, J.L., and S.W. Jacobson. 1994. Prenatal alcohol exposure and neurobehavioural development. *Alcohol Health and Research World* 18(1): 30–36.
Jayawickrama, N. 2002. *The judicial application of human rights law: National, regional and international jurisprudence*. Cambridge, UK: Cambridge University Press.
Kleinhaus, K., M. Perrin, Y. Friedlander, O. Paltiel, D. Malaspina, and S. Harlap. 2006. Paternal age and spontaneous abortion. *Obstetrics and Gynecology* 108(2): 369–377.
Koren, G., I. Roifman, and I. Nulman. 2004. Hypothetical framework: FASD and criminality—Causation or association? The limits of evidence-based knowledge. *Journal of Fetal Alcohol Syndrome International* 2(e6): 1–4. http://www.motherisk.org/JFAS_documents/FASD_Criminality.pdf.
Larcher, V., and J. Brierley. 2014. Fetal alcohol syndrome (FAS) and fetal alcohol spectrum disorder (FASD)—diagnosis and moral policing: An ethical dilemma for paediatricians. *Archives of Disease in Childhood* 10: 1136.
Lee, H.J., J.-S. Ryu, N.Y. Choi, Y.S. Park, Y.I. Kim, D.W. Hanm, K. Ko, C.Y. Shin, H.S. Hwang, K.-S. Kang, and K. Ko. 2014. Transgenerational effects of paternal alcohol exposure in mouse offspring. *Animal Cells and Systems* 17(6): 429.
Linder, E.N. 2005. Punishing prenatal *alcohol* abuse: The problems inherent in utilizing civil commitment to address addiction. *Illinois Law Review,* March 2005.
McNulty, M. 1987. Pregnancy police: The health policy and legal implications of punishing women for harm to their fetuses. *Review of Law and Social Change* 16: 277–319.
Merrick, J.C. 1993. Maternal substance abuse during pregnancy: Policy implications in the United States. *The Journal of Legal Medicine* 14: 57–71.
Michaelis, E.K., and M.L. Michaelis. 1994. Cellular and molecular bases of alcohol's teratogenic effects. *Alcohol Health & Research World* 18(1): 17–21.

Ministry of Justice, New Zealand. 2011. Youth Court, Issue 53. I, EP (1999). http://www.justice.govt.nz/courts/youth/publications-and-media/principal-youth-court ne. Accessed 17 June 2014.
Ministry of Justice, New Zealand. 2015. www.justice.govt.nz/publications. The NZ legal system.
Moore, T.E., and M. Green. 2004. Fetal alcohol spectrum disorder (FASD): A need for closer examination by the criminal justice system. *Criminal Reports* 19: 99–108.
Moss, K. 1990. Substance abuse during pregnancy. *Harvard Women's Law Journal 12*278–299.
Oaks, L. 2001. *Smoking and pregnancy: The politics of fetal protection*. NJ: Rutgers University Press.
Oster, E. 2013. *Expecting better: Why the conventional wisdom is wrong and what you really need to know*. New York: The Penguin Press.
Pollitt, K. 1990. 'Fetal rights', a new assault on feminism. *Nation* 250(12): 409–414.
Paltrow, L.M. 1990. When becoming pregnant is a crime. *Criminal Justice Ethics* 9.
Per Westman, H. 1993. Protection of the unborn child and the rights of parents. *SvenskJuristTidning s* 479.
Post, L.F. 1996. Bioethical consideration of maternal-fetal issues. *Fordham Urban Law Journal* 24(4): Article 10.
Priaulx, N. 2015. Pre(natal) crime: Pregnant women, substance abuse and the law. *Medico-Legal Journal* 83(1): 43–46.
Roebuck, T.M., S.N. Mattson, and E.P. Riley. 1999. Prenatal alcohol exposure again linked to delinquency, psychiatric woes. *Alcoholism, Clinical and Experimental Research* 23: 1070–1076.
Royal College of Obstetricians and Gynaecologists. 1996. *Guideline No.9: Alcohol consumption in pregnancy.* London: Royal College of Obstetricians and Gynaecologists.
S.A.F. Ocean Indien. 2014. http://www.safoceanindien.org./en/fetal-alcohol-syndrome/95-safoi-anglais/latest-news/200.
Salmon, J.V. 2007. *Fetal alcohol syndrome: New Zealand birth mothers' experiences*. Wellington, New Zealand: Dunmore Pubs.
SAMHSA [The Substance Abuse and Mental Health Services Administration]. 2006. *Fetal alcohol spectrum disorders and the criminal justice system.* Washington, DC: US Department of Health and Human Services. http://www.fasdcenter.samhsa.gov/documents/WYNK_Criminal_Justice5.pdf.
Savulescu, J., and L. de Crespigny. 2014. Should it be a crime to harm an unborn child? *The Conversation*, 20/3/2014. Accessed 10 July 2014.
Sayal, K., J. Heron, J. Golding, R. Alati, G.D. Smith, R. Gray, and A. Edmond. 2009. Binge pattern of alcohol consumption during pregnancy and childhood mental health outcomes: Longitudinal population-based study. *Pediatrics* 123: e289–e296.
Schmidt, C., et al. 1990. A long-term perspective of FAS. *Alcohol Health & Research World* 18(1): 74–81.
Schroedel, J.R. 2000. *Is the fetus a person: A comparison of policies across the fifty states*. Cornell: Cornell University Press.
Steinhausen, H.C., J. Williams, and H.L. Spohr. 1993. Long-term psychopathological and cognitive outcome of children with fetal alcohol syndrome. *Journal of the American Academy of Child and Adolescent Psychiatry* 32: 990–994.
Social Security & Child Support Commissioners. 2013. http://www.osscsc.gov.uk/Aspx/view.aspx?id=4073. Accessed 10 July 2014.
Stratton, K., C. Howe, and F. Battaglia (eds.). 1996. *Fetal alcohol syndrome: Diagnosis, epidemiology, prevention and treatment*. Washington, DC: National Academies Press.
Streissguth, A.P., F. Bookstein, H.M. Barr, P.D. Sampson, K. O'Malley, and J.K. Young. 2004. Risk factors for adverse life outcomes in fetal alcohol syndrome and fetal alcohol effects. *Journal of Developmental and Behavioural Pediatrics* 25: 228–238.
The Life Resources Charitable Trust. 2011. www.life.org.nz/rightsunbornchild.

Warren, M.A. 1989. The moral significance of birth. *Hypatia* 4(3): 46–65.
Webb, M.B. 2000. *Policing pregnancy: The case of foetal exposure to alcohol.* New Zealand: Litigation Dept., Russell McVeagh McKenzie Bartleet & Co.
Zhou, F. 2008. Fetal alcohol spectrum disorders study group (FASDSG). http://www.rsoa.org/fas.html. Accessed 7 Aug 2014.

Why People with FASD Fall for Manipulative Ploys: Ethical Limits of Interrogators' Use of Lies

Stephen Greenspan and John H. Driscoll

Introduction

People with FASD are easily manipulated by others, and such social vulnerability is a major reason why they are so likely to get into serious legal and other forms of difficulty. Police interrogations are also manipulative, in that a detective is attempting to use various ploys—including deceptive statements–to persuade someone to do something he is initially reluctant to do, namely admit to having committed a criminal act. In this paper, the authors use an audiotape of an actual interrogation of a young man with FASD who initially strongly denied guilt, to better understand the way interrogative manipulations (including lies) were used to break his will and get him to confess to a crime which he may or may not have committed. Cognitive and other impairments in people with FASD are discussed in relation to specific interrogative ploys—particularly the introduction of pseudo-factual and other forms of lies–that take advantage of those impairments. Ethical cautions regarding lie-infused interrogation, particularly when used with brain-impaired people who have trouble discerning factual-potential and the motivations of liars, are raised. A point that is strongly emphasized in this paper, and derived mainly from the writings of Jerome Bruner about narrative story-telling, is that for an adolescent or adult to stay out of jail requires a level of meta-cognition in which the intentions of a manipulator (whether a co-criminal or an interrogator) are thought about on a functional and abstract level. Unfortunately, people with FASD operate on a concrete and non-reflective level, which is why so many people with that disorder end up in jail or prison

S. Greenspan (✉) · J.H. Driscoll
University of Connecticut, Mansfield, USA
e-mail: stephen.greenspan@gmail.com

M. Nelson and M. Trussler (eds.), *Fetal Alcohol Spectrum Disorders in Adults: Ethical and Legal Perspectives*, International Library of Ethics, Law, and the New Medicine 63, DOI 10.1007/978-3-319-20866-4_2

Social Vulnerability of People with FASD

People with Fetal Alcohol Spectrum Disorder (FASD) do not, as a rule, show good social insight or judgment (Streissguth and Kanter 2002). Nor do they possess an ability to adequately apply their thoughts and intuitions to determine how to act safely in complicated or multifaceted situations. That is the main reason why they get in trouble with the law and "mess up" in other ways, such as getting fired from jobs (Edwards and Greenspan 2011). A big contributor to the poor social judgment of people with FASD and other brain-based disorders is their inability to see through, and resist, the risky manipulations, pressures and temptations offered by others (Greenspan and Woods 2014). These manipulations often involve pseudo-factual hypothetical statements that are in effect lies; a real life example (which resulted in a lengthy prison sentence) involved a young man being falsely told by a peer that his mother was being physically abused by a boyfriend, who needed to be beaten up in order to be taught a lesson. The assault took a severe turn and the targeted man died, which likely was the manipulator's intention all along. But manipulation can take many forms other than committing assault or murder, ranging from being persuaded to consume drugs or alcohol when one is trying to quit, to being pressured to confess to a crime when one is trying to maintain one's innocence.

Susceptibility of people with FASD to being manipulated by others is an obvious result of their neuro-cognitive limitations, but other factors enter in as well (Greenspan and Driscoll 1997a). These include their social neediness, which is a reflection of having few if any non-deviant friends, and their personality adaptations (a tendency to mimic and accede to the requests of others) which people with a history of failure tend to make (Zigler 2013). Suggestibility is a personality trait with a cognitive component that has been found to be high in people with FASD and which makes them susceptible to being manipulated in various contexts (Brown et al. 2011). Various studies have established that people with FASD and related neurodevelopmental disorders have a several-times greater probability of giving false confessions (Fast and Conry 2006), a reflection of the fact that interrogations, even those undertaken with the most honorable of intentions, are heavily reliant on the use of manipulative methods, typically involving some form of deception.

Although there is an acknowledged cognitive basis for the susceptibility of people with FASD to manipulation, that basis has generally been analyzed on a global and superficial level, such as by general reference to their impairments in "executive functioning" without linking it to specific actions. While executive functioning is a valid and useful construct, a more helpful analysis would delve deeper, by identifying specific ploys engaged in by a manipulator, and then seeing how those ploys tie in with specific cognitive processes, including executive functioning, and limitations. Such a more contextualized approach to understanding the poor judgment (with usual predictable bad outcomes) of people with FASD has not been widely attempted, to our knowledge.

An obstacle to conducting such an analysis is that we typically do not know, on a moment-to-moment basis, the deceptive tactics that are used in a particular incident. An exception to that generalization, however, can be found in recorded police interrogations, where the ebb and flow of a manipulative transaction can be minutely observed and analyzed. Police interrogation of people with FASD is an interesting subject for investigation in its own right, independent of its providing a window into the problems cognitively impaired individuals have in perceiving and resisting manipulative ploys. That is because people with brain-based impairments are notoriously susceptible to giving false confessions (Kassin et al. 2009), and are unable to resist the deceptive questioning ploys by interrogators, the use of which have been authorized by the US Supreme Court (Skolnick 1985) and prominently promoted in interview training manuals (Imbau et al. 2011). The current paper, therefore, is intended to shed light on how the cognitive limitations of people with FASD contribute to the ease with which they can be deceptively manipulated in general, and also to shed light on how this tendency contributes in particular to their vulnerability to giving false, as well as truthful but involuntarily-given, statements in police interrogation situations.

In pursuing these purposes, we shall first describe all of the deception ploys used in a successful (in the sense of producing a breakdown of will to resist) interrogation of a young man with FASD; examine the appropriateness of using a "big lie" to break down his resistance to incriminating himself; and then explain why it is that FASD and related modes of impaired cognitive functioning made him especially vulnerable to having his "will overborne" by a persistent interrogator. In a final section, we discuss ethical and legal implications flowing out of an enhanced understanding of the deceivability and gullibility of people with FASD and related disorders.

A Listing of Ploys Used to Break the Will of a Person with FASD

The case used in this paper involves "Juan," a homeless young man of 20 who was diagnosed with Fetal Alcohol Syndrome at a specialized FAS clinic as a pre-adolescent. Because Juan's legal case is still wending its way through the legal system, we shall slightly change some facts and take various other pains to disguise his identity. The basic crime of which he was charged was statutory rape, although the more serious but in this case, less provable charge of forcible rape was also alleged. Juan was on probation for an earlier relatively minor property crime. He has no history of violence or crimes against persons. He had failed to fulfill some requirement of his probation, as people with FASD have a difficult time remembering meetings with probation officer, paying fines, giving urine samples, etc. He consequently probably expected to be rearrested. The arresting person, "Officer Truax," delivered Juan to the presence of "Detective Smith," without telling him the reason. In the following paragraphs, we shall list each of the ploys told, and the response of Juan to each ploy. We shall save an analysis of each ploy and how it intersected with Juan's impairments, for a later section.

False Friendliness Ploy

Officer Truax brought Juan into a room in the police station, and introduced him to Detective Smith, who he described as "a really nice guy." Truax offered to fetch Juan a soft drink and told him that it was the only time he had ever offered such a kindness to anyone brought in for questioning. Juan was wary but seemed somewhat reassured.

Concealing True Purpose Ploy

Detective Smith told Juan that he was picked up for a probation violation but that he also wanted to question him about another matter. Juan was never told the nature of this other matter, and was asked not a single question about the probation violation. Juan made no audible response other than "okay."

Ploy to Create an Incentive to Cooperate

Juan was told that he was technically under arrest for the probation violation. Detective Smith told Juan that after they cleared up the probation matter, he could not be released right away but would have to be brought to the jail before he could be released. Likely, this caused Juan to think that by cooperating he would go free, while refusing to cooperate would land him in jail. It also minimized the seriousness of the situation and caused Juan to think that he was not in any real danger. Juan said "I understand" and asked no questions, such as the one that would occur to most people: "If I am not here for the probation violation, then just what do you want to question me about?"

Ploy of Reading Miranda Rights Quickly and Without Real Probes

Detective Smith read Juan the Miranda warning in a very cursory and formulaic manner. After every sentence was read quickly, the detective said to Juan "Do you understand that?" and Juan replied reflexively with "yes." Juan was never asked to repeat it in his own words, or to demonstrate what it really meant, such as that he could stop talking at any moment and walk out, which he probably felt he could not do in this case as he was told he was under arrest for the probation violation warrant. This ploy was intended to comply with the law by reading the rights without giving Juan any encouragement to actually assert those rights.

Ploy of Asking Many Irrelevant Questions

The first 30 min of the interrogation consisted of a string of questions having nothing to do with the alleged criminal offence, which still was not divulged. They had to do with Juan's life and touched on such things as his parents, siblings, extended family, friendships, past addresses, current living arrangements, school history, past court experiences, etc. Often, Juan was asked for specific dates about past events, such as when he attended a particular school. The effect was to create almost a trance-like question and answer sequence, where Juan would reply to each question with a truthful response, and no question stood out as particularly challenging. Juan replied without delay to every question and demonstrated good recall about various details of his life and its history.

Ploy of Easing Indirectly into Dangerous Territory

Detective Smith started to ask Juan about his sexual history, and also about his use of alcohol and other intoxicants. When asking about his female friends, Smith wanted to know their ages, the dates and frequency of sexual encounters, and the locations. He also asked about Juan's male friends and one young man in particular, a slightly older person nicknamed "Ratman," who lived in an abandoned house. Ratman purchased alcohol for under-aged youths and had video-watching parties where various young people would lounge on mattresses, drink and sometimes have sex, which Ratman apparently liked to watch. Juan answered every one of these intensely personal questions without hesitation or apparent evasion, and did not once say "none of your business." The one time Juan stiffened a little was when Smith mentioned a girl named "Caroline" and asked if Juan had ever had sex with her. Juan quickly denied it, saying he considered her like a younger sister and would never have sex with her.

Ploy of Seizing on a Minor Inaccuracy

At least a half dozen young women with whom Juan had been intimate were discussed in the course of the interview. None of them were under-age, with the exception of Caroline. One of the girls, "Brenda," apparently was several months older than the age that Juan said she was. The detective told Juan that he was sure he knew Brenda's real age and that he was intentionally lying. The detective said to Juan "I know the answer to most of the questions I am asking you, and if you tell a falsehood I will know it and I will think poorly of you. If you want me to trust you and think you have a good character, it is important that you answer truthfully to every question I ask you." Juan said he understood.

Ploy of Telling a Lie About Incriminating Evidence

Detective Smith returned to the topic of Caroline, with several questions about the nature of their relationship. Juan mentioned that Caroline had a huge crush on him, and had several times asked him to sleep with her but that he had always resisted the temptation, both because of her age and the fact that he had promised her brother he would look out for her. Juan indicated that he thought of Caroline as a kind of younger sister and he made it a point to never have sex with family. He even broke off a relationship with another young woman when he discovered they may have been distantly related. Smith asked about a particular evening at Ratman's house when Juan admitted being drunk to the point of blacking out and waking up later lying next to Caroline. The detective asked Juan if he could have had sex with Caroline that night and he replied by saying he was certain he did not. Smith then asked Juan "well how then can you explain how your DNA came to be found inside Caroline's vagina when she was checked at a medical clinic?" Juan replied to this totally fabricated assertion by saying "I have no idea."

Ploy of Mixing Truth or Hypothetical Truth with Fiction

A key to causing an emotional breakdown in Juan came when Detective Smith asked him if had ever had a sexually transmitted disease. Juan replied that he found out he had an STD when one of his sexual partners told him she was pretty sure she had contracted it from him. Smith then truthfully told Juan that Caroline had contracted the same disease, and that was how Juan's name came to the attention of the police, as Caroline was asked by a nurse to reveal the names of her sexual partners. Detective Smith again told Juan that they knew he had sex with Caroline, as he had gotten the information from others at the party, which was not exactly true. Detective Smith then repeated the big lie about finding Juan's DNA inside Caroline's vagina. At this point, Juan started to sob and said that his practice was to pull out when he felt himself about to come, but he supposed it did not always work perfectly.

Ploy of Suggesting a Culpability-Minimizing Theory

Once Juan started to cry, it became likely that Detective Smith would be able to "reel him in." The key to this lay in offering Juan a hypothetical story that Smith believed to be true, and then getting Juan to confirm and elaborate on that story. Three motivational components to Smith's success in "closing the deal" were: (a) convincing Juan that he would feel better if he came clean, (b) telling him repeatedly that what he did was "not the worst thing in the world," "most of the people I have interviewed have done much worse things", etc., thus minimizing the seriousness of the crime and implying a confession would likely would not bring serious consequences, (c) implying that it would be in Juan's interest in terms of a lesser consequence if he was truthful, and (d) providing Juan with a plausible explanation for his conduct on the night in question that reduced Juan's feelings of shame and guilt for an act of sex with a young girl he had promised to protect and felt bad about.

This hypothesized story, which Juan agreed to and even elaborated on, involved: (a) Juan being so drunk that his self-control and even consciousness was lessened, (b) Caroline being the sexual aggressor in initiating the sex and in carrying it to fruition, and (c) once the sexual encounter started, Juan got too aroused to be able to stop. The one place where Juan showed some resistance was when Detective Smith suggested that Caroline changed her mind and tried to end the encounter. Juan initially said it did not happen, but then admitted she might have weakly said "no" a couple of times, but then changed her mind again. Detective Smith later cleverly came back to this theme by asking Juan if he had ever completed a sex act with a different girl after she said "no" and he replied "yes, her name was Barbara, but she changed her mind and said to go ahead." In asking this question, Detective Smith was less interested in pursuing another crime as he was in getting Juan to indirectly strengthen the validity of his earlier quasi-admission that the sex with Caroline might have been involuntary.

Cognitive Deficiencies that Place People with FASD at Risk in an Interrogation Situation

Although there are personality (e.g., non-assertiveness) and state (e.g., inebriation) factors which might contribute to a foolish or risk-unaware act (Greenspan 2009a, b) such as adopting an interrogators story and agreeing or even embellishing on it, our basic focus in this paper is on cognitive explanations. In this section, we approach the phenomenon of Juan's caving into Detective Smith sequentially in terms of several cognitive theories. In each section, we relate the theory under discussion with what is known about FASD and the cognitive impairments associated with the disorder.

Categorization Perspective on Interrogative Deception Vulnerability

Jerome Bruner, one of the principal architects of the cognitive revolution in psychology, has opined that cognition is mainly a matter of categorization (Bruner et al. 1956) or procedural narrative knowledge (Bruner 1986). By that, he meant placing an event, object or situation into a self-constructed category, whose meaning comes from its resemblance to other phenomena which share some defining features. In the case of a manipulative situation such as a police interrogation, the most adaptive categorization would be "situations where somebody is trying to get one to do or say something seriously harmful to one's interests." In this sense, many of the narrative-enhancing ploys used by Detective Smith were intended to make it more difficult for Juan to categorize the situation in that manner and thus get into a "fight-or-flight" mind-set. Juan lacked a procedural awareness of the story telling techniques being used on him as well. Examples included initial friendliness, minimizing or obscuring the true purpose of the occasion, engineering a conversational quality which steered away from red-flag topics, mixing true, hypothetically true and false statements (thus obscuring the motives and deceitfulness of the interrogator) and effort to create the impression that keeping the esteem of the detective, rather than staying out of jail, was the real purpose of the interaction.

A key to categorizing a situation as manipulative is recognizing that there are lies being told and that hypothetical statements are not always true. That is because lying about the manipulator's intent, about potential benefits to the target, about the facts of the manipulator's persuasive arguments, etc., are always a central feature of any manipulation. In that sense, a particularly confusing challenge in the interrogation was the mixing of true (e.g., "Caroline contracted the same STD that you had") with untrue (e.g., "your DA was found in Caroline's vagina") statements. This requires—in a manner analogous to a Piagetian conservation task–an ability to hold onto two conflicting realities, one untruthful statement such as the DNA ploy, in the face of many truthful statements which by their number are more salient and tend to create an illusion of truthfulness. For Bruner (1961), who early-on was heavily influenced by Piaget, the inability to conserve a critical reality e.g. a big lie, in the face of salient and thus reality-distorting artifices such as many irrelevant truths or half-truths is a reflection of overreliance on what he termed enactive (or action-based) or iconic (internalized image based) representational modes of thinking. Moreover, it is an indication of great limits in the use of symbolic, word and formula-based, declarative and procedural knowledge particularly at the level of meta-cognition. People who have developmental disabilities that are impacting their social thinking may be capable of categorization on a concrete symbolic level, but they will likely have increasing difficulties with functional categories and extreme problems with abstraction of categories on a hypothetical level. For instance, if an interrogator in a case similar to Juan's were to lay the groundwork for his story version of what "may or may not" have happened by detailing actions and images, he or she is progressing towards the use of concrete categorization and potentially functional purposive description of what may have happened as fact. At the more abstract symbolic level what may be factual is readily then classified as known verified fact. The procedure is seemingly transparently exploratory and then at once totally confirmatory.

Bruner (1961) posited three modes of representing reality: enactive (action-based), iconic (image-based) and symbolic (abstract). These are in some ways analogous to Piaget's pre-operational (enactive), concrete operational (iconic) and formal operational (symbolic) stages, both in the mode of problem-solving and in that both formulations involve an invariant stage sequence. However, unlike Piaget, for Bruner the stages are not tied to specific ages, and it is thus possible for an adult, especially one with a neurodevelopmental disorder such as FASD to be operating on a very primitive level, where reality is basically a set of discrete concrete actions perhaps also functional or purposive in nature that arc not approached on any meta-cognitive level. That appears to be exactly what happened in the interrogation session with Juan, where he could not step back and readily categorize the meaning of the situation beyond a concrete or functional level, let alone figure out or abstract what was happening in his present situation and what his options were in that situation. He became the main character in the confirmatory story.

Bruner (1986) more recently draws a distinction between narrative thinking and paradigmatic thinking. The latter entails truth-finding or verifiable procedures. It is paradigmatic in nature, logical and scientific. On the other hand, the narrative mode of thought seeks similarity to truth through story telling. It is a process of shared reality through mutual transactional experience and not necessarily through "learning" in the traditional sense of seeking truth. Narrative thought can be defined by changes in intention. It creates a world of possibilities rather than certainties. It introduces doubt while scientific thought utilizes falsification and verification to create certainties and establish truth. It engenders performances rather than verification of truth. Stories can be true, not true or partially true.

Theory of Mind Perspective on Interrogative Deception Vulnerability

Part of the ability to survive a manipulative interrogation depends on social perspective-taking, an important aspect of social intelligence (Greenspan and Driscoll 1997b). Today, perspective-taking is subsumed under the rubric of "theory of mind," and refers to the ability to put oneself in the shoes of another person and understand their motives as well as how they see you. Such ability has obvious implications for surviving an interrogation session, as one cannot act purposively or skillfully in a manipulative situation without being able to understand what the other person is up to. Theory of mind has been studied mainly as a possible explanation for the inability of people with autism to exhibit social reciprocity (Baron-Cohen 1995), but deficiencies in this ability have been found in people with FASD (Kilchenbaum et al. 2012; Rasmussen et al. 2009) and other neurodevelopmental disorders which share some of the same symptoms as found in people with Autism Spectrum Disorder.

Communicative Processing Perspective on Interrogative Deception Vulnerability

FASD is known to produce damage to the prefrontal cortex, and such damage, according to Zalla et al. (2014) is associated with deficits in "encoding and retrieving complex events, such as action and narrative knowledge…[and in] processing different aspects of narrative representations, such as the figurative moral meaning, syntactic features, and inference generation [of story processing]." Persons with frontal lobe lesions "…showed a pattern of deficit at an early stage of story comprehension that specifically involved the ability to reconstruct the sequential links among events and to extract inferential knowledge from the text during encoding" (Zalla et al. 2014). Individuals with FASD have impairments in their ability to analyze story narratives, and cannot elaborate on stories well (Thorne et al. 2007). An interrogation is basically a situation in which an interrogator develops a story and tries to convince a subject to buy into and even elaborate on that story. In the absence of an ability to semantically process a story, people with FASD and related brain disorders are very likely to confabulate (Fotopoulou 2008; Gibbard et al. 2003), that is to pretend that things happened in a way they may not have.

Depletion of Cognitive Resources Perspective on Interrogative Deception Vulnerability

It is not an accident that the most critical piece of deceptive information is usually introduced late in an interrogation, at a point when the subject is beginning to get tired. Dragging an interrogation out is, in fact, an explicit tactic used almost universally by interrogators (Gudjonsson 2003), in outlawed form involving prolonged sleep deprivation, but in routine form going on for several hours. Psychologist Roy Baumeister and colleagues (Baumeister 2001; Baumeister et al. 1998) have shown conclusively that exhaustion is a powerful force that depletes a person's cognitive resources, causing a drop-off in intellectual functioning, and a loss of willpower (Baumeister and Tierney 2011). Given the limited cognitive resources which people with FASD have to begin with, and a general lack of confidence and self-efficacy (Cicerone and Azulay 2006) in the ability of brain-impaired people to assert their will and prevail in a challenging situation, the fatigue induced in an interrogation is likely to have particular impact on ability of someone with FASD to stick to his or her story, regardless of whether or not it is true.

Executive Functioning Perspective on Interrogative Deception Vulnerability

Ability to persevere when cognitive or volitional resources are depleted, otherwise known as "willpower," i.e., the ability to persist in thoughts feelings and actions, can be considered an overriding self-regulatory process (McCloskey and Perkins 2012) which, while not strictly cognitive in itself, has a major impact on the ability to function intelligently as in making good decisions in a manipulative interrogation situation. Other more specifically self-regulatory processes, which tap cognitive style as well as cognitive competence, and help determine how intelligently one functions in the world fit under the broad umbrella construct known as "executive functioning." In line with the organization leadership analogy implicit in the term, executive functioning is a "cognitive control," "cognitive supervision" or "cognitive management" system which coordinates the implementation and use of whatever intellectual capacities one may have.

While executive functions are not the same as intelligence, they determine how efficiently one can use the intelligence one possesses to develop effective use of declarative and procedural knowledge. Thus, one can function more poorly in school and in life than one could predict from one's IQ if one has impairments in executive functioning. Such impairments typically are associated with brain lesions, especially in the frontal lobes of the brain, and are so universally found in people with FASD that deficient executive functioning is in fact one of the main diagnostic indicators for FASD. This explains why people with FASD, in spite of having average IQ's in the 80s, almost universally function adaptively as if their IQs were in the 60s and 70s and, thus, often qualify for services for people with intellectual disability (Greenspan, Brown & Edwards, this volume).

In a comprehensive review of the executive functioning literature, Brown and Connnor (2014) observe that cognitive problems in FASD demonstrate "ineffective use of *higher-level* executive skills" that require conscious effort, and "especially when tasks are complex and require rapid processing." The authors go on to state that "adolescents and adults with FAS appear to have significant difficulty with tasks requiring complex sustained attention and shifting attention…, suggesting that attention disturbance in FASD may become noticeable when tasks involve complexity." All of these skills are required to survive, or even to cognitively process, a manipulative interrogation session.

Do the Ends Justify the Means in Use of Deceptive Interrogation?

Joseph Buckley, a co-author of books promoting the "Reid technique" of interrogation used by Detective Smith, has argued that the use of big lies (such as explicitly referring to false evidence) should only be used as a last resort (2012), although it begs the question whether smaller lies pervade the technique. By last resort, one assumes he meant situations where absence of lying does not attain the desired confession, but that of course is based on the notion, which he denies, that the purpose of an interrogation is not to obtain the truth but rather to obtain support for what the interrogator believes to be the truth. The fact that Juan eventually caved and admitted to having had sex with Caroline will likely be seen by some readers as justification for using the "big lie" (that his DNA was found inside her body) that undoubtedly played the most critical role in causing the crumbling of his will to continue asserting his innocence. There are three counter-arguments that could be made: (a) not all confessions are truthful and the confession that Juan gave could have been false; (b) lying or hypothetical factualizing is difficult to justify for use with people experiencing social cognitive disability, regardless of the purpose, and use of a big lie as an interrogative technique is intrinsically unfair, regardless of whether or not it appears to have worked; and (c) people with FASD, and other neuro-developmental disorders, lack the skills to be socially aware of potentially deceptive situations, utilize insight or apply judgment to determine action steps in their best interests, assert their rights, or resist giving an involuntary and possibly invalid confession. Their self-determination, in other words, is overborne.

Possibility of False Confession

There is today a very substantial scientific literature on the phenomenon of false confession (Kassin and Gudjonsson 2004), and many publicized cases of long prison sentences as well as death sentences being overturned when DNA or other evidence showed that confessions used to obtain their convictions were invalid. A great many of these since-overturned confessions were made by people with neurodevelopmental disorders (Perske 2011). In the most authoritative review of the research literature on false confessions, Kassin and Gudjonsson (2004) indicate that the two major risk factors for false confession are age (children and adolescents are at very high risk of false confession) and intellectual impairment. See also Mclachlan and Roesch (2012) on the vulnerability of youths to giving false confessions. The same factor is responsible for false confession in both populations: incomplete brain development, which affects one's ability to make good decisions under conditions of pressure and ambiguity (Clare and Gudjonsson 2010). Elizabeth Loftus, the leading authority on implantation of false memories, believes that in many cases, especially with vulnerable subjects, the giving of a false confession is made possible by convincing a subject that the interrogator's account of the person's behavior actually is what occurred. According to Leo and Drizin (2010), once a partial confession is obtained, an interrogator will, as was the case with Detective Smith and Juan, "pressure the suspect to provide a post-admission narrative that they jointly shape."

Some Lie-Based Techniques Cross an Ethical Line

Deontological/religiously-motivated ethicists such as Kant (1781), declare the use of categorical imperatives to generate universal maxims. In other words, if actions can be performed under all circumstances and conditions then they are justifiable. For instance, "it is never morally acceptable to lie". Alternatively, consequentialist/ utilitarian ethicists such as John Stuart Mill (Strasser 1991) assert that there are circumstances—particularly in the service of a socially desirable end—where lying is acceptable. Police interrogation authors such as Buckley (2012) assert that lying in a police interrogation is acceptable as long as it facilitates the process of getting at the truth, but acknowledge that on occasion it may result in something that is not the truth. The 1969 U.S. Supreme Court ruling, in *Frazier v. Cupp* (394 U.S. 731, 1969) decided that it is generally acceptable for an interrogator to use lies as an interrogative tactic, but within certain limits. The limit is invoked when an interrogator does something that "shocks the conscience of the community." An example of such a line-crossing would be telling a subject that confessing is the only way to keep her child from being taken away from her custody. Another example would be manufacturing evidence, such as a phony typed statement from an alleged confederate or a phony lab report on blood evidence tying the suspect to the crime. Courts have ruled that such a step is out of bounds, although the reason seems to be as much concern over the possibility that this phony evidence could inadvertently make its way into a trial record and be used years later in an appeal as if it were real, as it is over the likelihood that such evidence is extremely coercive. However, while phony reports about evidence or eyewitness statements are improper, it is acceptable to falsely make reference to the existence of such incriminating evidence. In the case of Juan, that would translate to a "stop" sign preventing Detective Smith showing Juan a phony lab report establishing his DNA being found inside Caroline, but a "proceed" sign for the tactic of claiming that such a report exists. Excuse us if we do not see much difference, especially when it is a ploy used against a guileless, non-reflective and naively trusting brain-impaired person such as Juan.

The basis of moral philosophy, according to Rawls (Nagel 1973) is justice, which he defines as fairness, or an absence of discrimination which advances the interests of one person; in other words, every person should be treated equally. Given that people with FASD are at a distinct disadvantage in even claiming, let alone exercising, their right to avoid self-incrimination in the face of manipulation, one can conclude that the use of interrogative ploys, especially when they involve introduction of grossly false facts, is inherently unfair or manipulation of vulnerability to suggestibility or deceivability. The most important skill needed to survive a manipulative interrogation, or any manipulative transaction for that matter, is the ability to recognize the event for the dangerous game that it is and if possible to walk or better yet run away without playing it. The US Constitution guarantees every citizen as well as (at least in theory) non-citizen resident the right to do just that, but the people most in need of protection from interrogation (those who are poor, uninformed, brain-damaged or otherwise overly trusting) generally fail to exercise that right. Once a person agrees to be interrogated, then he or she has entered into a dangerous place that requires some degree of cognitive wherewithal to keep from being led down the garden path to self-incrimination, regardless of whether or not the person has actually committed the act which he or she is suspected of having committed.

It is likely in many instances, such as the example reviewed in this paper, that the process of interrogation is a narrative story-telling technique which enlists its subject to perform his or her assigned role rather than a methodology which seeks truth through the use of the scientific method of verification and falsification. Is interrogation a means to give possible meaning to experience and/or possibly change someone's intentional view of what happened or a tool to derive real factual information? Oftentimes it is the former, namely a strategy to confirm a story as fact rather than the actual fiction it may be. In the end, people with a developmental disorder such as FASD are not equipped with the meta-cognitive and abstract thinking skills required to assess the actual situation they are in. They become caught up in the story as if it is an exercise in fact-finding rather than in story-confirming. Rather than engage in a mutual transaction, people with FASD such as Juan subjugate independent thinking to confirmation of a mutual experience carefully delineated by their interrogator/story teller.

A question to ask is: "Are interrogators story tellers with a plausible but potentially flawed story, enlisting their FASD or otherwise brain-impaired subjects into helping them factualize their hypotheses?" The answer appears in some contexts, and with some people, to be "yes."

References

Baron-Cohen, S. 1995. *Mindblindness: An essay on autism and theory of mind.* Cambridge: MIT Press.

Baumeister, R.F. 2001. Ego depletion, the executive function, and self-control: An energy model of the self in personality. In *Personality psychology in the workplace*, ed. B.R. Roberts, and R. Hogan, 299–316. Washington, DC: American Psychological Association.

Baumeister, R.F., and J. Tierney. 2011. *Willpower: Rediscovering the greatest human strength.* New York: Penguin group.

Baumeister, R.F., E. Bratlavsky, M. Muraven, and D.M. Tice. 1998. Ego depletion: is the self a limited resource? *Journal of Personality and Social Psychology* 74(5): 1252–1265.

Brown, N.N. and P. Connor. 2014. Impact of executive dysfunction on learning in children with fetal alcohol spectrum disorders (FASD). Seattle, Washington: Unpublished manuscript.

Brown, N.N., Gudjonsson, G., and Connor, P. 2011. Suggestibility and fetal alcohol spectrum disorders (FASD): I'll tell you anything you want to hear. *Journal of Psychiatry and Law* 39(1): 39–71.

Bruner, J.S.W., J.J. Goodnow, and G.A. Austin. 1956. *A study of thinking*. New York: Wiley.

Bruner, J.S.W. 1961. *The process of education*. Cambridge: Harvard University Press.

Bruner, J.S.W. 1986. *Actual minds, possible worlds*. Cambridge: Harvard University Press.

Buckley, J.P. 2012. The use of deception in interrogations. Paper presented as part of the Temple Law Review symposium on "deception and other strategic tactics used during police interrogations". www.temple.edu/lawreview/2012/11/14/the-use-of-deception-in-interrogation. Accessed 9 Nov 2012.

Cicerone, K.D., and J. Azulay. 2006. Perceived self-efficacy nd quality of life in traumatic brain injury. *Journal of Head Trauma Rehabilitation* 21(5): 408–409.

Clare, I.C.H., and G.H. Gudjonsson. 2010. The vulnerability of suspects with intellectual disabilities during police interviews: A review and experimental study of decision-making. *Mental Handicap Research* 8(2): 110–128.

Edwards, W., and S. Greenspan. 2011. Adaptive behavior and fetal alcohol spectrum disorders. *Journal of Psychiatry and Law.* 39(4): 419–447.

Fast, D., and J. Conry. 2006. The challenge of fetal alcohol syndrome in the criminal legal system. *Addiction Biology* 9(2): 161–166.

Foutopoulou, A. 2008. False selves in neuropsychological rehabilitation: The challenge of confabulation. *Neuropsychological Rehabilitation: An International Journal* 18(5–6): 541–565.

Greenspan, S. 2009a. Foolish action in adults with Intellectual Disabilities: The forgotten problem of risk-unawareness. In *International Review of Research in Mental Retardation*, ed. L.M. Glidden. New York: Elsevier.

Greenspan, S., and J. Driscoll. 1997a. The role of intelligence in a broad model of personal competence. In *Contemporary intellectual assessment: Theories, tests and issues*, ed. D.P. Flanagan, J. Genshaft, and P. Harrison. New York: Guilford.

Greenspan, S., and J. Driscoll. 1997b. Social intelligence and developmental disorder: Mental retardation, learning disabilities and autism. In *Ellis' handbook of mental deficiency*, ed. W.F. MacLean, Jr. 3rd edn. Mahwah, NJ: Erlbaum.

Greenspan, S., and Woods, G.W. 2014. Intellectual Disability as a disorder of reasoning and judgment: The gradual move away from IQ ceilings. *Current Opinion in Psychiatry* 27: 110–116.

Gibbard, W.B., P. Wass, and M.E. Clarke. 2003. The neuropsychogical implications of prenatal alcohol exposure. *Canadian Child and Adolescent Psychiatry Review* 12(3): 72–76.

Greenspan, S. (2009b). *Annals of Gullibility*. Westport: Praeger.

Greenspan, S., N.N. Brown, and W. Edwards. (this volume). FASD and the concept of "intellectual disability equivalence." In *Ethical and legal issues in fetal alcohol spectrum disorder*, eds. M. Nelson and M. Trussler. Amsterdam: Springer.

Gudjonsson, G.H. 2003. *The psychology of interrogations and confessions: A handbook.* New York: Wiley.

Imbau, F.E., J.E. Reid, J.P. Buckely, and B.C. Byrne. 2011. *Criminal interrogation and confessions*. Byurlington: Jones & Bartlett Learning.

Kant. E. 1781. *Critique of pure reason.* Indianapolis: Hackett Publishing.

Kassin, S.M., S.A. Drizin, T. Grisso, G.H. Gudjonsson, R.A. Leo, and A.D. Redlich. 2009. Police-induced confessions: Risk factors and recommendations. *Law and human behavior*, (access at http://ssm.com/abstract=1483878).

Kassin, S.M., and G. Gudjonsson. 2004. The psychology of confessions: A review of the literature and issues. *Psychological Science in the Public Interest* 5(2): 33–67. (A supplement of Psychological Science).

Kilchenbaum, N.M., K. Thomas, C.D. Molteno, N.C. Dodge, E.M. Meintjes, J.L. Jacobson, and S.W. Jacobson. 2012. Fetal alcohol spectrum disorders and theory of mind: An exploratory study of South African children. *Alcoholism, Clinical and Experimental Research* 36(205A): 0145–06008.

Leo, R.A., and S.A. Drizin. 2010. The three errors: Pathways to false confession and wrongful conviction. In *Police interrogations and false confession: current research, practices and policy recommendations*, eds. G.D. lassiter and C. Meissner. Washington, DC: American Psychological Association.

Loftus, E.F. (2004). The devil in confessions. Editorial in S.M. Kassin & G. Gudjonsson (2004). The psychology of confessions: A review of the literture and issues. *Psychological Science in the Public Interest*, *5* (2). (A supplement of Psychological Science).

Mclachlan, K., and R. Roesch. (2012). Police interrogation experiences and false confessions: Examining the vulnerabilities of juveniles with FASD. Paper presented at the annual meeting of the American Psychology-Law Society. San Juan, Puerto Rico. Accessed 14 Mar 2012.

McCloskey, G., and L.A. Perkins. 2012. *Essentials of executive functions assessment*. New York: Wiley.

Nagel, T. 1973. Rawls on justice. *The Philosophical Review* 82(2): 220–234.

Perske, R. 2011. Perske's list: False confessions from 75 persons with Intellectual Disability. *Intellectual and Developmental Disabilities* 49(5): 468–479.

Rasmussen, C., K. Wyper, and C. Talwar. 2009. The relation between theory of mind and executive functions in children wirh fetal alcohol spectrum disorders. *Canadian Journal of Clinical Pharmacology.* 16(2): e370–e380.

Skolnick, J.H. (1985). Deception by police. In *Moral Issues in Police Work*, eds. Elliston and Feldberg, 75–98. Savage, MD: Rowman & Littlefield.

Strasser, M.P. 1991. *The moral philosphy of John Stuart Mill*. Lakefield, NH: Longwood Academic.

Streissguth, A., and J. Kanter. 2002. *The challenge in fetal alcohol syndrome: Overcoming secondary disabilities*. Seattle: University of Washington Press.

Thorne, J.C., T.E. Coggins, O.H. Carmichael, and S.J. Astley. 2007. Exploring the utility of narrative analysis in diagnostic decision making: Picture-bound reference, elaboration, and fetal alcohol spectrum disorders. *Journal of Speech, Language and Hearing Research* 50(2): 459–474.

Zalla, T., M. Phipps, and J. Grafman. 2014. Story processing in patients with damage to the prefrontal cortex. *Cortex* 38(2): 215–231.

Zigler, E. (2013). Intelligence and cognitive exceptionality: The motivational perspective. In *Cognition and motivation: Forging an interdisciplinary perspective* (pp. 433–449). ed. S. Kreitler. New York: Cambridge University Press.

Neurodevelopmental Disorder Associated with Prenatal Alcohol Exposure: Consumer Protection and the Industry's Duty to Warn

Susan D. Rich and Laura J. Riley

Abstract In this chapter, we posit that manufacturers and distributors of alcohol (i.e., the alcohol industry) should participate in and be responsible for the health and human costs of Neurodevelopmental Disorder associated with Prenatal Alcohol Exposure (ND-PAE). Alcohol is a food and drug by federal definition in the United States and should be under FDA regulation, like pharmaceuticals and tobacco. More precise labeling and advertising, to include the 2005 revised U.S. Surgeon General's advisory (CDC 2005), will promote universal prevention through preconception approaches. Regulation and taxation of alcohol would finance targeted screening, intervention and treatment services.

Nomenclature Disclaimer: With the publication of the Diagnostic and Statistical Manual 5th Edition (American Psychiatric Association, 2013), the category of "Other Specified Neurodevelopmental Disorder" includes "Neurodevelopmental Disorder associated with Prenatal Alcohol Exposure (ND-PAE)," with proposed diagnostic criteria listed in Section 3 under "Conditions for Further Study." Because the topic presented in this paper has little to do with nomenclature and more to do with ethical responsibility versus legal obligation, we are opting to refer to all individuals affected by maternal alcohol use as ND-PAE and defer readers to the diagnostic criteria in DSM-5. To avoid confusion, we will use ND-PAE throughout this chapter in lieu of FASD or other prior nomenclature, with the exception of Fetal Alcohol Syndrome (FAS), where necessary.

S.D. Rich (✉)
Potomac, Maryland, USA
e-mail: dr.sdrich@gmail.com

L.J. Riley
Los Angeles, California, USA
e-mail: laura.riley.2010@lawmail.usc.edu

M. Nelson and M. Trussler (eds.), *Fetal Alcohol Spectrum Disorders in Adults: Ethical and Legal Perspectives*, International Library of Ethics, Law, and the New Medicine 63, DOI 10.1007/978-3-319-20866-4_3

Why Prevent Neurodevelopmental Disorder Associated with Prenatal Alcohol Exposure?

There is no denying over 40 years of evidence that alcohol is a potent neurodevelopmental teratogen and may be the most common cause of non-inherited intellectual disability (Merck Manual for Healthcare Professionals, online). With prevalence estimates of 2–5 % of school aged American children (May et al. 2014), ND-PAE is at least as concerning as HIV/AIDS or autism, each individually affecting less than 1 % of the U.S. population (CDC 2013a, b). The following statistical comparison provides insight into the magnitude of various neurodevelopmental conditions:

1. Neurodevelopmental Disorder associated with Prenatal Alcohol Exposure (ND-PAE): "CDC studies have shown that 0.2–1.5 cases of Fetal Alcohol Syndrome (FAS) occur for every 1,000 live births in certain areas of the United States. Other studies using different methods have estimated the rate of FAS at 0.5–2.0 cases per 1,000 live births. Scientists believe that there are at least three times as many cases of FASDs as FAS." (http://www.cdc.gov/ncbddd/fasd/data.html) Worldwide research puts ND-PAE as high as 2–5 % of school aged children (May et al. 2009, 2014).
2. Autism: "About 1 in 88 children [just over 1 %] (11.3 per 1000 in 2008) has been identified with an autism spectrum disorder (ASD), according to estimates from CDC's Autism and Developmental Disabilities Monitoring (ADDM) Network." (http://www.cdc.gov/ncbddd/autism/data.html)
3. Cerebral Palsy (CP): "prevalence varied by site, ranging from 2.9 per 1,000 8-year-olds in Wisconsin to 3.8 per 1,000 8-year-olds in Georgia. The average prevalence of CP across the four sites was approximately 3.3 per 1,000 or 1 in 303 8-year-old children in the United States." (http://www.cdc.gov/ncbddd/cp/data.html)
4. Down's Syndrome: each year about 6,000 babies (about 1 of every 691 babies) in the United States are born with Down syndrome. (http://www.cdc.gov/ncbddd/birthdefects/downsyndrome.html)
5. Spina Bifida: "…each year, about 1,500 babies are born with spina bifida." (http://www.cdc.gov/ncbddd/spinabifida/data.html)

The Substance Abuse and Mental Health Services Administration (SAMHSA) FASD Center for Excellence reports prevention to be "cost effective." An ND-PAE birth carries lifetime health costs of $860,000, which can be as high as $4.2 million (Lupton et al. 2004). The U.S. spends at least $5.4 billion per year on services for individuals with Fetal Alcohol Syndrome, or FAS (Harwood 2003). This includes neonatal care, management of developmental delays and birth defects, special education, a range of developmental disabilities services, mental and general health care costs, and supported living costs. These amounts do not include criminal justice services or substance abuse treatment for the 60–80 % of individuals with ND-PAE who end up incarcerated by age 18 and/or with substance use disorders (Streissguth and O'Malley 2000). Since FAS accounts for only 10–20 % of all ND-PAE cases, the extrapolated estimate for ND-PAE would be $21.6 billion.

Legislation to Regulate and Accurately Label Alcohol

The Pure Food and Drug Act in 1906 allowed regulation of foods, pharmaceuticals, and beverages. A 35 year old debate between the Food and Drug Administration (FDA) and the Bureau of Alcohol, Tobacco and Firearms (ATF) has focused on which agency governs the labeling of beverages containing alcohol. According to the United States Code of Federal Regulations, by definition, alcohol is both a food and a drug, underscoring the government's duty to warn and responsibility to protect the public from harm. The Federal Food, Drug, and Cosmetic Act (FFDCA), §201. [301] defines "food" as:

(1) articles used for food or drink for man or other animals,
(2) chewing gum, and
(3) articles used for components of any such article.

The FFDCA does not make an exception for or separately define alcohol—an article used for drink for man. Alcohol has been used for centuries as a beverage made by fermenting (wine) grapes, other berries, and fruits; by brewing (beer) or distilling (liquor) hops, barley, and other grains. In a hearing before a Congressional Subcommittee on Alcoholism and Drug Abuse of the Committee on Human Resources of the 95th Congress (January 31, 1978), Commissioner Kennedy testified,

> Since 1938 when the Federal Food, Drug, and Cosmetic Act was enacted, FDA has always considered alcoholic beverages to fall within the definition of 'food' and has taken enforcement actions against such products on charges of adulteration and, in at least one case, for misbranding.

The Food Packaging and Labeling Act (FPLA) of 1967 gave authority to the Federal Trade Commission and the FDA over labeling of "consumer commodities" for net contents, identity of commodity, and name and place of business of the product's manufacturer, packer, or distributor. Additional regulations in the act "prevent consumer deception (or to facilitate value comparisons) with respect to descriptions of ingredients, slack fill of packages, lower price labeling, or characterization of package sizes" (FPLA 1967). Further, the Nutrition Labeling and Education Act (NLEA) of 1990 provides FDA with specific authority to require nutrition labeling of most foods regulated by the Agency; and to require that all nutrient content claims (i.e., 'high fiber', 'low fat', etc.) and health claims be consistent with agency regulations. Essentially, the "truth in labeling law" of 1990 now means that if orange juice is not fresh squeezed, the label cannot read "fresh squeezed."

Drugs are defined in the same section of the FFDCA as:

(A) articles recognized in the official United States Pharmacopoeia, official Homoeopathic Pharmacopoeia of the United States, or official National Formulary, or any supplement to any of them; and
(B) articles intended for use in the diagnosis, cure, mitigation, treatment, or prevention of disease in man or other animals; and
(C) articles (other than food) intended to affect the structure or any function of the body of man or other animals; and
(D) articles intended for use as a component of any article specified in clause (A), (B), or (C). A food or dietary supplement for which a claim, subject to Sections 403(r)(1)(B) and 403(r)(3) of this title or Sections 403(r)(1)(B) and 403(r)(5)(D) of this title, is made in accordance with the requirements of Section 403(r) of this title is not a drug solely because the label or the labeling contains such a claim. A food, dietary ingredient, or dietary supplement for which a truthful and not misleading statement is made in accordance with Section 403(r)(6) of this title is not a drug under clause (C) solely because the label or the labeling contains such a statement.

As a drug, alcohol is used as an ingredient in a variety of over-the-counter and prescription medications, including cough syrups, sleeping medicine, and others. Despite its medicinal purposes, it has a propensity to be habit forming yet is not included among "Habit Forming Substances" listed in §352, *Misbranded Drugs and Devices* of the FFDCA. The label, according to the regulation, must bear *the name and quantity or proportion of such substance or derivative and in juxtaposition therewith the statement 'Warning - May be habit forming.'* Following the "Tylenol scare" of the 1980s, the FDA began mandating sealed safety containers for over-the-counter medications (Rehak 2002). The debate about FDA guidelines for risks and benefits of pharmaceutical products and the protection of the public from harm is an ongoing concern of Federal regulatory agencies (IOM 2014).

Though tobacco is neither a food nor a drug, the Family Smoking Prevention and Tobacco Control (FSPTC) Act (signed into law by President Barack Obama on June 22, 2009) gave the FDA the ability to regulate tobacco products (FSPTC 1996). Ten years earlier, the Federal government successfully sued one of the largest manufacturers of tobacco, Phillip Morris, for fraudulent and unlawful conduct and reimbursement of tobacco-related medical expenses:

> Judge Kessler explains that the evidence shows that the Defendants knew for fifty years or more that cigarette smoking caused disease, but repeatedly denied that smoking caused adverse health effects. Judge Kessler describes the Defendants' efforts during this time to attack and discredit the scientific evidence of a link between cigarette smoking and disease (Freiberg 2006).

For 50 years leading up to the FSPTC Act, the public health, medical, and legal campaign against societal harm due to tobacco products led to wider restrictions on smokers, eventually even bans on smoking in public places. The public outcry from resentful tobacco consumers was met with a larger demand for smoke-free restaurants, shopping areas, and recreational centers.

Surgeon General's Warning and Alcohol Labeling

In the United States, the Alcohol and Tobacco Tax and Trade Bureau is responsible for the regulation of alcohol labeling and advertising, including a government health warning statement to appear on all alcohol beverages. However, this responsibility has not seemed to impose any real legal, enforceable duty to date. In 1981, the initial US Surgeon General's Advisory on alcohol use in pregnancy recommended that pregnant women avoid exposure to alcohol (CDC 1981). In 1989, following attempted legislation in the United States, alcoholic beverages began carrying a briefer version of the original US Surgeon General's Warning.

> The first part of the warning reads: "Government Warning: According to the Surgeon General, women should not drink alcoholic beverages during pregnancy because of the risk of birth defects." Various researchers have examined exposure to the warning label and its impact on drinking during pregnancy. In general, the studies concluded that although awareness of the alcohol beverage warning label increased after the implementation of the law, this awareness has attenuated over time. Furthermore, the warning label's impact on drinking during pregnancy has been modest (Hankin 2002).

However, decreased publicity and public health interest in the issue led to a dramatic (4-fold) increase in alcohol consumption by pregnant women from 1991–95 (CDC 1997). A follow up study to the original study in 2001–05 showed a drop overall in the numbers of pregnant women drinking; however, the rate of moderate to heavy alcohol use did not go down (Denny et al. 2009). In 2005 the Office of the Surgeon General released an updated advisory on alcohol use in pregnancy stating that:

> We do not know what, if any, amount of alcohol is safe. But we do know that the risk of a baby being born with any of the fetal alcohol spectrum disorders increases with the amount of alcohol a pregnant woman drinks, as does the likely severity of the condition. And when a pregnant woman drinks alcohol, so does her baby. Therefore, it's in the child's best interest for a pregnant woman to simply not drink alcohol (CDC 2005).

The advisory also recommended that all childbearing age women be adequately informed about the dangers of alcohol use before (not just during) pregnancy and implied that women using alcohol during the reproductive years should use reliable contraception and plan their pregnancies carefully to avoid unintentional prenatal alcohol exposure (HHS 2005). The advisory suggests that any amount of alcohol consumed by a mother during pregnancy leaves her at risk for causing undue harm in her child.

Other government actions include an August 5, 2010 resolution passed in the United States Senate designating September 9, 2010 "National Fetal Alcohol Spectrum Disorders Awareness Day" (S.Res.612, 111th Cong. 2010). The Healthy People 2020 Report sets a target of 98.3 % of pregnant females aged 15–44 years abstaining from alcohol in the past 30 days (this would be 10 % improvement from 89.4 % in 2007–08) (Healthy People 2020). At the same time, the 2010 Dietary Guidelines Report somewhat contradicts the Surgeon General's warning:

> Heavy drinking during pregnancy can produce a range of behavioral and psychosocial problems, malformations, and cognitive dysfunction in the offspring (NIAAA 2003; NIAAA 2009). Even daily moderate drinking during pregnancy, especially in the first few months or before the pregnancy is recognized, may have behavioral or neurocognitive consequences in the offspring. This effect may be from the direct toxic effects of alcohol or its metabolites or the effect that alcohol has on suppressing folate status—a known determinant of neural tube defects. …heavy drinking during pregnancy can cause serious health consequences to the unborn infant. Even daily moderate alcohol consumption among pregnant women may not be without risk and should be avoided (DGAC 2010).

This comprehensive government report (the Section on Alcohol is 24 pages in length) does not clearly state the dangers of *any level* of prenatal alcohol consumption. Some pregnant women may conclude that it is safe to drink moderately, or to simply avoid episodes of heavy (binge) drinking.

Given that alcohol could technically qualify as both a food and a drug, its labeling should meet the more rigorous standards than those set out specifically for alcohol now. Alcoholic Beverage Labeling is currently addressed in the United States Code of Federal Regulations. Section 213 states Congressional findings that *the American public should be informed about the health hazards that may result from the consumption or abuse of alcoholic beverages.* And while a more detailed labeling requirement section exists, it only requires the same statement that was written in 1988 be located in a conspicuous and prominent location. At the same time, the warning label on most alcoholic beverages is printed in small font with a color that makes it less conspicuous. There is neither a requirement that the text of the label be re-examined to reflect updates in the medical field nor that it comply with alcohol's FFDCA definition as a food, or a drug.

There was some movement to improve labeling for alcohol in the form of a Bill in 1997 (Truth in Alcohol Beverage Labeling Act 1997). It made suggestions like disclosing the number of drinks each beverage contains rounded to the nearest quarter drink. Notably, it did not include any additional warning for pregnant women or those of a childbearing age. In any event, the Bill did not move forward from the House subcommittee on Health and Environment, leaving the 1988 Alcoholic Beverage Labeling Act with no updates. Nor has anything been updated since the 1999 statement by the ATF that:

> The hazards of heavy alcohol consumption are clear and substantial and have far-reaching health and social consequences. Alcohol is the second leading cause of preventable deaths in the United States as well as most industrialized countries, second only to cigarette smoking. In pregnant women, heavy alcohol consumption is associated with fetal alcohol syndrome (CFR Parts 4, 5, 7; Notice No. 884).

This statement was included in a report proposing to amend regulations to prohibit the appearance on labels of any statement making a substantive claim regarding health *benefits* associated with the consumption of alcohol. Of course it is important not to advertise false health benefits of alcohol, but this report heads in the wrong direction by attempting to avoid a regression in the labeling of alcohol as opposed to adding further cautionary notices to the label. Another government agency, also does not give proper attention to the harmful effect of alcohol for pregnant women.

As written, the current message focuses only on the pregnant woman without warning populations at highest risk of causing unintentional damage before they know they are pregnant: social and binge drinkers. It is also completely lacking for directions for use (e.g., to be consumed with meals or on a full stomach in no more than one ounce of absolute alcohol per 30 min). There is no warning for the chronic alcoholic, persons with diabetes, hepatitis, or other contraindicated conditions, as well as no warning for childbearing age consumers at risk of pregnancy. Likewise, there is no warning against use by lactating mothers, children, or adolescents—populations which risk disproportionate damage to the developing central nervous system. Finally, there are no guidelines for unsafe doses or duration of administration (i.e., not to exceed X amount of alcohol within a given period of time for an average size adult) as is required for medically-beneficial drugs.

To that end, as both a food and a drug, alcohol is mislabeled in that it fails to reveal a variety of facts about the consequences which may result from its customary use. Notwithstanding the devastating sequelae (intellectual disability and physical birth defects) associated with heavy drinking throughout pregnancy, the warning current warning label fails to warn social drinkers: (1) that *a majority of birth defects occur in the first 8 weeks of pregnancy* (before women know they are pregnant); (2) that *moderate amounts of alcohol* (five to seven or more drinks per week) consumed throughout pregnancy can cause attention deficits and hyperactivity, learning disabilities, problem solving dysfunction, and other minimal brain damage described as *neurodevelopmental disorders*; (3) that *reliable contraception should be utilized* to prevent these problems.

The Alcohol Industry's Duty to Warn

Like the tobacco and pharmaceutical industries, alcohol manufacturers and distributors have a responsibility for the impact of their product on human health (i.e., the loss of intellectual potential of children exposed prenatally). One way to reduce and/or account for the financial and human costs of ND-PAE is to impose a greater responsibility through government regulation and taxation of the alcohol industry. This approach is similar to consumer protection of truth in labeling laws and oversight of the tobacco and pharmaceutical industries by the FDA. Government agencies and professional medical organizations can use the "ND-PAE tax" funds to appropriately educate and intervene through screening, targeted information dissemination, and referral to services.

Understanding the complex etiology of ND-PAE and the difficulty reaching all consumers uniformly, we suggest that marketing and packaging of alcoholic beverages should include the following elements from the revised 2005 Surgeon General's Warning (CDC 2005):

> WARNING: **AVOID PREGNANCY** - Maternal alcohol use is the leading known cause of intellectual disability and preventable birth defects. A majority of physical birth defects and serious cognitive problems occur when drinking alcohol as early as the third week

after conception. These problems can include: learning disabilities, memory and problem solving difficulties, attention deficits, mood issues, hyperactivity and other neurodevelopmental disorders. There is no safe amount or time to use alcohol in pregnancy. Sexually active childbearing age alcohol consumers should either **avoid pregnancy** or not use alcohol if pregnant or planning a pregnancy.

Summary

ND-PAE is preventable, and unfortunately still very prevalent. Clearly, alcohol manufacturers carry a duty to inform the public about the risk of alcohol use before pregnancy recognition (not just during pregnancy), given the likelihood of unplanned pregnancies in childbearing age alcohol users. Taxing the alcohol industry will fund grass roots local and state partnerships between condom manufacturers, medical associations, and nonprofits directed at ND-PAE awareness, such as the National Organization on Fetal Alcohol Syndrome (NOFAS) and Planned Parenthood. Recognizing that information alone does not change behavior, we believe this approach will at least hold the industry accountable for the lifelong physical and neurodevelopmental problems caused by prenatal alcohol exposure.

References

American Heart Association. 2014. Alcohol and heart health. http://www.heart.org/HEARTORG/GettingHealthy/NutritionCenter/HealthyEating/Alcohol-and-Heart-Health_UCM_305173_Article.jsp.

American association of pediatrics. Joint call to action on alcohol and pregnancy: Eliminating the risk for fetal alcohol spectrum disorders. http://www.aap.org/en-us/advocacy-and-policy/aap-health-initiatives/fetal-alcohol-spectrum-disorders-toolkit/Pages/Joint-Call-to-Action-on-Alcohol-and-Pregnancy.aspx#sthash.uEDjb3cO.dpuf.

Baby center, How can I safely have an occasional drink? Expert advice. http://www.babycenter.com/0_alcohol-and-nursing-moms_3547.bc.

Centers for disease control and prevention (CDC). 1981. Surgeon general's advisory on alcohol and pregnancy. *FDA Drug Bulletin* 11(2): 9–10.

Centers for disease control and prevention (CDC). 1997. Alcohol consumption among pregnant and childbearing-aged women—United States, 1991 and 1995. *Morbidity and Mortality Weekly* 46(16): 346–350.

Centers for disease control and prevention (CDC). 2005. Notice to readers: Surgeon general's advisory on alcohol use in pregnancy. *Morbidity and Mortality Weekly* 54(09): 229.

Centers for disease control and prevention (CDC). 2013a. Vital signs: Binge drinking among women and high school girls—United States, 2011. *Morbidity and Mortality Weekly* 62(01): 9–13.

Centers for disease control and prevention (CDC). 2013b. Monitoring selected national HIV prevention and care objectives by using HIV surveillance data—United States and 6 U.S. dependent areas—2011. *HIV Surveillance Supplemental Report* 2013;18(No. 5). Published October 2013.

Denny, C.H., J. Tsai, R.L. Floyd, and P.P. Green. 2009. Alcohol use among pregnant and non-pregnant women of childbearing age—United States, 1991–2005. *Morbidity and Mortality Weekly Report* 58(19): 529–532.

Department of the Treasury, 27 CFR Parts 4, 5, and 7, Bureau of alcohol, tobacco and firearms [Notice No. 884' *Health claims and other health-related statements in the labeling and advertising of alcohol beverages*. http://www.atf.gov/regulations-rulings/rulemakings/notices/notice-884.pdf.

Fair Packaging and Labeling Act (FPLA). 15 U.S.C. §§ 1451–1461: 16 C.F.R. Parts 500, 501, 502, 503.

Freiberg, M. 2006. Tobacco control legal consortium. The verdict is in: Findings from United States v. Philip Morris, ed. K. Cork, and M. Mahoney. The hazards of smoking. http://publichealthlawcenter.org/sites/default/files/resources/tclc-verdict-is-in.pdf.

Hankin, J.R. 2002. Fetal alcohol syndrome prevention research. National Institute on Alcohol Abuse and Alcoholism. (http://pubs.niaaa.nih.gov/publications/arh26-1/58-65.htm).

Harwood, H. 2003. The Lewin group, economic costs of fetal alcohol syndrome. Fairfax, Virginia; SAMHSA's FASD Center for Excellence website. http://fasdcenter.samhsa.gov/publications/cost.aspx.

Healthy people 2020, Maternal infant and child health. Increase abstinence from alcohol among pregnant women. www.HealthyPeople.gov. MICH-11.1. http://www.healthypeople.gov/2020/topics-objectives/topic/maternal-infant-and-child-health/objectives.

House of Representatives proposed bill 1977 (H.R. 1977), introduced by Rep. Joseph P. Kennedy, II at the 105th Congress, 1st Session, 19 June 1997.

Institute of Medicine (IOM). 2014. Characterizing and communicating uncertainty in the assessment of benefits and risks of pharmaceutical products, 12–13 Feb 2014.

Lupton, C., L. Burd, and R. Harwood. 2004. Cost of fetal alcohol spectrum disorders. *American Journal of Medical Genetics Part C: Seminars in Medical Genetics* 127C(1): 42–50.

May, P.A., J.P. Gossage, W.O. Kalberg, L.K. Robinson, D. Buckley, M. Manning, and H.E. Hoyme. 2009. Prevalence and epidemiologic characteristics of FASD from various research methods with an emphasis on recent in-school studies. *Developmental Disabilities Research Reviews* 15(1): 176–192.

May, P.A., A. Baete, J. Russo, A.J. Elliott, J. Blankenship, W.O. Kalberg, D. Buckley, M. Brooks, J. Hasken, O. Abdul-Rahman, M.P. Adam, L.K. Robinson, M. Manning, and H. Eugene Hoyme. 2014. Prevalence and characteristics of fetal alcohol spectrum disorders. *Pediatrics* 134:5 855–866. Published ahead of print October 27, 2014.

Merck pharmaceuticals. The Merck manual for healthcare professionals (Merck). Metabolic, electrolyte, and toxic disorders in neonates: Prenatal drug exposure, alcohol. http://www.merckmanuals.com/professional/pediatrics/metabolic_electrolyte_and_toxic_disorders_in_neonates/prenatal_drug_exposure.html?qt=fetal%20alcohol%20syndrome&alt=sh#v1088405.

Oster, E. 2013. Take back your pregnancy. *Wall Street Journal*, 9 Aug 2013.

Report of the DGAC on the dietary guidelines for Americans. 2010. Part D. Section 7: Alcohol. http://www.cnpp.usda.gov/Publications/DietaryGuidelines/2010/DGAC/Report/D-7-Alcohol.pdf.

Rehak, J. 2002. Tylenol made a hero of Johnson & Johnson: The recall that started them all, *New York Times*. 23 March 2002.

Senate resolution 612, 111th Congress, 2010.

Streissguth, A.P., and K. O'Malley. 2000. Neuropsychiatric implications and long-term consequences of fetal alcohol spectrum disorders. *Seminars in Clinical Neuropsychiatry* 5(3): 177–190.

U.S. Code of Federal Regulations (21 CFR 211.132), 2014.

U.S. Code of Federal Regulations (U.S.C. 215), 2012.

U.S. Code of Federal Regulations (U.S.C. 213), 2010.

U.S. Dept. of Health & Human Services (HHS). 2005. Office of the Surgeon General, news release: "U.S. surgeon general releases advisory on alcohol use in pregnancy," 21 February 2005. Accessed http://www.surgeongeneral.gov/pressreleases/sg02222005.html.

Experience in Involuntary Hospitalisation of Drug and Alcohol Addicted Mothers During Their Pregnancies at Borgestadklinikken, Norway

Egil Nordlie

This chapter deals with coercive measures in relation to pregnant substance abusers in Norway and experiences from Borgestadklinikken—Blue Cross South, where many of these women are treated. The legal basis is described and ethical problems to do with the women's loss of liberty compared with the benefits for the child are discussed. In order to throw light on some of the problems, a hypothetical letter has been included in which summarizes what many of these women have said. The chapter ends with results from the author's own study on the birth weight

E. Nordlie (✉)
Borgestadklinikken—Blue Cross South, Skien, Norway
e-mail: egil.nordlie@borgestadklinikken.no

M. Nelson and M. Trussler (eds.), *Fetal Alcohol Spectrum Disorders in Adults: Ethical and Legal Perspectives*, International Library of Ethics, Law, and the New Medicine 63, DOI 10.1007/978-3-319-20866-4_4

of the children. It has positive effects on children and mothers when involuntarily hospitalization is made as early as possible during pregnancy. Consequently this will reduce the risk of developing FASD and other negative consequences for the children. The reasons why, relatively speaking, so few women with alcohol problems are hospitalized involuntarily will be discussed.

It is however important for readers from other countries, such as those from North America, to understand that some European governments often have a different kind of relationship with its people than North American governments. The Norwegian government wants to help women to have healthy babies. In North America, governments want mothers to take personal responsibility for their actions. The Norwegian approach might perhaps seem like a form of punishment to a North American, when it is intended as assistance so a mother has a healthy baby.

Legal Basis

Proposing legislation that results in a loss of liberty for individual members of the population will always be a dilemma. In 1994–95, there was a great deal written in the Norwegian media about children born to women who use drugs and alcohol during pregnancy. In 1994, one of Norway's biggest television channels broadcast a documentary about the severe postnatal symptoms suffered by infants when the mother had used heroin during pregnancy. The programme also looked at alcohol use in pregnancy and the knowledge available on this problem at the time, with one of Norway's leading paediatricians appearing and shedding tears over the fate of these infants. Experts begged the politicians to do something. Although the Minister of Children at the time did not believe there to be any need for coercive legislation, the opposition was so strong that an addition to the existing Norwegian Social Services Act of 1993 was passed. Counterarguments that it was a backward step in the battle for women's rights to deprive women of their rights, that women could go underground or that there was a risk that the frequency of abortions would increase had no effect. The interests of the child were perceived to be more important.

The first two subsections in section 10.3 of the Social Services Act read as follows:

> It can be decided that a pregnant drug or alcohol user should be admitted to an institution designated by a regional health authority without her consent (…) and detained there throughout the pregnancy if the abuse is of such a nature that it is highly probable that the child will be born with defects, and if voluntary assistance is insufficient. The County Social Welfare Board shall also decide whether there are grounds for taking urine samples from the pregnant woman during her stay in the institution.
>
> The purpose of admission is to prevent or limit the probability of the child being harmed. Importance shall be attached to the woman being offered during her stay in the institution satisfactory help with her drug or alcohol abuse and to enable her to care for the child.

Norway is the only country to have such a statutory provision, but similar legislative proposals are also being debated in some other countries. The original wording proposed by the health authorities was "*…that the child will be born with severe defects…*", but the Norwegian Parliament omitted the word "*severe*" in order to strengthen legal protection for the foetus. The intention of the provision is to prevent negative consequences for the foetus, not to punish the woman. The provision also makes it a requirement that "voluntary measures are insufficient". However, voluntary measures do not necessarily have to been *tried* before coercion is decided on. Social services also sometimes threaten the woman with coercion if she does not admit herself voluntarily, which is a form of passive coercion (Søvig 2007).

I have, on several occasions given lectures in other countries on our coercive treatment and found that the same counterarguments are put forward. Although the battle for women's rights is important, of course, it seems strange in my opinion that there is not a greater understanding of the interests of the child being given priority. A Danish newspaper ran an article on our sheltered unit for these women in May 2014, and said that it perhaps took a paediatrician in tears for these infants to receive the protection of such a law. My impression is that in parts of the world where not as much progress has been made with regard to women's and children's rights there is considerable opposition to coercive measures against women.

Generally speaking, Norway does not stand out as a country with many statutory provisions that permit coercive measures over and above criminal measures (Søvig 2007). As a comparison, the mental health, child protection and communicable diseases legislation in many countries contains provisions permitting coercive measures if the person in question is a serious danger to himself or others. However, under section 10.2 of the Social Services Act, drug and alcohol users can also be hospitalised involuntarily if their life and health are at serious risks.

Our clinic, Borgestadklinikken, which specialises in the treatment of pregnant drug and alcohol users and families with young children, set up a small unit for pregnant women with drug and alcohol problems and mothers with young children in 1991. Since 1993 the clinic has had a larger family unit, which also admits fathers. Borgestadklinikken has had provision for pregnant women who are hospitalised involuntarily since the coercion legislation came into force in 1996. A separate sheltered unit for them was set up at the institution in 1998.

Between 1996 and 2013 there were 608 temporary coercion orders against pregnant drug and alcohol users, with 362 being upheld by an ordinary ruling of the *County Social Welfare Boards*. Borgestadklinikken has had more admissions than any other institution under this coercion provision and can document 197 admissions between 1996 and 2013. The age range was 16–45 with an average of 28, the same as for the rest of the country (Wiig and Myrholt 2012). The average age for women hospitalised involuntarily with alcohol as their primary diagnosis was 35 on a national basis. There were between three and five admissions to Borgestadklinikken annually in the first few years, but the number started increasing in 2001, and every year since 2003 between eight and 14 women hospitalised involuntarily have completed their pregnancy at Borgestadklinikken. In 2012 and 2013 there were 14 admissions each year.

Duty to Inform

Section 32 of the Norwegian Health Personnel Act contains another important statutory provision:

> Notwithstanding the duty of confidentiality (…), the health personnel shall of their own accord provide information to the social welfare service, when there is reason to believe that a pregnant woman is abusing intoxicating substances in such a way that it is highly probable that the child will be born with defects, cf. the Social Services Act section 10.3.

When it comes to the duty of health personnel (doctors, psychologists, midwives, health visitors, nurses, etc.) to inform social services, issues can arise. On several occasions the author has provided training for fellow doctors and been told the following: "We would rather reach a preventive agreement with the woman than report her directly to social services. We could lose her trust and she might go underground. That would be worse." But what the duty-to-inform provision says is indisputable. The number of reports of concern has increased in recent years as health personnel have gradually come to understand how important it is to speak out. There are few reports of women leaving the country or going underground, but isolated cases do occur. According to an evaluation report (Lundeberg et al. 2010), the figures from the County Social Welfare Boards also show that it was the woman herself who contacted social services in up to 41 % of cases.

Further Proceedings

Social services must obtain all the necessary information, after which coercive measures may become relevant if voluntary measures are insufficient. Social services can issue a *temporary order* (98 % of all cases), in which case the woman will be admitted to a suitable institution as soon as possible. Institutions that can take involuntary patients are specially designated by the health authorities (Søvig 2007).

An average of 55 days passed between social services receiving the report of concern and issuing the emergency order (Lundeberg et al. 2010). In terms of the safety of the foetus this seems too long, but timing has recently improved. Social services must submit the temporary order to the chairperson of the regional County Social Welfare Board within 48 hours. The woman can appeal against the emergency order and is entitled to a ruling within a week. Social services must then prepare and send the final case to the County Social Welfare Board within two weeks. If they exceed this deadline, the case is dropped. It then takes between four and seven weeks for the cases to be considered by the County Social Welfare Board (Søvig 2007). The women are still hospitalised during this period, which is a long time to wait for a final ruling. When social services and the County Social Welfare Boards assess the risk to the child, they base their decisions on expert knowledge, the observations and reports of concern they have received regarding the pregnant woman through social services, and statements by the woman and her lawyer. The County Social Welfare Board consists of a chairperson (who is a jurist), a specialist and a layperson. In difficult cases the County Social Welfare Board can be enlarged with two more members. It is also possible to use *simplified* proceedings if the woman and social services agree. The County Social Welfare Boards can rescind the order, but found in favour of social services in 92 % of cases. The woman can appeal the coercion order to the District Court.

Social services monitors the woman while she is hospitalised involuntarily and cooperates with the institution. Social services can rescind the order during the pregnancy, but rarely does so. Her case must be reviewed every 3 months. The institution's medical director cannot rescind the order, but he can decide to release her if, for example, she is a risk to others or the institution is unsuitable for her. This must be done in consultation with social services.

Under the coercion provision the woman must be offered treatment for her drug or alcohol problem and she must receive help with taking care of her child. One complication is that some of the women do not want treatment and deny that they have a drug or alcohol problem. There is a strong emphasis on helping the woman to care for her child during the treatment phase and through cooperation with child protection while the woman is hospitalised. According to one of Borgestadklinikken's reports (Wiig and Myrholt 2012), 51 % were allowed to care for their child after the birth (1996–2013).

Deliberations by the County Social Welfare Boards

The coercion order is dependent on the use of drugs or alcohol during pregnancy being such that it is highly probable that the child may have defects, i.e. the probability of defects occurring must be greater than 50 %. Defects in this context are considered to mean serious reactions and resulting conditions in the child such as teratogenic defects, withdrawal symptoms, growth retardation, risk of functional abnormalities, etc. When social services and the regional County Social Welfare Boards assess the risk to the child, they base their decisions on expert knowledge, the observations and reports of concern they have received regarding the pregnant woman through social services, statements by the woman and her lawyer, and witnesses. There is some debate as to whether the order is based on a 'precautionary principle', although there is no mention of this in the coercion provision. The experts and other advisers in these coercion cases provide details of the harmful effects that the individual substances can have on a foetus and the fact that using a variety of substances can amplify the negative effects of each. The use of alcohol at any stage of pregnancy is given great weight.

Stress factors, use of tobacco, nutritional state and other attendant health conditions also affect foetal development. However, smoking and snuff use are not taken into account in assessing coercive measures. The use of *legally* prescribed substitute preparations (methadone and buprenorphine) does not constitute grounds for coercive measures if the pregnant woman is not also using drugs and alcohol. On the other hand, *illegal* consumption of the same preparations would be significant in assessing the need for involuntary hospitalization, because it is considered to be more dangerous for the foetus when the mother has a variable consumption of opioids than a constant level of opioids during the pregnancy.

A Woman's Own Description of Being Hospitalised Involuntarily

To throw more light on how the women fare during their involuntary hospitalisation, the following story is a description of a hypothetical woman's 'own experiences' of being hospitalised involuntarily. In this hypothetical case, she writes a letter to a friend. The description is based on my own experience as a doctor to these women over a period of 18 years and represents my understanding of the reactions, thoughts and challenges often shared by them.

Hi Vicky!

Elisabeth here! What has happened to me in recent weeks has set off all sorts of thoughts and emotions that I just have to write about.

I still don't understand how they could hospitalise me against my will four weeks ago. While Petter was in town, the drug and alcohol abuse counsellor from the municipal social services came to the house and read out an order they had written, saying that there was a strong suspicion that I used too much alcohol and pills, and that I'd twice tested positive for cannabis in my urine. I'd also phoned up and asked to postpone a check-up with the midwife because I was in Denmark. Surely you're allowed to go to Denmark when you get a free ferry ticket. They also said they had received a report of concern from my midwife and GP, who said I smelled of alcohol and had been a bit of a mess when I went for a check-up. I'd only drunk some beer at a party the day before. I know I shouldn't drink alcohol when I'm pregnant, so I didn't do it again.

By the way, I'm already 18 weeks pregnant. I'm having a boy. Hooray! Cathrine, who is nearly 3, will have a baby brother. She's so excited! I told her when I last visited her in the foster home. I think it's shameful that she has a mother who has been hospitalised against her will.

I probably drank more when I was pregnant with Cathrine than I did this time. Of course I read in the leaflets the midwife gave me that alcohol can damage the baby in all sorts of ways, but I'm sure that's just if the mother drinks a lot. But I cut down on my alcohol consumption a lot when I thought I was pregnant again. I was 11 or 12 weeks gone by then, I think. But you know yourself how bad you feel when you cut back. I got myself some pills, Valium and Rivotril, to help with the cravings. And then I smoked a bit of cannabis some evenings. I didn't think it was as dangerous as alcohol. It's not as if I'm a pill abuser or drug addict, and I'd virtually stopped drinking, so I think it was really unfair to hospitalise me against my will. But I'd had a couple of warnings about the possibility of coercion if my urine samples were positive or I didn't attend check-ups. Maybe I should have gone to my auntie in Spain. They wouldn't have found me there. It's easy to buy pills there too.

It was worse for one of the other girls here. She was actually caught red-handed. When they came to get her, the amphetamines were on the bedside table and she was totally high already. Both she and her boyfriend became angry, and tried to resist. The police were called and he was carted off to the police station, while she was taken to the clinic. She initially made a bit of trouble here, including setting the fire alarm off, but soon calmed down. That was a couple of months ago. But, believe it or not, she consoles me and says that we would never have managed on our own. Strange – she thinks that this can turn our lives around and is glad she was coerced. I'm not though, if they'd just given me a chance, I could have showed them I can cope. – But I may get that chance! Because involuntary hospitalisation is a temporary measure, my whole case is going before the County Social Welfare Board next week and my lawyer will try to have the coercion order overturned. But social services are adamant and think I should be hospitalised for my whole pregnancy. Idiots! They don't understand a thing! My case officer and doctor here will be giving evidence in the case. They'll have to say how I've behaved here of course and whether they think I'll get drunk. I hope they know what they're talking about.

I understand that everyone wants to protect the baby of course. But what about my rights, and the fact that I want to protect my baby too and have a right to show that I can manage that? No mother wants her baby to come to any harm! You know, Vicky, I also thought about having an abortion for the first few days after coming here. I couldn't bear the horrible thought of having to stay here, of all places, in a drug rehabilitation clinic for five or six months. Of course I knew I was entitled to an abortion even if I was under a coercion order. I brooded night and day, I was restless and felt pretty bad. I had a few withdrawal problems from the pills and was actually given Sobril for eight to 10 days so I could come off them gradually, which helped a bit. I talked to the midwife and doctor and other members of staff, who said it was up to me to decide whether to have an abortion. But when I went to the hospital to find out how far gone I was and got to see the baby on the ultrasound, I was in no doubt that it was my baby and I was going to have it. I slept more peacefully that night.

The Sheltered Unit is completely different from the Detoxification Unit. Everyone is pregnant, apart from a couple of blokes, the partners of two of the girls. The staff are very strict and inflexible, but easy to get along with too. They actually show they care about us and show understanding for our situation. But the doors and windows are locked, and there's a high fence outside so we can't get out and drugs can't get in. I still believe that they think too much about protecting the baby from drugs and not enough about it not being good for the baby if the mother gets stressed out. But one of the others, who has been here for five months, told me yesterday that when I've been here two or three months, I might understand why it's so strict. Someone managed to get out and got really high. But she was sorry and came back of her own accord before the police or social services found her.

I've started attending group meetings with the other girls. They talk about all sorts of things, social and financial stuff, and their relationship with their partners. We also talk about our relationship with the babies inside us, they call it the 'mothering process'. It's actually quite nice. The midwife is supposed to teach us about our pregnancy and health, while the doctor is supposed to explain about possible harm to the foetus from drug and alcohol use during pregnancy. The others says that the doctor doesn't talk much about the worst effects, but says instead that we need to be clean for the rest of our pregnancy, regardless of how late in our pregnancy we stopped using drugs and alcohol. I hope I can get out of some of the group meetings. We also do crafts and hobbies. Among other things, we get to make a plaster copy of our own tummy in our eighth month. We also get to go on trips in the clinic's minibus and visit shopping centres and other places, but accompanied by staff.

I've already had a few chats with the woman who's my case officer. She's a social worker and is going to help me with a lot of the things I struggle with and advise me on how to deal with social services, child protection, my lawyer, my family and the baby's father. I didn't want to talk to her to begin with, because she couldn't help me get out of here. But she told me that, if things go well and I'm not seen as an escape risk, I may be allowed out on my own, for an hour at first and maybe building up to several hours. I also know that some people get leave to go home overnight even though they have been hospitalised against their will, and I hope I will too eventually if I have to stay here.

The others are also struggling. The woman in the room next to mine used to do a lot of heroin, but has been on Subutex for a year now. But then she got high on cannabis and amphetamines and was hospitalised against her will. She knows there is a high risk of the baby having a lot of withdrawal symptoms after it is born, but doesn't know if she dares to gradually come off the Subutex because of withdrawal symptoms. The Norwegian health authorities recommend staying on Subutex during pregnancy if the woman is in the medication-assisted rehabilitation programme. But it is also possible to come off it gradually if she wants to and both she and the foetus receive proper monitoring. Although she's been hospitalised against her will, she can decide for herself. I haven't told her, but I think it's totally unfair and infuriating that she gets to use medicines here that can cause the same problems for the baby as the drugs the rest of us have been hospitalised against our will for.

And now for the hardest thing I want to say to you, Vicky. I don't really know what I should do about my relationship with Petter. He's a great guy and I'm sure he's the baby's father. But he's on the move a lot and uses cannabis and amphetamines. He said he doesn't want to give up drugs and alcohol, "because that's life". But I have dreams of me and Petter and the wee chap being a little family together. The baby's father can also be admitted to the Sheltered Unit, but he needs to be motivated to do it. My common sense tells me I should drop him, but my heart says I should give him a chance. He's allowed to come and visit me, but then he has to give a negative urine sample on the same day. If he can manage that, I'll say that he has to see his GP and get a referral for admission here. I've asked the others here, the staff and my case officer what they think, but they advise me to wait a bit.

I also asked mum. She doesn't think me and Petter enjoying ourselves with a drop of beer now and again is that dangerous. "I drank when I was pregnant with you too and everything was fine," she told me. But no, everything wasn't fine! In the last few days I've been thinking about why I wasn't very good at school, why I was bullied for being so small and sickly, and because I had a squint in my left eye. They called me the 'ugly duckling'. Adults too. So recently I've been wondering if things might have gone better for me if mum had been hospitalised against her will when she was pregnant. Then I might have managed to write this letter myself. I just have to come clean, Vicky, and tell you that one of the other girls is writing down the words in this letter for me. She was hospitalised against her will because of heavy tablet use early in her pregnancy. She's a teacher, 41 years old and a mother of three already. Imagine what a shock it was for her to be hospitalised against her will. But the coercion order was rescinded by the County Social Welfare Board and she decided to stay in here of her own free will after all. She hasn't taken any pills even though she's in here voluntarily, and that is great.

Now I have to get this letter in the post, Vicky. Social services phoned and said my lawyer wants to talk to me about the coercion case that's going before the County Social Welfare Board. What's more, child protection also phoned earlier today and wants to make an appointment to visit my three year old. My case officer has told me that it's a good idea to cooperate with child protection before the birth so that I get to know what they think about me being allowed to keep the baby when it's born. I don't understand it! Why shouldn't I be allowed to keep the baby? After all, I'm going to be clean for the rest of my pregnancy. Anyway, they're not allowed to make any decisions before the child is born.

What a long letter, Vicky. Lots about myself, but you'll appreciate that I'm in a difficult situation. How are you by the way? You've got your life straight now, haven't you?

Love Elisabeth

Sheltered Unit for Pregnant Drug and Alcohol Abusers Hospitalised Involuntarily at Borgestadklinikken and Voluntary Admissions Unit

The "Sheltered Unit" for pregnant women hospitalised involuntarily has seven places and is staffed by nurses, social educators, activity therapists, social workers, a midwife and a doctor (Myrholt et al. 2010). It is also possible to speak to a pastor and receive treatment from a physiotherapist or dentist. The women are usually initially admitted for a short stay in the clinic's detoxification unit. It is important for their actual reception to be as kind and considerate as possible. How quickly they are transferred to Shielded Unit depends on their condition, but between 3 and 8 days is normal. They all have their own case officer and the offer of a lawyer. The women take part in individual sessions, assessment, education and group meetings. They have an appointment with the doctor once a week and with the midwife every couple of weeks. Importance is attached to good relations with the woman, factual information on the harmful effects of drugs and alcohol, and not feeling guilty unnecessarily. It is also vital to stress that the foetus is just as important a patient as the woman. There is also space for fathers, who can be admitted together with the woman towards the end of the pregnancy. The men often attend sessions and follow the women's regime.

Many of the women have very complex medical problems, both mental and physical, and are treated for them with their pregnancy in mind.

They attend group meetings which address the following topics: foetal development, childcare, music for the baby and hearing development, making a plaster tummy model, maternity preparation, physical changes, exercise, health and diet, effects of drugs and alcohol on a foetus, effects of drugs and alcohol on adults, addiction treatment, smoking and the offer of a stop-smoking course, childbirth, pain relief, breathing exercises, the postpregnancy period, sexuality, sexually transmitted diseases, prevention, parenting, the father's role, violence, aggression, shame, self-respect, dignity, values, attitudes, religious belief, and networks.

They also take part in expressive arts and music therapy and can go on joint activities outside the institution.

Some are allowed out unattended, with this being assessed on an individual basis. It is important that the woman should have as much control as possible within the constraints that exist. The women also take part in the open handover, i.e. they can be present when one staff shifts hands over to another.

They have ultrasound scans at the local hospital in Skien, a 20-min drive from the clinic, at least twice (17 and 32 weeks). But most have more frequent scans. There is a low threshold for having extra scans if unexpected conditions arise.

Most of the women have a dignified leaving ceremony shortly before their due date. The majority of births happen at the local hospital. There is a close working relationship with the hospital. Many women also take up the offer of a six-week stay after the birth, in some cases without the child as a result of it being taken into care at birth.

The pregnant women hospitalised voluntarily in the Family Unit follow a similar regime, but are free to go out when not doing regular activities. They share a communal area with other families with children who are in the clinic voluntarily.

Differing Practice in Institutions

Other institutions in Norway have different regimes for involuntary patients with regard to treatment, rules and sanctions for breaking the rules. There is a broad spectrum of treatment offered, from places with very little provision to places with all sorts of opportunities for assessment, treatment and development of individual plans and treatment plans (Lundeberg et al. 2010). Some institutions have locked doors for patients throughout their stay, while in others the patients live on an open ward together with voluntary patients of both sexes. Statistically, there are more escapes from institutions with locked doors. Patients do not receive good enough care on mixed-sex wards (Lundeberg et al. 2010). They should be treated in institutions with well-organised wards designated by the regional health authorities.

Some institutions only want to treat these patients in accordance with the special voluntary section (10.4) in the Social Services Act. This is based on the woman admitting herself for treatment voluntarily but entering into a written agreement with the institution that she can be kept in for 3 weeks if she wants to leave. This can be done three times during her stay in the institution. Other countries have similar statutory provisions for drug and alcohol users in general.

Results from a Study of Birth Weight in Children Born to Women Hospitalised Involuntarily and Voluntarily

The author of this chapter carried out a study for the period 1996–2009 in order to elucidate whether involuntary treatment has a positive effect on children in accordance with the primary purpose of the coercion provision. (Nordlie 2012). They were compared with children born to voluntary patients. A total of 110 involuntary and 54 voluntary patients (three twin pregnancies) completed their pregnancy at Borgestadklinikken in this period.

It is important to obtain data on whether the coercion provision creates favourable conditions for the child. The study therefore aimed to elucidate whether coercive measures against pregnant women with drug and alcohol problems produce a different outcome in the child's birth parameters, primarily birth weight, compared with children born to voluntary patients in the same institution during the same period. It also aimed to establish whether length of stay affected birth weight.

In accordance with the coercion order, the women were monitored by means of random urine tests for drugs and alcohol at least once a week throughout their stay.

Owing to good monitoring by staff, there is very little risk of drug and alcohol use during their stay. The voluntary patients also provide urine samples once a week.

Serious alcohol abuse was given as the sole reason for the coercion order in five of the cases. Otherwise it was for the most part polysubstance abuse, usually involving amphetamines, cannabis, habit-forming medication and opiates, but with alcohol as part of the abuse in 25 cases.

Among the voluntary patients, two specified alcohol as their only substance. Otherwise alcohol was used together with drugs in most of the other cases. The consumption of and combination of different drugs was not as extensive as among the involuntary patients.

Eight of the women hospitalised involuntarily also used methadone legally during pregnancy. The coercion order was issued purely on the basis of their illegal drug use, however. Eight of the 12 children where the mothers (involuntary and voluntary patients) used methadone legally during pregnancy suffered neonatal withdrawal symptoms.

Roughly as many boys as girls were born in both groups (54 girls/56 boys for involuntary patients and 31 girls/26 boys for voluntary patients, including three sets of twins) and the birth weights were not differentiated between the sexes. They were all live births.

The mean birth weight, length and head circumference for children born to women hospitalised involuntarily until the birth were 3,165 g, 48.9 and 34.0 cm ($n = 110$), while the same figures for children born to women hospitalised voluntarily in the same period were 3,362 g, 49.6 and 34.5 cm ($n = 57$). According to the Norwegian register of births, the mean weight, length and head circumference for live births in Norway were 3,531 g, 50.1 and 35.2 cm in the same period ($n = 760{,}943$).

The mean birth weight for children born to mothers hospitalised involuntarily was 197 g ($p = 0.051$) lower than for children of voluntary patients and 366 g ($p < 0.001$) lower than for live births in Norway in the same period. The primary and secondary risks of actual drug use may be assumed to be a significant cause of the birth weights in children born to involuntary patients being lower than the mean for all children born in Norway in the same period. Even a borderline significant difference in birth weight is an indication that mothers hospitalised involuntarily are more at risk than voluntary patients, which is in accordance with the intention of the Act.

The mean birth weight was 3,512 g ($n = 31$) when the mothers had been hospitalised involuntarily for *more than* 140 days *(half their pregnancy) and* 3,029 g ($n = 79$) when the mothers had been hospitalised involuntarily for *less than* 140 days ($p = 0.001$). The birth weights were 3,439 g ($n = 25$) and 3,301 g ($n = 32$) when the mothers had been hospitalised voluntarily for *more and less than* 140 days respectively.

Birth weight, length and head circumference are an indicator of what the health-related conditions and conditions for growth were like for the foetus during the pregnancy. The results of this study were most unequivocal for birth weight, so the decision was taken to focus the results on birth weight as an indicator. Birth weight is also commonly used in similar studies internationally. However, genetic factors, smoking, diet, sex and disease also affect birth weight.

On average, the children of mothers hospitalised involuntarily who were in the clinic for more than half their pregnancy had a birth weight on a par with other Norwegian babies despite their mothers smoking heavily. The mean birth weight for children born to mothers involuntarily hospitalised for more than half their pregnancy was all of 483 g higher than for children of mothers involuntarily hospitalised for less than half their pregnancy ($p = 0.001$).

FASD babies have a much lower birth weight that average. Thirty of the 110 women had used alcohol during pregnancy. Although this number is small, the results may indicate that there is not a high incidence of FASD among the children whose mothers were hospitalised in the institution for more than half their pregnancy. On average, children born to mothers hospitalised for less than half their pregnancy weighed almost half a kilogram less. It is difficult to say whether this is due to alcohol consumption. Other signs of FASD were not investigated as closely after birth.

In my experience, over the 18 years for which the coercion provision for pregnant women has been in existence, the attitude to coercive measures has changed in social services, in primary healthcare and among the women themselves. It is no longer just in the most at-risk cases that coercive measures are implemented. At Borgestadklinikken this patient group has also received more satisfactory provision owing to better staffing of the unit, close pregnancy monitoring and changed attitudes to the women in the form of more freedom within a secure framework. There is less stress. Fewer weeks of substance use and births closer to the due date are also among the reasons for birth weight increasing (Fig. 1).

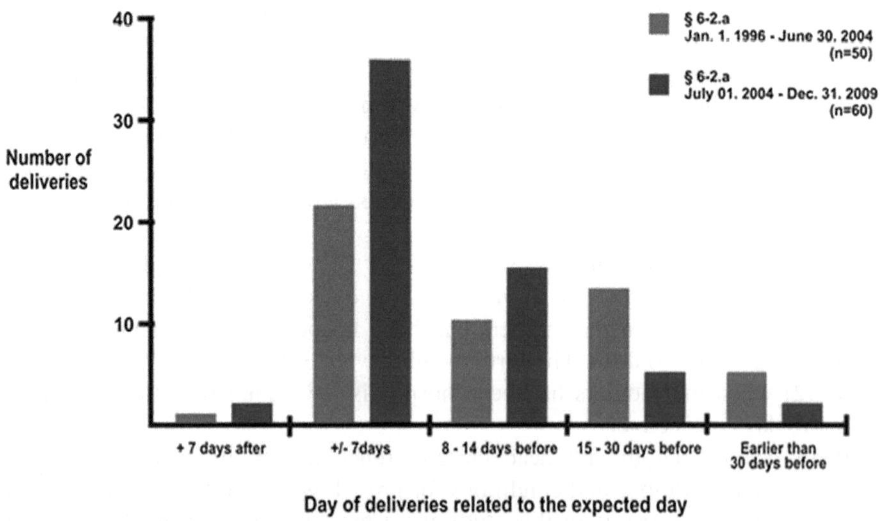

Fig. 1 Delivery statistic for involuntarily hospitalized pregnant woman

The main conclusion of this study must be that there are clear indications that measures under section 10.3 of the Social Services Act are effective and absolutely necessary, and that it is best for mother and baby for the measures to be implemented as early as possible (Nordlie 2012).

Figure 1 shows births in relation to due date for the periods 1996–2004 and 2004–2009. As a result of better health and less stress for the mother, the births occurred closer to due date in the second period than in the first for the involuntary patients.

Other Reports

Unfortunately, there are no national reports focusing on the incidence of FASD among these children. However, Borgestadklinikken has produced several other reports on coercive measures for pregnant women. Some factual information from the reports is mentioned in the rest of the text. Here are some of the other important results:

Who informed social services in 1996–2005 (Hansen and Svenkerud 2006) compared with 2006–2010 (Wiig and Myrholt 2012):

	1996–2005: (%)	2006–2010: (%)
Child protection service	19	16
The women themselves	19	11
Drug treatment institutions	18	11
Already known to social services	14	11
Relatives	3	10

On average, the women were hospitalised 6 weeks earlier in their pregnancy in the period 2006–08 (mean 19 weeks pregnant) compared with 1996–99 (mean 25 weeks pregnant) (Grimstad and Kristiansen 2009).

In total, 13 % of the pregnant women opted for an abortion in 1996–2010 (Wiig and Myrholt 2012). There was a slight increase in the number of abortions as a result of the women being hospitalised earlier in their pregnancy. It is impossible to know whether the women would have chosen to have an abortion if they had not been hospitalised. All the women were given the opportunity to return to Shielded Unit voluntarily to stay for a few days following a spontaneous or induced abortion. There are, incidentally, very few spontaneous abortions.

As far as the condition of the babies after birth is concerned, a report (Hansen and Svenkerud 2006) shows that of the 26 children born in 1996–2005 for which they had good data, 18 had defects that could have been due to the effects of drugs and alcohol in pregnancy. This included premature birth, low birth weight, withdrawal problems, deformities (e.g. gastroschisis and oesophageal atresia), delayed motor development, etc.

Alcohol and Coercion

It is estimated that at least 450 children a year are born in Norway with FASD and possibly many more. Alcohol sales in Norway increased by 48 % between 1993 and 2008, mainly in the form of wine, and alcohol consumption in fertile women is on the increase. Nowadays, young women drink twice as much alcohol as their mothers did at the same age. It is therefore a cause for concern that relatively few women are hospitalised involuntarily with alcohol abuse as their main problem when we know that alcohol is the most toxic substance for the foetus and that coercive measures can prevent FASD.

Why is that the case?

Based on the aforementioned report (Lundeberg et al. 2010) and my own experiences, the reasons might be as follows:

1. Alcohol consumption is easier to conceal than the use of illegal substances.
2. Alcohol is more socially acceptable than the use of other substances.
3. The duty to inform has been applied less rigorously than for illegal substances. There are also ethical aspects to where the line should be drawn for reporting concern when a woman uses relatively moderate amounts of alcohol regularly, but where the foetus is nevertheless in the danger zone for defects. It is particularly challenging if the woman engages in moderate binge drinking.
4. Pregnant women with alcohol problems have less contact with social services.
5. Some pregnant women actually reduce their alcohol consumption to avoid a reaction by the heath service/social services, but are nevertheless in the danger zone.
6. The health service/social services are not always good enough at avoiding moralising and telling off, and reassuring the woman that voluntary measures and even coercion would be best for both her and the child.
7. Several women with alcohol problems have had their coercion order rescinded.

Closing Comments

When it comes to pregnant drug and alcohol users, we are talking about the vulnerability of two individuals. The coercion provision is based on the child being the weaker and more vulnerable part. In cases where it is not possible to guarantee the child the necessary conditions for its own development and health by means of recommended voluntary provision for the mother, it is therefore necessary to protect the child by implementing measures against the mother's will. Because the consequences for a new human life are so serious and the measures cause the woman to be deprived of her freedom, it is important to have a sound legal framework, well-trained experts on the County Social Welfare Boards and an adequate number of treatment places with good therapeutic provision. It is also important to be aware that the mother's well-being during the pregnancy is also best for the child.

In recent years, positive steps have been taken in many countries in the prevention of FASD and other foetal defects caused by substance abuse. In Norway the health authorities have run several information campaigns on alcohol and pregnancy. The KoRuS-Sør resource centre, based at Borgestadklinikken, has conducted many information and cooperation projects on substance abuse and pregnancy throughout the country over a period of many years.

Norwegian law professor Karl Harald Søvig writes in his book that the foetus has *relative* legal protection under Article 2 of the European Convention on Human Rights, depending on the context (Søvig 2007). Under the abortion laws in many countries the foetus has *graduated* legal protection, all depending on how advanced the pregnancy is. Some countries with the death penalty actually prohibit the execution of pregnant women. Under the UN Convention on the Rights of the Child, a foetus has the same rights as child after birth as long as the intention is that it should be born. It ought, therefore, to be quite natural to have a law that protects the child against substance-related defects, especially FASD, even if it results in loss of liberty for the woman, as long as it produces good results.

Some women do not discover they are pregnant until well into their pregnancy, and some do not manage to give up substance abuse despite being pregnant. Many of these women believe that it was involuntary hospitalisation that made them turn their lives around and become clean. They would never have managed it on their own and recognise that it was necessary for other people to intervene. Several have said that coercion saved both their own life and that of their child. One of the women even wrote to the Minister of Health, thanking him for there being such a statutory provision. It is important to listen to what the women themselves have to say!

There is no longer much discussion about Norway keeping its special coercion provision for pregnant drug and alcohol users. But there is definitely room for improvement when it comes to carrying out the interventions, something the health authorities are aware of and want to do something about through regulations, for example. Defining the duty to inform in more detail and having well-established, effective chains of action in social services for substance users, especially in the case of alcohol use, is important. Consistency of assessments and processing times at the County Social Welfare Boards, uniform, evidence-based treatment regimes in the institutions, and better procedures for cooperation with social and child protection services are also areas for improvement.

The Norwegian Health Authorities will publish revised guidelines for coercion in 2015. However, based on the experience and knowledge we have in Norway, we feel that other countries should consider the introduction of similar statutory provision due to the positive benefits to the children, the women, and society.

References

Grimstad, V., and R. Kristiansen. 2009. Gravide innlagt på Borgestadklinikken etter LOST § 6.2a i perioden 2006–2008 (Pregnant women admitted at Borgestadklinikken in the period 2006–2008).

Hansen, F., and S. Svenkerud. 2006. Gravide innlagt på Borgestadklinikken etter LOST § 6.2a i perioden 1996–2005. ISBN 82-90354-64-9. (Pregnant women admitted at Borgestadklinikken in the period 1996–2005).

Lundeberg, I.R., et al. 2010. Tvang overfor rusmiddelavhengige - Evaluering av Lov om sosiale tjenester §§ 6-2, 6-2a, og 6-3. UNI Rokkansenteret (Coercive measures against substance addictives—an evaluation).

Myrholt, R., et al. 2010. Behandlingstilbudet ved Skjermet enhet. Statusrapport. Borgestadklinikken (Teatment at shielded unit).

Nordlie, E. 2012. Fødselsvekt hos barn født av tvangsinnlagte gravide rusmiddelmisbrukere. Rusfag nr.1-2012. Pg 126–137 (Edit) (Birth weights of babies born of mothers who were involuntarily hospitalised because of substance abuse).

Søvig, K.H. 2007. Tvang overfor rusmiddelavhengige (Coercive measures against substance additives). Abstract in English.

Wiig, E., and R. Myrholt. 2012. Gravide innlagt på Borgestadklinikken etter §6.2a i perioden 2009–2010 (Pregnant women admitted at Borgestadklinikken in the period 2009–2010).

Is It Ethical to Condone Low Levels of Drinking in Pregnancy?

Janni Niclasen

Introduction

Today, it is widely agreed that exposure to very large doses of alcohol during pregnancy acts as a teratogen (Randall 2001) that can cause structural damage to specific parts of the brain including the corpus callosum, cerebellum and hippocampal areas (Abel 1998). It is also well established that such damage can have a wide range of deleterious effects on children's cognitive, behavioural and physical development (NIAAA: National Institute of Alcohol Abuse and Alcoholism 2000; Jones et al. 1973; O'Leary 2004; Stratton et al. 1996). However, studies investigating prenatal exposure to low doses of alcohol in relation to neurodevelopmental outcomes in childhood have been far less conclusive (Gray and Henderson 2007a; Henderson et al. 2007a; Abel 1998; Gray et al. 2009; Sayal 2007). Whereas some studies have reported negative associations with a wide range of neurodevelopmental outcomes (Sayal et al. 2007; Fried and Watkinson 1988; Testa et al. 2007), others have not been able to establish such associations (Kelly et al. 2010), and even others have reported a J-shaped (i.e. curvilinear) function, suggesting that exposure to a little alcohol apparently acts as a protective factor for the developing foetus (Kelly et al. 2009).

From a scientific perspective it therefore cannot be concluded whether it is ethical to condone low levels of drinking in pregnancy. Some studies have reported that it is beneficial to the children that the mothers abstain from alcohol during pregnancy. However, other studies report that it is, by contrast, beneficial to the child that the mother drinks small doses of alcohol. In the latter case it might indeed therefore be considered unethical to abstain. In what follows I will begin by giving a brief historical perspective to the subject of prenatal exposure to alcohol and subsequent child development. I will then proceed

J. Niclasen (✉)
Department of Psychology, University of Copenhagen, Copenhagen, Denmark
e-mail: Janni.Niclasen@psy.ku.dk

M. Nelson and M. Trussler (eds.), *Fetal Alcohol Spectrum Disorders in Adults: Ethical and Legal Perspectives*, International Library of Ethics, Law, and the New Medicine 63, DOI 10.1007/978-3-319-20866-4_5

to introduce the findings, in rather more detail, from the current human literature looking at prenatal exposure to low doses of alcohol and child neurodevelopment. I will then discuss why the results are so conflicting. This will be done by discussing a number of methodological limitations of the human studies. Thereafter I will consider findings from animal studies and discuss the limitations of extrapolating such results to humans. Finally, I will return to discussing whether it is ethical to condone low levels of drinking in pregnancy. My conclusion is that it is 'better to be safe than sorry' and recommend complete abstinence.

Fetal Alcohol Syndrome (FAS): A Historical Perspective

The term 'Fetal Alcohol Syndrome' (FAS) was coined by Jones and Smith in their pioneering article from 1973, "*Recognition of the fetal alcohol syndrome*" (Jones and Smith 1973). In this, and two later articles, they systematically delineated the association between maternal alcohol abuse in pregnancy and a specific pattern of growth retardation, dysmorphic facial features and central nervous system dysfunctions in the child (Jones et al. 1973, 1974; Jones and Smith 1973). These were the first scientific articles for many years in English to have been published on the deleterious effects of alcohol on the developing foetus (Randall 2001; Calhoun and Warren 2007; George 1951; Warner and Rosett 1975).

However, the suspicion of alcohol as a culprit of dysfunctions and abnormalities related to human reproduction and child development was not new. In the first half of the 1700s, during the "*Gin epidemic*" in England, several reports documented the adverse effects on the developing foetus of maternal drinking in pregnancy (Randall 2001; Calhoun and Warren 2007; Warner and Rosett 1975; Mattson and Riley 1998).

In 1725 James Sedgewick, a London apothecary, noticed that there was a relation between the "*mothers ill-spent life during her pregnancy and consequences on infants*" (Warner and Rosett 1975). A year later the College of Physicians petitioned the British parliament to control the distilling trade and called gin "*a cause of weak, feeble distempered children … born weak and silly … shrivelled and old, as though they had numbered many years*" (Calhoun and Warren 2007). Novelist and anti-gin campaigner Henry Fielding blamed gin consumption for the "*increased crime and increased ill-health among children*" (Randall 2001; Calhoun and Warren 2007; Warner and Rosett 1975; Mattson and Riley 1998). However, because alcohol along with opium was the only anaesthetic available in the 18th century it could not be prohibited by obstetricians (Warner and Rosett 1975).

The first epidemiological study of women consuming alcohol in pregnancy was carried out by William Sullivan, a deputy medical officer of the convict prison in Parkhurst, England (Randall 2001; Calhoun and Warren 2007; Warner and Rosett 1975). In 1899 he followed 600 children born to 120 imprisoned, alcoholic women and 28 non-drinking relatives as controls. Among the alcoholic women he observed twice as many infant mortalities, he observed that 80 women had three or more infant deaths, that 56 % died at birth or before the age of two and that the children of the alcoholic women were not productive members of society later in life (Randall 2001; Calhoun and Warren 2007; Warner and Rosett 1975). Around the same time, in 1905, in the USA, MacNicholl surveyed alcohol as a cause of mental retardation among school children in the city of New York. Among the 6624 children of drinking parents he found 53 % to be "*dullards*", compared to 10 % "*dullards*" among the 13,523 children of abstainers (MacNicholl 1905). It was also around this time that several researchers started using animal models to investigate the deleterious effects of prenatal alcohol exposure and showed that offspring of alcohol-exposed parents often had physical defects (Randall 2001).

From the beginning of the 1900s to the 1960s, however, interest on the topic virtually disappeared. There was a general paradigm shift from a focus on prenatal and hereditary factors, to an approach emphasizing the importance of early childhood factors as important for child development (Warner and Rosett 1975). In the early 1970s interest in the adverse effects of alcohol was renewed (Jones and Smith 1973; Mattson and Riley 1998; Jones 2012). In particular, the introduction of the FAS term made the topic of prenatal alcohol exposure and neurodevelopment find its way back onto the scientific agenda. Whereas the early studies mostly described single cases severely affected by FAS, researchers gradually began to investigate the effects of much lower doses of alcohol. As a consequence, considerable interest began to be focussed on identifying whether there exists a safe, lower level below which drinking is not associated with any harm to the developing foetus (Kelly et al. 2010; Henderson et al. 2007b; Sayal et al. 2013; Underbjerg et al. 2012)—that is, whether it is safe to condone low level drinking in pregnancy?

Findings from the Human Literature

Whereas few professionals today would disagree that prenatal exposure to large doses of alcohol causes irreversible brain damage (NIAAA: National Institute of Alcohol Abuse and Alcoholism 2000; Rasmussen et al. 2008), an on-going debate prevails as to whether exposure to lower doses of alcohol is also damaging for the developing foetus. As mentioned above, studies investigating prenatal exposure to low doses of alcohol and neurodevelopmental outcomes in childhood has not been conclusive and no clear dose-response associations have been established (i.e. the higher dose, the more damage) (Gray and Henderson 2007a; Henderson et al. 2007a; Abel 1998; Gray et al. 2009; Sayal 2007). Some studies have found associations in childhood with externalising and aggressive behaviour (Sood et al. 2001), poor mental health (Sayal et al. 2007; Fried and Watkinson 1988; Testa et al. 2007), low IQ (Streissguth et al. 1990; Larroque and Kaminski 1998), hyperactivity (NIAAA: National Institute of Alcohol Abuse and Alcoholism 2000; Olson et al. 1997), impulsivity (NIAAA: National Institute of Alcohol Abuse and Alcoholism 2000), inattention (Streissguth et al. 1986, 1994), learning and memory difficulties (NIAAA: National Institute of Alcohol Abuse and Alcoholism 2000; Olson et al. 1997), lack of coordination (NIAAA: National Institute of Alcohol Abuse and Alcoholism 2000), dysexecutive functioning (NIAAA: National Institute of Alcohol Abuse and Alcoholism 2000) and deficient social skills (NIAAA: National Institute of Alcohol Abuse and Alcoholism 2000). Other studies have reported no such associations in childhood with mental health (Kelly 2010), IQ (Eriksen et al. 2012), academic achievement (Kelly 2010), hyperactivity/inattention (Rodriguez et al. 2009), language delay (O'Leary et al. 2009), attention (Underbjerg et al. 2012) and executive functions (Skogerbo et al. 2012). Some other studies have even reported a J-alcohol shape, indicating that exposure to a little alcohol apparently acts as a protective factor for the developing foetus. One such study found that the worst mental health and cognitive outcomes at age three were apparent in the offspring of both total abstainers and heavy drinkers (Kelly et al. 2009). Another study showed that light and moderate drinking in the first three months of pregnancy was positively associated with mental health scores at age 14 (Robinson et al. 2010). Along the same line, one thorough systematic review of the literature concluded that there is no convincing evidence that prenatal exposure to low-moderate doses of alcohol is negatively associated with neurobehavioural outcomes (Gray and Henderson 2007b).

Considering the ambiguity of results, why is it that most Westernised countries recommend abstinence when, in fact, there is no scientific evidence that low level drinking is harmful to the developing foetus? Around half of all the studies actually report the most beneficial outcomes for the children of mothers who report drinking a little alcohol in pregnancy. On that basis it may seem unethical to abstain in pregnancy. Why, therefore, is it that we do not recommend women to drink low doses in pregnancy? Is it perhaps time to revise our guidelines and to recommend the intake of a little alcohol during pregnancy?

Methodological Issues

The systematic review acknowledged that many of the human studies had methodological weaknesses (Henderson et al. 2007a). Most of the studies presented above are based on data from prospective pregnancy cohorts. Cohort studies are a type of observational study mainly used in medicine and the social sciences. In such studies a group of individuals are followed from pregnancy where they are asked about their alcohol intake in pregnancy. Then at a later point in time their children are assessed. In the statistical analyses the information from the mother's pregnancy about the alcohol intake is then related to the information about their child's developmental stage (e.g. hyperactivity, learning difficulties or executive functions). Typically, in the statistical analyses the children of women with different intakes (i.e. those who abstain, with those who report an average of 1 drink/week, with 2 drinks/week etc.) are compared.

There are several strengths in such a design, including the prospective nature of the data collection. However, there might also be problems as well. What if there are differences between women who report drinking and those who do not drink in pregnancy on other parameters, i.e., not just on their alcohol intake? What if the women differ in relation to other pregnancy-related factors including their general lifestyle and socio-demographics? Furthermore, what if they also differ on post-natal conditions, including psychological and home-environmental factors? Such factors, known as (potential) confounders and mediators, respectively, might insufficiently or inappropriately be taken into account in the statistical analyses. In this case the results might reflect these variables rather than the alcohol intake per se. Furthermore, the studies might also be characterised by a problematic, imprecise definition of alcohol categories. Finally, there might be problems in relation to how and when child's neurodevelopment is assessed. There seems to be a tendency in research to apply conveniently brief screening instruments to measure children's neurodevelopment. It also appears that children are often assessed at too early an age when the potential developmental consequences for the child of maternal alcohol consumption during pregnancy have not yet manifested themselves. Below, I will consider each of these points in turn. The point I will make is that all of these factors can have consequences for the results obtained in the statistical analyses, and thus for the conclusions that can be drawn as to whether it is ethical to condone low level drinking in pregnancy.

Confounding

Confounding means 'to mix together', and can be defined in the present context as the mixing together of the effect of an exposure with a factor that is statistically associated with the exposure and causally associated with the outcome (Juul 2008). Figure 1 shows that alcohol exposure is hypothesized to be causally associated with neurodevelopment in the foetus. However, smoking (a potential confounder) is also hypothesized to be causally associated with the outcome (foetal

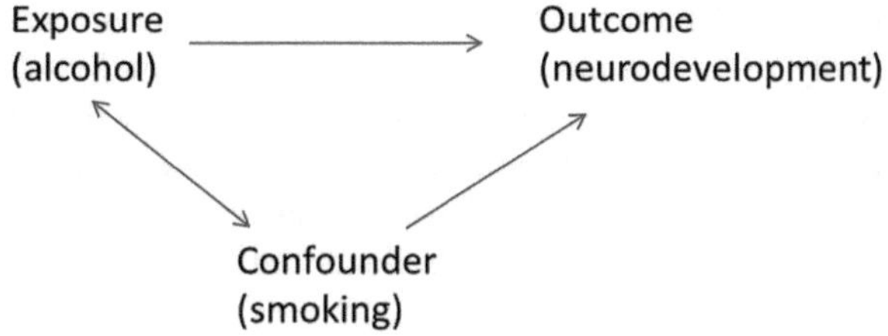

Fig. 1 The association between the exposure to alcohol and the outcome, namely child neurodevelopment and the influence of smoking as a confounder

neurodevelopment) but is statistically associated with alcohol intake. In this situation smoking is therefore considered a confounder that should be controlled for in the statistical analyses, i.e. include the variation between the different exposure groups of that factor. Controlling for smoking will remove some of the apparent association between alcohol and neurodevelopment. If we conversely do not do that then the effect of alcohol will be exaggerated. This is known as 'residual confounding'.

Residual confounding refers to the confounding that remains after attempting to adjust completely for confounding (Juul 2008; Rothman 2002). Residual confounding remains if important confounders have not been controlled for or if a confounder has not been recorded correctly and hence lead to erroneous data. The effects of residual confounding can either mask a true association or, indeed, create a spurious association, i.e. a false or non-causal association (Porta 2008).

A wide range of potential confounders could (and ideally should) be controlled for in human studies. These include prenatal exposure to smoking, caffeine, marijuana and other illicit drugs, psychotropic medication, vitamin, physical activity, education, income and other measures of socioeconomic status (SES), psychiatric problems and age. These factors can all be considered potential confounders as they have all been found to be causally associated with child neurodevelopment and statistically associated with maternal alcohol intake in pregnancy (Rodriguez et al. 2008, 2009; Braun et al. 2006; Latimer et al. 2012; Linnet et al. 2003; Linnet 2009; Mick et al. 1996; Obel et al. 2008; Thomsen and Skovgaard 2007).

In large scale cohort studies researchers usually control for a few of these factors including maternal age, smoking and educational level. However, what if large differences are observed on a wide range of other factors and these are not controlled for in the statistical analyses? Then this can potentially lead to residual and unmeasured confounding, i.e. it will bias the results of the statistical analyses. In order to investigate this one study set out to describe those women who drink in pregnancy and those women who do not drink in pregnancy on a large number of background characteristics, i.e. on a large number of potential confounding factors. The aim was to investigate whether alcohol-related differences could be observed on such factors. A selection of the variables is presented in Table 1.

The women are subdivided into exposure categories according to their self-reported intake throughout pregnancy. The '0' group has reported abstaining throughout pregnancy, the following group between >0–10 throughout pregnancy and so on. The high-intakers in this study includes women who have reported drinking >90 units in pregnancy. The high intakers (defined as those women who reported drinking more than 90 units of alcohol in pregnancy) were older, more likely to have a university degree, eat fish and have a pre-pregnancy Body Mass Index within the normal range. They were on the other hand less likely to watch television and drink cola. Conversely, the abstainers (0 units in pregnancy) were younger, more likely to have mandatory education only, drink cola, watch television, smoke cigarettes, live alone and have psychiatric problems. They were less

Table 1 Differences on a number of background characteristics between women with different self-reported intake of alcohol. Adapted from Niclasen (2014)

Alcohol units consumed during pregnancy	Full sample	0	>0–10	>10–30	>30–90	>90
N	63,464	7,204	19,111	15,054	16,619	5,476
Age	30.5	29.5	29.8	30.5	31.1	32.5
Mandatory education only	8.4 %	17.8 %	9.3 %	6.6 %	5.4 %	6.5 %
University degree	14.6 %	5.8 %	11.3 %	15.4 %	19.2 %	21.2 %
Psychiatric diagnoses prior to pregnancy (yes)	2.9 %	4.8 %	3.0 %	2.4 %	2.4 %	3.0 %
Smoking (yes)	25.2 %	31.5 %	24.1 %	23.3 %	23.5 %	30.9 %
Pre-pregnancy BMI within the normal range (yes)	67.6 %	56.8 %	64.0 %	68.4 %	72.8 %	76.0 %
Exercise in the last part of pregnancy (yes)	24.3 %	17.5 %	22.3 %	25.2 %	27.8 %	27.2 %
TV (2½ h or more per day)	21.1 %	29.6 %	23.4 %	20.1 %	17.4 %	15.7 %

likely to do exercise and eat fish. What happens if we do not control for these factors in our statistical analyses? The described factors might be mentally protective for the high-exposed children, but disadvantageous for the unexposed children. What would happen if we were conducting a study looking at low doses of alcohol and child neurodevelopment?

One study did this. The starting point of the study was the same sample of women and the children were assessed behaviourally and emotionally with a questionnaire at age seven (Niclasen et al. 2013). In the statistical analyses the researchers controlled for maternal and paternal education, smoking and psychiatric disorders and maternal psychological well-being in pregnancy. The study indeed reported the most favourable outcomes for the high intakers (over 90 units of alcohol in pregnancy) and the least favourable outcomes for the children of the abstainers. Does this result imply that exposure to over 90 units of alcohol in pregnancy has a positive effect on a developing foetus? Or does it indicate a positive influence of other factors that are not controlled for in the statistical analyses?

Mediation

In order to understand the lack of consistency in the literature, and to make progress, we need not only to draw on knowledge from epidemiology regarding prenatal factors but also on knowledge from psychology regarding post-natal, childhood-related factors. Within alcohol epidemiology there has been a great focus on the prenatal environment, and an almost complete lack of focus on post-natal conditions. However, it is well known from psychology that childhood-related factors and the home environment have a very substantial impact on child development. The point that I will try to make here is that such psychologically-oriented early childhood factors should be taken into account when interpreting the results of the epidemiological studies investigating prenatal exposure to alcohol in relation to child development. That there is too little focus on the post-natal environment has been recognised in the literature (Sayal 2007). Abel noted that "*that a child's postnatal environment affects his/her behaviour is hardly surprising, but it has not received the attention it deserves in the present context (i.e. in the alcohol literature)*" (Abel 1998, p. 127). Further, a comprehensive review concluded that "*the two most important types of confounding of effects on neurodevelopmental outcome are failure to control for the postnatal environment and failure to control for factors which are strongly genetically influenced and which may be related to both prenatal alcohol exposure and the outcome*" (Gray and Henderson 2007a, p. 20).

These childhood-related factors are known as mediators. A mediator (Fig. 2) can be defined as "*a variable that occurs in a causal pathway from a causal (independent) variable to an outcome (dependent) variable. It causes variation in the outcome variable and itself is caused to vary by the original causal variable. Such a variable will be associated with both the causal and the outcome variable*" (Porta 2008, p. 131). A wide range of mediating factors could be relevant to control for in the statistical analyses including attachment, the child's IQ (Streissguth 1996), parental mental health in childhood (Sayal 2007), home environment (Abel 1998; Streissguth 1996), quality of parenting (Sayal 2007), changes in living arrangement (Streissguth 1996), violence (Streissguth 1996) and social support (Abel 1998). These factors are especially important in the first 2 years of life when the child's brain is particularly plastic (Abel 1998). Therefore, a child's behaviour and cognitive functioning is a product not only of prenatal factors (such as alcohol) but also of postnatal exposures (such as attachment).

Attachment can be defined as "an affectional tie that one person or animal forms between himself and another specific one—a tie that binds them together in space and endures over time" (Ainsworth and Bell 1970). In the 1950s Bowlby was the first to demonstrate the lasting consequences of the mother-child relationship on a wide range of developmental cognitive and mental health outcomes (Bowlby 1950). He concluded that infants who develop a *secure* attachment style are those with a history of sensitive and responsive maternal care. This style is associated with better emotional regulation, higher self-esteem, and more

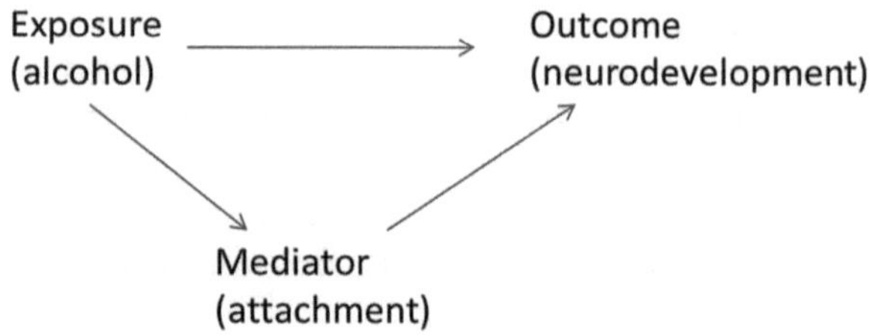

Fig. 2 The association between the exposure to alcohol and the outcome, namely child neurodevelopment and the influence of attachment as a mediator

develop coping skills. In turn this makes the children better able to handle stressful or challenging situations and conversely lowers their risk for poorer mental health outcomes later in life. Secure attachment is also found to be associated with better academic performance in adolescents and better cognitive performance in childhood (Mikulincer and Shaver 2007a, b). On the other hand, children with an *insecure* attachment are by contrast at greater risk for poor mental health outcomes (Ainsworth and Bell 1970; Sroufe 2005). Therefore, attachment has a "large" effect on mental health development, whereas exposure to low doses of alcohol all other things being equal will have a "small" effect on child development. Therefore, the potential negative effect of alcohol will most likely be masked by the relatively larger effect of attachment and other mediators. Moving back to the model in Fig. 2 it can be said that alcohol is hypothesized to be causally associated with child neurodevelopment, but also causally associated with attachment. Attachment is also hypothesized to be causally associated with child neurodevelopment. The mediator (attachment) thus causes variation in the outcome (neurodevelopment) but is itself caused to vary by the original causal variable (alcohol).

In summary, mediating psychological factors such as attachment are rarely, if ever, controlled for in the statistical analyses of the large-scale epidemiological cohort studies. Because they are strong predictors of mental health they most likely mask the potential, probably small, effect of low dose exposure. Because mediating psychological factors are not included in large scale cohorts we do not know whether large differences exist between exposure groups (as was the case with the confounding factors described above). But it is surely fair to postulate that large differences do exist and that such mediating factors in turn bias the results of the statistical analyses in the studies investigating prenatal exposure to alcohol and child development.

Imprecise Definitions of the Exposures

Above I have focused on problems concerning factors that ideally should be included in the statistical analyses, i.e. confounders and mediators. Below I will focus on problems related to the exposures (i.e., the mother's alcohol intake). What pieces of information do we use to classify the women? Firstly, I will focus on what is known as "*dose*", "*pattern*" and "*timing*", after which I will examine the validity of the self-reported measures of alcohol intakes generally used in large-scale cohort studies.

It is well recognised by researchers that exposure to larger *doses* of alcohol causes more damage than exposure to smaller doses. The literature presented above does so by distinguishing between exposure to different average doses of alcohol, for example comparing women who abstain in pregnancy, with those who drink one unit per week, with two units per week etc. Thus this is one factor that is always considered important and included in the definitions of the exposure categories. It may further complicate matters that, while scientists may have some consensus on the size of a drink of alcohol, the public may generally not be aware of what the typical size of a dose is.

Apart from accepting the importance of dose, the most recent literature also recognised the importance of considering the *pattern* of the exposure, i.e. the quantity consumed on a typical occasion (O'Leary et al. 2010). Most studies today therefore distinguish between exposure to averaged-over-time (lower) doses of alcohol and to 'binge' drinking (Gray and Henderson 2007a; Henderson et al. 2007a, c; Underbjerg et al. 2012; Eriksen et al. 2012; O'Leary et al. 2009; Skogerbo et al. 2012; Kesmodel et al. 2012; Flak et al. 2013). Binge drinking is most often defined as an intake of a minimum of five units of alcohol on a single occasion (Gray and Henderson2007a; Henderson et al. 2007c). The rationale for this distinction is that it has been established that exposure to binge drinking is more devastating for the developing central nervous system because it is the peak blood alcohol concentration that determines the level of the damage (Gray and Henderson 2007a; Henderson et al. 2007c). In other words, binge drinking causes greater harm than exposure to an equivalent amount of alcohol spread over several days, weeks or months (Henderson et al. 2007c; Kesmodel 2001). Generally, there seems to be evidence of a very deleterious effect of being exposed to binge drinking. One review concluded (Gray and Henderson 2007b) that children exposed to binge drinking during their mothers' pregnancies consistently showed poorer neurodevelopmental outcomes.

Apart from dose and pattern, a third factor, namely *timing,* seem to be of particular importance to understanding the effects of prenatal exposure to alcohol on child neurodevelopment. Most human studies have only investigated alcohol exposure during early pregnancy, (Sayal et al. 2007, 2009, 2013; Kelly 2010; Underbjerg et al. 2012; Skogerbo et al. 2012; Robinson et al. 2010) despite the fact that there appear to be two critical periods in human fetal development, when the brain is especially vulnerable to insult (Abel 1998). The first period begins during the first trimester, from gestational weeks 12–20. This period is characterised by a rapid rate of nerve-cell proliferation. The second period begins during the third trimester and continues until age 18 months of the child and is characterised by a brain growth spurt. Unfortunately, very few human studies have attempted to investigate possible effects of timing. One study that did so found that exposure to moderate-high levels of alcohol in the third trimester was negatively associated with language delay at age two (O'Leary et al. 2009). Another recent study compared binge drinking in the early and late part of pregnancy (not taking average intake into account) (Niclasen et al. 2013). This study reported that a small but significant association was observed on externalising scores (defined as behaviours that actions direct problematic energy outward) at age seven for the early exposure. However, the effect for the late exposure was much greater.

At least two conclusions can be drawn from the above evidence. One concerns the size of the effect of the exposure. Virtually all human studies focus on early exposure and report small effects of alcohol. But it may be that if the researchers instead focused on late exposure we might generally see stronger effects of alcohol. The second conclusion is related to the combination of the three factors dose, pattern and timing. Most researchers do not investigate all three factors. However, they all have a great impact on the results (small versus large doses; average versus binge drinking; and early versus late pregnancy exposure).

Apart from the questions of dose, pattern and timing, data on exposure can also suffer from reporting bias whereby women with actually a high intake may report only a zero or a very low intake. The former would thus be misclassified into the abstaining group (Abel 1998). It has been demonstrated that mothers who drink often deny their drinking habits (Garcia-Algar et al. 2012), that mothers who admit to drinking do not remember details correctly (Garcia-Algar et al. 2012), and that other women, in order to avoid disapproval, purposely under-report their alcohol intake (Kesmodel 2001; Kesmodel and Frydenberg 2004). Particularly women with a high intake tend to under-report alcohol consumption during pregnancy, especially if the information is not collected in a careful and sensitive manner (Kesmodel 2001). Because the under-reporting probably varies between groups this will also lead to misclassification that varies between intake groups. This, in turn, will inevitably have an impact on the results.

Problematic Use of Outcome Measures

Whereas the sections above were concerned with issues related to confounders and mediators, and to definitions of the exposure categories, this third section is focused on problems related to the outcome measures used in the large-scale cohort studies. The related critique can be subdivided into two aspects. The first part is concerned with the use of specific outcome measures. Those often consist of brief screening tools rather than extensive cognitive and neuropsychological test batteries. The second part is related to the ages at which the children are assessed. Here, the point that I will put forward is that the children are often assessed at too early ages when the potential neurobehavioural symptoms are not yet present.

It has been reported that people prenatally exposed to alcohol exhibit impairments on the performance of relatively complex and novel tasks (Kodituwakku et al. 2001a, b). Often researchers in the large-scale cohorts rely on information about child development from brief screening tools (i.e. short questionnaires) filled in by the children's mothers. On the positive side such screening tools often show excellent psychometric properties, particularly when used in large-scale studies. However, they may simply not be sensitive enough to identify potential damages from alcohol exposure. However, neuropsychological tasks are very difficult to incorporate in most large-scale cohorts because this is both very expensive and time-consuming have participants undergoing such assessments.

In practice, most often information on child development is only available from parental ratings although it is known from the literature that teachers are actually better at identifying for example externalising symptoms, like ADHD. In order to get a diagnoses of ADHD the problem behaviour has to be present in at least two settings, typically home and school. Because it is a particular problem for many children to 'stay still and keep focus' in a school setting this is typically where the symptoms are present. Parent raters are also known to give a conservative estimate of behavioural problems in their children and parents' ratings are generally found to be lower than those of teachers (Obel et al. 2004; Niclasen 2012). Therefore, it can be argued that teacher raters are needed in order to get better ratings of the children—at least concerning externalising behaviours.

The second element of critique regarding outcome measures is concerned with the age at which the child was assessed. Most often the timing of follow-ups of children is chosen for practical and pragmatic reasons rather than theoretical ones. This is problematic. When studying neurodevelopmental outcomes in children, the ages at which the follow-ups take place should be evidence and theory driven, rather than practicality-driven. Cohort studies have thus been criticised for assessing the children at inappropriate ages (Powell 2012; Parker and Brennan 2012; Astley and Grant 2012). For example, one study among 2,600 children with Fetal Alcohol Syndrome reported that about half had normal developmental scores at pre-school whereas all had severe brain dysfunction at age ten. In the same group of children 10 % were found to have attention problems at age five, whereas 60 % had attention problems at age ten (Astley 2010). The lack of consistency in the literature may thus in part be a consequence of this—that the children are assessed at inappropriate ages, in particular often too early, thus missing behavioural effects that may only manifest themselves later in childhood. Assessing the children at too early ages will imply two things. Firstly, that a potential effect of alcohol is not found in a study. Secondly, and perhaps more importantly, if there are many such studies it will bias the overall evidence of the association between low level drinking and child neurodevelopment.

Findings from the Animal Literature

I have described above how human observational studies are prone to methodological limitations. These include problems with variables that are not sufficiently controlled for in the statistical analyses (i.e. confounding and mediation), impressions in relation to the exposures and a problematic use of outcome measures. The advantage of animal studies, as oppose to observational human studies, is that all of these factors can be experimentally manipulated to disentangle the effects of dose, pattern and timing. In animal studies the subjects can be assigned into precisely defined exposure groups (Obviously, this would be out of the question in human studies). In turn, the problems of confounding and mediation can be kept to a minimum. Animal studies may thus shed light on the extent to which prenatal exposure to alcohol act as a teratogen on the developing foetus (Randall 2001).

Experimental animal studies have indeed demonstrated that the neuroteratogenic effects depend on dose, pattern and timing (Guerri et al. 2009; Plant 1999). Regarding *dose*, animal studies have concluded that exposure to low-moderate doses of alcohol is negatively associated with the anatomical development of the forebrain in rats (Ikonomidou et al. 2000), and attention and neuromotor functioning in rhesus monkeys (Schneider et al. 2001). Dose-response associations, which in human studies have been virtually absent, have also been demonstrated in numerous rat studies investigating spontaneous alternation, learning and conditioned taste aversion learning tasks (Stratton et al. 1996; Gray and Henderson 2007a; Gray et al. 2009; Riley et al. 1979).

In relation to *patterns*, animal studies have consistently shown negative effects on offspring exposed to binge-like patterns on discrimination tasks and association learning (as measured in mazes), motor skills (as measured by poor reflexes, coordination and balances), hyperactivity/inattention (as measured by preservation on reversal tasks and heightened exploratory behaviour), and social ability (as measured by play tasks, mating abilities and corporation) (Gray and Henderson 2007a; Gray et al. 2009; Driscoll et al. 1990).

The *timing* aspect has in the animal literature received much more attention compared to the human literature (Gray and Henderson 2007a; Gray et al. 2009). First trimester exposure has been found in rhesus monkeys to significantly decrease scores on infant neurobehavioural tests, whereas mid- to late gestation exposure results in reduced motor maturity (Schneider et al. 2001). Third trimester human brain development is the period of greatest brain growths and is equivalent to the early post-natal periods in rats. Numerous rat studies have reported that heavy alcohol exposure in late pregnancy results in reduced brain weight and volume, particularly in the forebrain, brainstem, cerebellum and corpus callosum (Ikonomidou et al. 2000; Chen et al. 2003; West 1987). A rhesus monkey study reported early gestation exposure as being negatively associated with infant neurobehavioural test scores whereas mid- to late exposure resulted in reduced motor maturity (Schneider et al. 2001). A study investigating binge-like drinking during the third trimester in vervet monkeys found significantly fewer neurons in the frontal cortex of the exposed offspring. The authors concluded that the entire dorsolateral prefrontal cortex was affected and that this provides an anatomical basis for the behavioural and cognitive deficits observed in children exposed to alcohol prenatally (Burke et al. 2009).

The primary advantage of studying prenatal exposure to alcohol by means of animal models is that it makes possible research that could for obvious ethical reasons not be carried out in humans. Thus, exposure and confounding and mediation can be controlled for in a more rigorous manner (Guerri et al. 2009). However, although animal studies have contributed greatly to the understanding of the effects of prenatal exposure to alcohol on neuroanatomical and neurodevelopmental outcomes they do have their limitations. The complex pre- and postnatal environments in humans cannot be replicated in animal models. Similarly, the complexity of human social behaviour in interaction with others, and the higher cognitive functioning in humans bear little resemblance to that of animals. Furthermore, the rapid third trimester brain development in humans does not take place until the early post-partum period in rodents. These factors may all limit the applicability of the results from the animal studies (Guerri et al. 2009).

Is It Ethical to Condone Low Levels of Drinking in Pregnancy?

Based on the evidence from the human literature I think the answer to this question is that we really do not know. The literature is far too contradictory. I would further assert that, with our current research practices, we will never know whether there is a safe lower level below which drinking is not associated with any harm to the developing foetus. This is because (a) the results of the state-of-the-art large-scale cohort studies probably reflect residual and unmeasured confounding and mediating factors; (b) it also seems that misclassification of alcohol consumption may impact upon the results, and problematic use of outcome measures. As long as we continue to use self-reported measures of consumption and as long as we do not sufficiently incorporate dose, pattern and timing into well-defined exposure categories we will be faced with the problem of misclassification; (c) finally, as long as the outcome measures consist of short screening questionnaires, rather than, for example, extensive and sophisticated neuropsychological tests, and the children are assessed at too early ages, potential associations will most likely not be unravelled.

From animal studies there appears to be somewhat more evidence that alcohol act as a teratogen on the developing foetus. However, most animal studies model exposure that corresponds to very high doses—not the usual 'one-drink-on-a-Saturday-night-light drinking patters that we are otherwise concerned with here. Such studies are very sparse. This might be because it is difficult to model such drinking behaviours in animals. Alternatively, it might be due to publication bias. It might be that animal studies focusing on low level drinking are particularly prone to publication. Whether or not this is actually the case is hard to know. Also, it is not without problems to extrapolate findings from the animal literature to humans: firstly because the complex pre- and post-natal human environments and complex social interactions are poorly approximated in animals and secondly because the third trimester brain development in humans takes place post natally in rodents. Therefore the applicability of the results is limited.

One way forward would be to revise our current research methods. I agree with Abel who concludes that "*until research progresses beyond simple group comparisons, our understanding of the nature of the cognitive deficits associated with the uniqueness of Fetal Alcohol Syndrome/Alcohol Related Birth Defects will never progress beyond superficiality*" (Abel 1998, p. 134). We could make use of a whole range of natural experiments and research designs including a mother/father design, a sibling design and a cultural comparison design. Each of these study designs will have its own strengths and weaknesses (Gray et al. 2009; Rutter et al. 2001). In the 'cultural design', data is compared between different cultural settings in which drinking in pregnancy is observed within different socio-economic groups. One study compared results from Denmark, where the affluent often drink in pregnancy, and Finland where it is more women with low socioeconomic status who drink in pregnancy (Rodriguez et al. 2009). This study did not find low doses of alcohol to be related to child inattention/hyperactivity once social adversity and smoking were taken into account.

In the 'sibling design' only mothers who changed their drinking habits between pregnancies are included. This way the effects of confounding and mediation are minimized. One study that did so did find that high exposed children exhibited more conduct problems than their low exposed siblings, thus suggesting a causal association. However, they did not find any associations with attention and impulsivity (D'Onofrio et al. 2007). Studies using mothers and children from extended families and adopted children would also be welcomed.

It can also be recommended to compare neurodevelopmental outcomes in children whose fathers drank in pregnancy but whose mothers did not drink in pregnancy, with children whose fathers did not drink but whose mothers did. If it is assumed that mothers drinking act as a biological teratogen, then similar effects in the two groups would imply that maternal drinking is not directly causative (Gray et al. 2009).

Conclusion

Because we really do not know the effect of low exposure to alcohol, and because we do know that exposure to large doses of alcohol act as a teratogen on the developing foetus, I believe that the safest thing is to recommend abstinence. It is, after all, 'better to be safe than sorry', and under these circumstances even low level drinking in pregnancy should be discouraged.

References

Abel, E.L. 1998a. Neurodevelopmental abnormalities. Structural and sensory. *Fetal alcohol abuse syndrome*, 101–110. New York: Plenum

Abel, E.L. 1998b. Neurodevelopmental abnormalities. Behavioural and cognitive. *Fetal alcohol abuse syndrome*, 111–138. New York: Plenum.

Ainsworth, M.D., and S.M. Bell. 1970. Attachment, exploration, and separation: Illustrated by the behavior of one-year-olds in a strange situation. *Child Development* 41(1): 49–67.

Astley, S.J. 2010. Profile of the first 1,400 patients receiving diagnostic evaluations for fetal alcohol spectrum disorder at the Washington State Fetal Alcohol Syndrome Diagnostic & Prevention Network. *The Canadian Journal of Clinical Pharmacology* 17(1): e132–e164.

Astley, S., and T. Grant. 2012. Another perspective on 'the effect of different alcohol drinking patterns in early to mid pregnancy on the child's intelligence, attention, and executive function'. *BJOG* 119(13): 1672–1675.

Bowlby, J. 1950. Research into the Origins of Delinquent Behaviour. *British Medical Journal* 1(4653): 570–573.

Braun, J.M., R.S. Kahn, T. Froehlich, P. Auinger, and B.P. Lanphear. 2006. Exposures to environmental toxicants and attention deficit hyperactivity disorder in U.S. children. *Environmental Health Perspectives* 114(12): 1904–1909.

Burke, M.W., R.M. Palmour, F.R. Ervin, and M. Ptito. 2009. Neuronal reduction in frontal cortex of primates after prenatal alcohol exposure. *NeuroReport* 20(1): 13–17.

Calhoun, F., and K. Warren. 2007. Fetal alcohol syndrome: Historical perspectives. *Neuroscience and Biobehavioral Reviews* 31(2): 168–171.

Chen, W.J., S.E. Maier, S.E. Parnell, and J.R. West. 2003. Alcohol and the developing brain: neuroanatomical studies. *Alcohol Research and Health* 27(2): 174–180.

D'Onofrio, B.M., C.A. Van Hulle, I.D. Waldman, J.L. Rodgers, P.J. Rathouz, and B.B. Lahey. 2007. Causal inferences regarding prenatal alcohol exposure and childhood externalizing problems. *Archives of General Psychiatry* 64(11): 1296–1304.

Driscoll, C.D., A.P. Streissguth, and E.P. Riley. 1990. Prenatal alcohol exposure: comparability of effects in humans and animal models. *Neurotoxicology and Teratology* 12(3): 231–237.

Eriksen, H.L.F., E.L. Mortensen, T. Kilburn, et al. 2012. The effects of low to moderate prenatal alcohol exposure in early pregnancy on IQ in 5-year-old children. *BJOG* 119(10): 1191–1200.

Flak, A.L., S. Su, J. Bertrand, C.H. Denny, U.S. Kesmodel, and M.E. Cogswell. 2013. The association of mild, moderate, and binge prenatal alcohol exposure and child neuropsychological outcomes: A meta-analysis. *Alcoholism: Clinical and Experimental Research*.

Fried, P.A., and B. Watkinson. 1988. 12- and 24-month neurobehavioural follow-up of children prenatally exposed to marihuana, cigarettes and alcohol. *Neurotoxicology and Teratology* 10(4): 305–313.

George, M.D. 1951. *London life in the eighteen century*.

Gray, R., R.A. Mukherjee, and M. Rutter. 2009. Alcohol consumption during pregnancy and its effects on neurodevelopment: What is known and what remains uncertain. *Addiction* 104(8): 1270–1273.

Garcia-Algar, O., D. Black, C. Guerri, and S. Pichini. 2012. The effect of different alcohol drinking patterns in early to mid-pregnancy. *BJOG* 119(13): 1670–1671.

Gray, R., and J. Henderson. 2007b. *Review of the fetal effects of prenatal alcohol exposure*. Oxford: National Perinatal Epidemiology Unit, University of Oxford.

Gray, R., and J. Henderson. 2007a. *Rewiev of the fetal effects of prenatal alcohol exposure*. Oxford: National Perinatal Epidemiology Unit, University of Oxford.

Guerri, C., A. Bazinet, and E.P. Riley. 2009. Foetal alcohol spectrum disorders and alterations in brain and behaviour. *Alcohol and Alcoholism* 44(2): 108–114.

Henderson, J., R. Gray, and P. Brocklehurst. 2007a. Systematic review of effects of low-moderate prenatal alcohol exposure on pregnancy outcome. *BJOG* 114(3): 243–252.

Henderson, J., R. Gray, P. Brocklehurst. 2007b. *Systematic review of effects of low–moderate prenatal alcohol exposure on pregnancy outcome*.

Henderson, J., U. Kesmodel, and R. Gray. 2007c. Systematic review of the fetal effects of prenatal binge drinking. *Journal of Epidemiology and Community Health* 61: 1069–1073.

Ikonomidou, C., P. Bittigau, M.J. Ishimaru, et al. 2000. Ethanol-induced apoptotic neurodegeneration and fetal alcohol syndrome. *Science* 287(5455): 1056–1060.

Jones, K.L. 2012. *FASD from a historical perspective*. Barcelona, Spain.

Jones, K.L., and D.W. Smith. 1973. Recognition of the fetal alcohol syndrome in early infancy. *Lancet* 302(7836): 999–1001.

Jones, K.L., D.W. Smith, C.N. Ulleland, and P. Streissguth. 1973. Pattern of malformation in offspring of chronic alcoholic mothers. *Lancet* 1(7815): 1267–1271.

Jones, K.L., D.W. Smith, A.P. Streissguth, and N.C. Myrianthopoulos. 1974. Outcome in offspring of chronic alcoholic women. *Lancet* 1(7866): 1076–1078.

Juul, S. 2008. *Epidemiologi og evidens*, 1st edn. Munksgaard.

Kelly, Y., A. Sacker, R. Gray, J. Kelly, D. Wolke, and M.A. Quigley. 2009. Light drinking in pregnancy, a risk for behavioural problems and cognitive deficits at 3 years of age? *International Journal of Epidemiology* 38(1): 129–140.

Kelly, Y.J., A. Sacker, R. Gray, et al. 2010. Light drinking during pregnancy: Still no increased risk for socioemotional difficulties or cognitive deficits at 5 years of age? *Journal of epidemiology and* community health.

Kesmodel, U. 2001. Binge drinking in pregnancy–frequency and methodology. *American Journal of Epidemiology* 154(8): 777–782.

Kesmodel, U., and M. Frydenberg. 2004. Binge drinking during pregnancy—is it possible to obtain valid information on a weekly basis? *American Journal of Epidemiology* 159(8): 803–808.

Kesmodel, U.S., H.L. Eriksen, M. Underbjerg, et al. 2012. The effect of alcohol binge drinking in early pregnancy on general intelligence in children. *BJOG* 119(10): 1222–1231.

Kodituwakku, P.W., W. Kalberg, and P.A. May. 2001a. The effects of prenatal alcohol exposure on executive functioning. *Alcohol Research and Health* 25(3): 192–198.

Kodituwakku, P.W., P.A. May, C.L. Clericuzio, and D. Weers. 2001b. Emotion-related learning in individuals prenatally exposed to alcohol: an investigation of the relation between set shifting, extinction of responses, and behavior. *Neuropsychologia* 39(7): 699–708.

Larroque, B., and M. Kaminski. 1998. Prenatal alcohol exposure and development at preschool age: Main results of a French study. *Alcoholism, Clinical and Experimental Research* 22(2): 295–303.

Latimer, K., P. Wilson, J. Kemp, et al. 2012. Disruptive behaviour disorders: a systematic review of environmental antenatal and early years risk factors. *Child: Care, Health and Development* 38(5): 611–628.

Linnet, K.M., K. Wisborg, NJ Secher, et al. 2009. Coffee consumption during pregnancy and the risk of hyperkinetic disorder and ADHD: A prospective cohort study. *Acta Paediatrica* 98(1): 173–179.

Linnet, K.M., S. Dalsgaard, C. Obel, et al. 2003. Maternal lifestyle factors in pregnancy risk of attention deficit hyperactivity disorder and associated behaviors: review of the current evidence. *American Journal of Psychiatry* 160(6): 1028–1040.

MacNicholl, T.A. 1905. A study of the effects of alcohol on school children. *Quarterly Journal of Inebriety* 27: 113–117.

Mattson, S.N., and E.P. Riley. 1998. A review of the neurobehavioral deficits in children with fetal alcohol syndrome or prenatal exposure to alcohol. *Alcoholism, Clinical and Experimental Research* 22(2): 279–294.

Mick, E., J. Biederman, and S.V. Faraone. 1996. Is season of birth a risk factor for attention-deficit hyperactivity disorder? *Journal of the American Academy of Child and Adolescent Psychiatry* 35(11): 1470–1476.

Mikulincer M., P.R. Shaver. 2007a. *Attachment in adulthood. Structure, dynamics and change*, 1st edn. New York: The Guildford Press.

Mikulincer M., P.R. Shaver. 2007b. Attachment orientations, behavioural self-regulation, and personal growth. *Attachment in adulthood. Structure, dynamics and change*, 1st edn., 219–250. New York: The Guildford Press.

NIAAA: National Institute of Alcohol Abuse and Alcoholism. 2000. 10th special report to the US congress on alcohol and health. Highlights from current research. Bethesda: Department of health and human services; Report No.: 00-1583.

Niclasen, J. 2012. SDQ Danish cut off scores: 5–7-year-olds. 18-12-2012.

Niclasen, J. 2014. Drinking or not drinking in pregnancy: The multiplicity of confounding influences. *Alcohol and Alcoholism* 49(3): 349–355.

Niclasen, J., A.M. Andersen, T.W. Teasdale, and K. Strandberg-Larsen. 2013. Binge drinking and cumulated alcohol exposure in pregnancy: behavioural and emotional development at age seven. *Journal Epidemiology and Community Health.*

Niclasen J, T.W. Teasdale, K. Strandberg-Larsen, and A.M. Nybo Andersen. 2013. Is alcohol binge drinking in early and late pregnancy associated with Behavioural and Emotional Development at Age Seven? *European Journal of Child and Adolescent Psychiatry.*

O'Leary, C.M. 2004. Fetal alcohol syndrome: Diagnosis, epidemiology, and developmental outcomes. *Journal of Paediatrics and Child Health* 40(1–2): 2–7.

Obel, C., E. Heiervang, A. Rodriguez, et al. 2004. The strengths and difficulties questionnaire in the Nordic countries. *European Child* and *Adolescent Psychiatry* 13(Suppl 2): II32-9:II32-II39.

Obel, C., K.M. Linnet, TB Henriksen, et al. 2008. Smoking during pregnancy and hyperactivity-inattention in the offspring—comparing results from three Nordic cohorts. *The International Journal of Epidemiology.*

O'Leary, C.M., S.R. Zubrick, C.L. Taylor, G. Dixon, and C. Bower. 2009. Prenatal alcohol exposure and language delay in 2-year-old children: the importance of dose and timing on risk. *Pediatrics* 123(2): 547–554.

O'Leary, C.M., C. Bower, S.R. Zubrick, E. Geelhoed, J.J. Kurinczuk, and N. Nassar. 2010. A new method of prenatal alcohol classification accounting for dose, pattern and timing of exposure: Improving our ability to examine fetal effects from low to moderate alcohol. *Journal of Epidemiology and Community Health* 64(11): 956–962.

Olson, H.C., A.P. Streissguth, P.D. Sampson, H.M. Barr, F.L. Bookstein, and K. Thiede. 1997. Association of prenatal alcohol exposure with behavioral and learning problems in early adolescence. *Journal of the American Academy of Child and Adolescent Psychiatry* 36(9): 1187–1194.

Parker, M.O., and C.H. Brennan. 2012. Low and moderate alcohol consumption during pregnancy: Effects on social behaviour and propensity to develop substance abuse in later life. *BJOG* 119(13): 1671–1672.

Plant, M.L., E.L. Abel, and C. Guerri. 1999. Alcohol and pregnancy. In *Health issues related to alcohol consumption,* ed. I. MacDonald, 2nd edn. ILSI Europe; 1999. 181–214.

Porta, M. 2008. *A dictionary of epidemiology*. Oxford: Oxford University Press.

Powell, G. 2012. Danish studies suggesting low and moderate prenatal alcohol exposure has no adverse effects on children aged 5 years did not use appropriate or effective measures of executive functioning. *BJOG* 119(13): 1669–1670.

Randall, C.L. 2001. Alcohol and pregnancy: Highlights from three decades of research. *Journal of Studies on Alcohol* 62(5): 554–561.

Rasmussen, C., G. Andrew, L. Zwaigenbaum, and S. Tough. 2008. Neurobehavioural outcomes of children with fetal alcohol spectrum disorders: A Canadian perspective. *Paediatrics and Child Health* 13(3): 185–191.

Riley, E.P., E.A. Lochry, and N.R. Shapiro. 1979. Lack of response inhibition in rats prenatally exposed to alcohol. *Psychopharmacology (Berl)* 62(1): 47–52.

Robinson, M., W.H. Oddy, N.J. McLean, et al. 2010. Low-moderate prenatal alcohol exposure and risk to child behavioural development: A prospective cohort study. *BJOG* 117(9): 1139–1150.

Rodriguez, A., J. Miettunen, T.B. Henriksen, et al. 2008. Maternal adiposity prior to pregnancy is associated with ADHD symptoms in offspring: Evidence from three prospective pregnancy cohorts. *The International Journal of Obesity (London)* 32(3): 550–557.

Rodriguez, A., J. Olsen, A.J. Kotimaa, et al. 2009. Is prenatal alcohol exposure related to inattention and hyperactivity symptoms in children? Disentangling the effects of social adversity. *Journal of Child Psychology and Psychiatry* 50(9): 1073–1083.

Rothman, K.J. 2002. *Epidemiology. An intoduction*. Oxford: Oxford University Press.

Rutter, M., A. Pickles, R. Murray, and L. Eaves. 2001. Testing hypotheses on specific environmental causal effects on behavior. *Psychological Bulletin* 127(3): 291–324.

Sayal, K. 2007. Alcohol consumption in pregnancy as a risk factor for later mental health problems. *Evidence-Based Mental Health* 10(4): 98–100.

Sayal, K., J. Heron, J. Golding, and A. Emond. 2007. Prenatal alcohol exposure and gender differences in childhood mental health problems: A longitudinal population-based study. *Pediatrics* 119(2): e426–e434.

Sayal, K., J. Heron, J. Golding, et al. 2009. Binge pattern of alcohol consumption during pregnancy and childhood mental health outcomes: Longitudinal population-based study. *Pediatrics* 123(2): e289–e296.

Sayal, K., E.S. Draper, R. Fraser, M. Barrow, S.G. Davey, and R. Gray. 2013. Light drinking in pregnancy and mid-childhood mental health and learning outcomes. *Archives of Disease in Childhood* 98(2): 107–111.

Schneider, M.L., C.F. Moore, and E.F. Becker. 2001. Timing of moderate alcohol exposure during pregnancy and neonatal outcome in rhesus monkeys (Macaca mulatta). *Alcoholism, Clinical and Experimental Research* 25(8): 1238–1245.

Skogerbo, A., U.S. Kesmodel, T. Wimberley, et al. 2012. The effects of low to moderate alcohol consumption and binge drinking in early pregnancy on executive function in 5-year-old children. *BJOG* 119(10): 1201–1210.

Sood, B., V. Aney-Black, C. Covington, et al. 2001. Prenatal alcohol exposure and childhood behavior at age 6 to 7 years: I. Dose-response effect. *Pediatrics* 108(2): E34.

Sroufe, L.A. 2005. Attachment and development: a prospective, longitudinal study from birth to adulthood. *Attachment & Human Development* 7(4): 349–367.

Stratton, K., C. Howe, F. Battaglia. 1996. *Fetal alcohol syndrome. Diagnosis, epidemiology, prevention and treatment*. Washington, D. C.: National Academy Press.

Streissguth, A.P., H.M. Barr, J. Kogan, F.L. Bookstein. 1996. Understanding the occurence of secondary disabilities in clients with fetal alcohol syndrome (FAS) and fetal alcohol effcts (FAE): Final report. Seattle: University of Washington school of medicine. Department of Psychiatry and behavioural sciences.

Streissguth, A.P., H.M. Barr, P.D. Sampson, J.C. Parrish-Johnson, G.L. Kirchner, and D.C. Martin. 1986. Attention, distraction and reaction time at age 7 years and prenatal alcohol exposure. *Neurobehavioral Toxicology and Teratology* 8(6): 717–725.

Streissguth, A.P., H.M. Barr, and P.D. Sampson. 1990. Moderate prenatal alcohol exposure: Effects on child IQ and learning problems at age 7½ years. *Alcoholism, Clinical and Experimental Research* 14(5): 662–669.

Streissguth, A.P., P.D. Sampson, H.C. Olson, et al. 1994. Maternal drinking during pregnancy: attention and short-term memory in 14-year-old offspring—a longitudinal prospective study. *Alcoholism, Clinical and Experimental Research* 18(1): 202–218.

Testa, M., B.M. Quigley, and R.D. Eiden. 2007. The effects of prenatal alcohol exposure on infant mental development: A meta-analytical review. *Alcohol and Alcoholism* 38(4): 295–304.

Thomsen P.H., A.M. Skovgaard. 2007. *Børne- og Ungdoms psykiatri. En lærebog om børn og unges psykiske sygdomme*. FADL Forlag.

Underbjerg, M., U.S. Kesmodel, N.I. Landro, et al. 2012. The effects of low to moderate alcohol consumption and binge drinking in early pregnancy on selective and sustained attention in 5-year-old children. *BJOG* 119(10): 1211–1221.

Warner, R.H., and H.L. Rosett. 1975. The effects of drinking on offspring: An historical survey of the American and British literature. *Journal of Studies on Alcohol* 36(11): 1395–1420.

West, J.R. 1987. Fetal alcohol-induced brain damage and the problem of determining temporal vulnerability: A review. *Alcohol and Drug Research* 7(5–6): 423–441.

Ethical Challenges When Screening for and Diagnosing FASD in Adults

Ilona Autti-Rämö

Knowledge of the long-term effects of FASD has increased in recent years, and special attention has been focused on adolescents and young adults with neurobehavioural and psychosocial problems or problems with the law without a prior FASD diagnosis. However, there has been little discussion about the ethical issues concerning the implementation of a screening programme for diagnosing FASD in adolescents and young adults. With the help of an in-depth framework, it is possible to capture the broad consequences that implementing a screening programme for the diagnosis of FASD holds for individual citizens, the health care system and society. Several critical ethical issues transferable from country to country can be identified that should be carefully discussed prior to implementing a screening programme for the diagnosis of FASD.

Research on FASD has, so far, been focused on the follow-up of children born to mothers with identified alcohol or drug abuse during pregnancy. According to these studies the identified large continuum of teratogenic effects of prenatal alcohol exposure has been as FASD (Fetal Alcohol Spectrum Disorder) which has been further divided into 4 groups: Fetal Alcohol Syndrome (FAS, Partial Fetal Alcohol Syndrome (PFAS), Alcohol Related Neurobehavioural Disorder (ARND) and Alcohol Related Birth Defect (ARBD) (Hoyme et al. 2005; Chudley et al. 2005; Astley 2006). However, there is so far no worldwide consensus on the specific criteria for the various subgroups or FASD.

The risk of being permanently damaged due to prenatal exposure to alcohol is not a rarity but a major concern, as was stated by the WHO (http://www.who.int/substance_abuse/msbalcstragegy.pdf). The best estimates on the prevalence of FASD are based on so called in-school studies (May et al. 2009). In these studies one age cohort is screened for symptoms suggesting FASD (growth retardation, typical facial features, complex neurobehavioural problems) and those screened

I. Autti-Rämö (✉)
The Social Insurance Institution, Research Department and Adjunct Professor,
Department of Pediatric Neurology, University of Helsinki, Helsinki, Finland
e-mail: Ilona.autti-ramo@kela.fi

M. Nelson and M. Trussler (eds.), *Fetal Alcohol Spectrum Disorders in Adults: Ethical and Legal Perspectives*, International Library of Ethics, Law, and the New Medicine 63, DOI 10.1007/978-3-319-20866-4_6

positive are further examined to receive the diagnosis FAS, PFAS, ARND or deferred. The lowest prevalence rates are based on national registers indicating either poor diagnostic resources or an unwillingness to give the diagnosis of FASD. The current estimate for the prevalence of FASD in western countries is 1/100 but it may be as high as 1 in 20. In Finland alcohol use among fertile-aged women has increased almost six-fold during the past 40 years and the share of abstaining fertile-aged women has decreased to about 10 % (Mäkelä et al. 2010). A similar increase has been reported in many Western countries. It is thus probable that the prevalence of FASD is higher among children and adolescents than among adults. In many Eastern countries the prevalence of alcohol consumption among fertile-aged women is still very low and it can be forecasted that with the increase of alcohol abuse among women in these countries the worldwide prevalence of FASD will also increase.

Knowledge of the long-term effects of FASD has increased in recent years, and special attention has been paid to adolescents and young adults with neurobehavioural and psychosocial problems or problems with the law without a prior FASD diagnosis. Currently in Canada and the United States new diagnostic facilities for diagnosing FASD in adolescents and adults are being founded and many juvenile courts both in Canada and the United States are sending young adults to FASD diagnostic centers. Some of the juvenile courts have developed screening protocols for FASD.

There has been very little discussion about the ethical issues concerning the implementation of a screening programme for diagnosing FASD. With the help of an in-depth framework, it is possible to capture the broad consequences that implementing a screening programme for the diagnosis of FASD holds for individual citizens, the health care system and society. Several critical ethical issues transferable from one country to another can be identified that should be carefully discussed prior to implementing a screening programme for the diagnosis of FASD.

Criteria for a Screening Programme

Within health care, screening for a disease is a societal decision. The aim is to improve the outcome and thus treatment and care that improve the outcome or wellbeing of the person need to be available. The person is not actively looking for a diagnosis him/herself and participation is based on informed consent. Screening can be population based or targeted at risk groups. Diagnostic procedures are conducted due to symptoms and based on individual decision: the patient wants to have a diagnosis and treatment to help overcome the symptoms. The professional tries to find a reason for the symptoms with the goal of starting a specific treatment. However, treatment may not always be available.

Every society creates norms that reflect its values. These norms form the basis for laws and moral rules that protect the individuals from harming each other.

Some rules are made explicit as regulations; others remain implicit but are nonetheless followed. In this article I use the Finnish criteria for the evaluation of a screening programme. These were originally based on the criteria presented by Wilson and Jungner (1968) combined with additions by The Danish Council of Ethics (2001). This set of criteria have been compiled and used by the Working Group for Screening at the Ministry of Social Welfare and Health to evaluate existing and proposed screening programmes (Table 1 and Autti-Rämö et al. 2005).

Table 1 Criteria for the evaluation of screening programmes in Finland. The origin of the criteria are marked with WHO (Wilson and Junger 1968) and D (The Danish Council of Ethics 2001) with the respective number of the original criteria. The existence of available evidence for each criteria when screening for FASD in adults is marked by Yes, Unclear or No

Criteria	Origin	Criteria fulfilled
1. The condition sought should be an important health problem both to an individual and society	WHO 1	Yes
2. The disease can be identified with screening at a latent or early symptomatic stage (WHO4)	WHO 4	No
3. The natural history of the condition, including development from latent to declared disease, should be adequately understood	WHO 7	No
4. There should be a suitable test or examination for screening for which following characters are known	WHO 5	
- validity of the testing system	D 11a	Unclear
- sensitivity and specificity	D 11b	No
- predictive value of tests	D 11 c	No
5. The test is acceptable to population	WHO 6	Unclear
6. An evaluation has been made of		
- stigmatization	D 12b	No
- the ethical and psychological consequences for the examinees	D 12a	No
- the consequences of "false positive" and "false negative" test results	D 12c	No
7. An agreed policy exists on whom to treat as patients	WHO 8	Unclear
8. An accepted and effective treatment exists	WHO 2	No
9. Facilities for diagnosis and treatment are available	WHO 3	No
10. The screening organization has been described in detail	D 14	
- At national and local levels	D 14a	Unclear
- Steering, quality control and registration system	D 14b,c	Unclear
- provision of information to target group	D 14e	Unclear
- triage	D 14d	Unclear
- staff training and	D 14f	Unclear
- test result dissemination and consultation	D 14 g	Unclear
11. The costs of casefinding (including diagnosis and treatment of patients diagnosed) need to be described	WHO 9	No
12. The effect of screening is evaluated	D 13	No
13. Screening is a continuing process	Who 10	Unclear
14. The costs of screening are economically balanced in relation to possible expenditure on medical care as a whole	Who 9	Unclear

With the help of these criteria I will first evaluate, whether screening for FASD among adolescents or adults, especially by court order, fulfils the criteria set for a screening programme within the health care system.

When going through the various criteria for a successful screening programme within the health care system (see Table 1), it is clear that a screening for FASD ordered by a juvenile court does not fulfill a majority of the criteria. There is thus not enough evidence that this screening programme can and will be of benefit for the person or the society from the use of health resources perspective.

Ethical Evaluation of Screening for and Diagnosing FASD in Adolescents and Adults

The European Union has funded a multinational project called EUnetHTA that aims to facilitate the efficient use of resources available for technology-assisted decision making about health in Europe (see http://www.eunethta.eu). The EUnetHTA collaboration has so far produced three core models to identify important, clinically relevant issues transferable from one country to another, which need to be answered when deciding on the implementation of a specific intervention, diagnostic method or screening programme. The adopted core models are divided into nine domains of equal importance: 1. Health problems and current use of technology; 2. Description and technical characteristics of technology; 3. Safety; 4. Clinical effectiveness; 5. Costs and economic evaluation; 6. Ethical aspects; 7. Organisational aspects; 8. Social aspects and 9. Legal aspects.

The EUnetHTA core models are an attempt to define and standardise elements of a health technology assessment. It is based on the idea that in spite of differences in health care systems, specific issues are of importance worldwide and the answers to these issues are completely or partially transferable from country to country. Thus the identified core issues should always be evaluated before implementing a new technology in any country (Saarni et al. 2008). In trying to identify the ethical challenges of screening for and making or excluding an FASD diagnosis in young adults I have used the EUnetHTA core models for screening and diagnostics. Several of the ethically important issues are of concern not only for the person with FASD but also for other stakeholders: parents, siblings, professionals and the society. As screening for FASD is a new programme, its consequences are likely to be more difficult to predict than the consequences of changing methods within an existing screening programme. Individual values, attitudes and expectations within and for the health care systems may be different in various cultures. Novel screening programmes can also have far-reaching and unexpected consequences on health and social care.

In this article I will focus on identifying the most relevant ethical issues faced when deciding on the implementation of screening for and diagnosing FASD in young adults by using the EUnetHTA core models for diagnostics and screening programmes. I do not attempt to present an ethical analysis of the various issues at stake but rather to reveal the possible unintended consequences of ethical relevance of which people involved in screening for and diagnosing FASD in adults should be aware. Special emphasis is given to screening for FASD in juvenile court.

1. Principal questions about the social aspects.

Identifying the conceptions behind the beliefs of the value of the screening programme may help to put them in perspective, when considering their overall acceptability. It is also important to identify those groups within the society for whom the screening programme and diagnostic tests will be used as the expert resources are sparse. Currently screening programmes are mainly used either to estimate the prevalence of FASD (in-school studies) or in juvenile court to define FASD as a mitigating factor. Both these programmes challenge traditional conceptions of a screening programme (e.g. voluntary basis and effective treatment is available). The health care system has to be prepared to handle moral issues raised by true positive and false negative findings.

2. Hidden or unintended consequences of the technology and its applications for different stakeholders.

Screening for FASD can have side-effects in addition to those following from its intended use. Unintended consequences may not always be predicted, though many are. Screening positive and being diagnosed with FASD may have effects on siblings, and one has to consider whether to offer diagnostic procedures for all siblings. To adopt coherent and just health policies one has to consider how relevantly similar situations are treated in a health care system. If a person was diagnosed with a hereditary disorder, one would usually offer the diagnostic procedure also to the siblings. In addition, mothers with alcohol abuse during pregnancy have been identified to have both increased morbidity and mortality (Kahila et al. 2010). When is it morally acceptable to neglect the need to examine the mother and siblings and provide them with the necessary treatment as well?

FASD can be a welcome diagnosis in some situations (e.g. foster care). The diagnosis may, however, not always be beneficial to the person but can lead to social stigma depending on the values and beliefs of the society as attitudes play a major role. One has to consider what the value of the diagnostic information is and what kind of information does the diagnose give to the person him/herself. The primary information from the diagnose is prenatal alcohol exposure that labels the mother in a way that no other disorder of prenatal origin does.

In screening programmes, diagnostic information by definition always necessitates further action, so all new screening programmes do have an impact on the health care processes and systems. Screening for and diagnosing FASD in a person also necessitates the evaluation of the needs of the mother and the siblings. Screening for FASD in adults may also change the concepts of the disorder if the definition of whom to treat as a patient is unclear (see point 4).

Screening for and diagnosing FASD by order of the juvenile court will require resources that are, so far, sparse. This will inevitably lead to a situation where a majority of persons diagnosed with FASD in their adulthood will have problems with the law. This may lead to a conception among laypersons that all persons with FASD have a high susceptibility to criminal behaviour which can lead to discrimination and stigmatization.

3. The value of a screening program and diagnostic testing

The ethical analysis of a screening programme should acknowledge the value-ladenness of the whole process from identifying difficulties in individual performance to organizing the required treatment and support. Thus diagnosing a disorder is morally acceptable so far as it has practical consequences to the person. Diagnosis as the only intervention is rarely possible. So far no studies are available on the value diagnosing FASD in adolescents or adults has for themselves. The effectiveness of screening and diagnostic programmes is mainly described either from the perspective of the legal system or foster care. The context-specificity of diagnosing FASD is clear; in juvenile court the value is related to minimizing legal consequences and in foster care it is about identifying both primary and secondary behavioural problems. But what is the value of the diagnosis for the person him/herself, other family members or the social and health care sectors?

4. Are the accuracy measures decided upon and balanced in a transparent and acceptable way?

Diagnosing FASD is based on a variety of cut-off limits (growth, palpebral fissure), more consensus and less evidence based definitions of affected brain domains and risk-levels on prenatal alcohol exposure. Diagnosing FASD requires extensive interpretative skills and the practical consequences of the accepted test "battery" and cut-off limits will depend on the population tested. FASD is not a dichotomous diagnosis and the children are inevitably exposed to a double burden: the effects of prenatal exposure and traumatizing postnatal living surroundings. If the problems arise first in adolescence it may be impossible to define whether they were due to prenatal alcohol exposure or due to a cascade of harmful events during childhood (Fig. 1).

No international consensus on the various cut-off limits, the definition of CNS involvement, or the level of risk exposure exist. The decision on cut-off values defines the sensitivity and specificity of the test battery and if the cut-offs are arbitrary, who does the balancing between true and false positives? This should be done considering the moral value of different results. In general, high specificity is required if false positives have serious consequences. The moral value of different results—true and false positive, true and false negative—may be different depending on their values in the society.

Several diagnostic challenges exist and one has to question whether the accuracy measures are balanced in a transparent and acceptable way. We do not yet have a reliable and objective method to identify prenatal alcohol exposure or to define the criteria for harmful exposure. There is so far no international consensus on what kind of neurobehavioural profile is typical and specific for FASD. Even the requirements for facial features differ in various diagnostic batteries and a lack of racial standards for palpebral fissure makes this criterion less valid than head

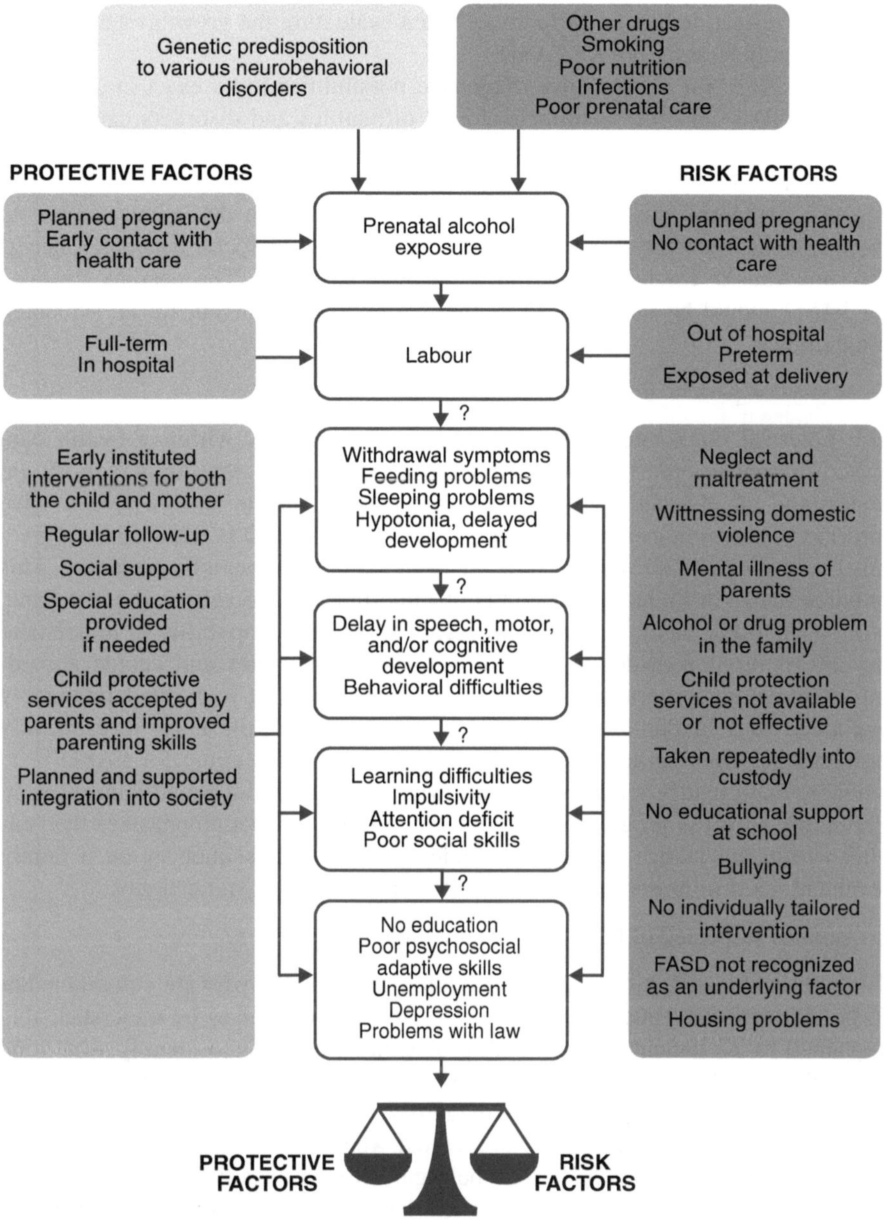

Fig. 1 The outcome in adulthood after prenatal alcohol exposure is affected by other risk and protective factors of pre- and postnatal origin

circumference. Criteria for growth retardation are based solely on growth curves. In clinical practice the growth of the parents is always taken into account when evaluating individual growth trajectories. There is no objective basis for neglecting hereditary effect and racial differences when evaluating the growth retardation or length of palpebral fissure for FASD.

There are so far no objective diagnostic possibilities to be exact in differentiating ARND from other neurobehavioural difficulties and disorders caused by a traumatized childhood. Thus there is always the possibility for both false positive and false negative diagnoses. The consequences of a false positive and a false negative diagnosis may be country specific depending on the value attached to the diagnosis in society—e.g., whether it is a "ticket" to social benefits, a necessary prerequisite for judicial action, a negative label, etc. The persons screened for FASD should be informed about these possibilities inherent for all screening programmes.

5. Effects on patient autonomy

The practical challenge with all screening programmes within a health care system is that in order to be fully autonomous, the participating person should understand all alternatives following different test results and be able to give informed consent at every step. When screening for FASD is required by a juvenile court, it can be questioned how the right to autonomy can be preserved. This requires the right to decide on things of importance to oneself on the one hand, but, on the other hand, also relevant information and a capability to understand the information, consider it in relation to personal values and decide accordingly. Screening programmes in general can be difficult to be understandably explained to a layperson. Screening for FASD in juvenile court is especially problematic. One has to be prepared to answer why screening for FASD is so valuable as to justify its use on people who cannot give informed consent to it. Especially so if the program is implemented in manner that emphasizes the benefits within the justice system and undervalues the consequences on a family level and leads to minor or no health related interventions and benefits.

6. Persons suspected of having FASD are vulnerable

The right and justification to use a technology for persons who are vulnerable has to be clarified. Any intervention that does not allow itself to be understandably explained to the patient is potentially problematic. One has to identify what kind of special care do persons suspected of having FASD need even when merely discussing the possibility of an FASD diagnosis. One has to be prepared to discuss why they are being examined for FASD and what it will mean for them if diagnosed. In some cultures FASD may be regarded as a negative label blaming the mother, a situation where the persons involved have no chance to defend themselves. In others "the system" may require an FASD diagnosis as it is considered a mitigating factor or a requirement to receive benefits, support or treatment.

Persons with FASD often have diminished decision-making capacity and understanding of long term consequences. Who has the right to balance the benefit against possible harm in these situations? If it is the legal system, on what grounds can these decisions be made? Is this specific, etiology-given diagnosis so valuable that it justifies its use on people who cannot give informed consent to it? It has to be remembered that FASD is a very heterogenous disorder and the diagnosis as such does not inform anything about the neurobehavioural profile. There are a large number of persons with various other neurobehavioural disorders of pre-, peri- or postnatal origin that can lead to difficulties in understanding the rules and regulations of the society. Would it be more justified to screen for and diagnose specific neurobehavioural profiles that diminish a person's capacity to understand the laws and norms of the society?

7. Informing the person

Within health care, the professional practice is to discuss the content and possible consequences of a screening programme and diagnostic test with patients. How is this performed in a juvenile court system? Are the persons explicitly informed about false positive and false negative screening results and that the diagnosis is based on expert knowledge and somewhat artificially defined cut-off criteria?

8. Effect on professional values, ethics and traditional roles

The patient–physician relationship is traditionally based on mutual trust, confidentiality and professional autonomy so that individual treatment decisions can be made in the best interest of the patient. In a patient–physician relationship, outside requirements cannot override the need for acknowledging individual patient values. As an FASD diagnosis gives more weight on etiology than the actual needs of the person it may interfere with core virtues and principles of medical and professional ethics. Interventions that align with professional ethics are more likely to be implemented successfully. For professional ethics it may be challenging to give the diagnosis of FASD to a person with a severely traumatizing childhood in spite of informed prenatal alcohol exposure. In such situations FASD may be an underlying disorder but it does not explain the difficulties encountered in adulthood.

9. Effect on human dignity and integrity

Technologies that are applied for persons with reduced autonomy may violate a person's dignity, i.e. challenge the idea that all human beings have intrinsic moral value, and should thus not be seen as means to others' ends. Labelling people may also threaten their dignity or define them as less worthy. Does an FASD diagnosis lead to better acceptance of one's problems or will it affect the individual child–parent relationship and if so, for better or worse? What does it mean for the person when the diagnostic procedure is required by the authorities and not by the person him/herself?

Integrity can be seen as a coherent image or identity of oneself. Institutions that tempt to dishonesty are detrimental to integrity; for example, systems where lying about one's state of health might lead to better treatment than being honest. There is no possibility to objectively verify or exclude prenatal alcohol exposure in a person with neurobehavioural difficulties and thus a risk exists that mothers tell about prenatal alcohol exposure in the hope of minor sentencing for their child.

10. Balancing the benefits and harms when implementing and not implementing the screening programme for FASD

The decision to implement a screening programme for FASD requires a careful decision on the balance between benefit and harm, cost-effectiveness, reallocation of resources etc. When this decision has been made on the system level, the decision on the individual patient level rests on both the professionals who offer the programme and the person who autonomously accepts to participate at every possible step. The individual decision has to be based on objective information on possible benefits and risks. Risks are only justified to the extent they are needed to create benefits. If not proven otherwise, the individual patient is generally to be seen as the best judge of risks and benefits for him/herself. The main questions are, whether we can foresee the possible benefits and risks of an FASD diagnosis for an individual and who will weigh them?

11. Effects on any other stakeholders

Some screening programmes have the potential to unfold unwanted or harmful effects not only on the persons being screened for but also indirectly on other stakeholders. Screening for FASD can have consequences for mothers, siblings, health care organizations, society etc. Diagnosing FASD may negatively interfere with the wellbeing of the whole family. Benefits and harms to individuals must be balanced with benefits and harms that can affect the society as a whole (maximizing public health). Changes in the requirements of diagnostic possibilities and treatment facilities may significantly alter also the requirements placed on the health care system.

12. Justice in the health care system

When deciding on implementing a new screening programme one has to consider the consequences implementing and not implementing have on justice in the health care system. Screening for and diagnosing FASD requires reallocation of human resources, funding and training. A large reallocation of resources may seriously jeopardize other patient groups with similar needs. If FASD is to be screened for, how can one guarantee that the programme is implemented in a way that there is equal access to those in equal need and benefit of the programme? Potential discrimination or other inequalities (geographic, gender, ethnic, employment, insurance) may prevent access to diagnostic procedures and potential inequalities and discrimination should be justified. Screening programmes can sometimes acquire significant symbolic value that may create demands that are not justified on health grounds. Is this the case in screening for FASD and not for neurobehavioural disorders in general in juvenile court? For diagnostic tests, issues of access to tests and treatments as well as labelling and potential discrimination of diagnosed persons may be relevant issues when considering the justness of the system.

13. Fair legislation and regulation

Screening and diagnostic technologies are commonly differently regulated than treatments, especially medications. Ethical reflection is needed when considering what kind of regulation is needed. Emphasis should be put on considering the ethically relevant aspects. For example, who has a right to get the results of the diagnostic procedure and for what purposes? Is legislation needed to ensure equal access or is it justified that the juvenile court has a priority line for the diagnostic procedures? As there is no international consensus on diagnostic criteria, it may also be necessary to define rules and regulations to ensure good quality of diagnostic procedures.

In Summary

Social life is grounded on prevalent morals, values and behavioural models as well as on national laws. The moral rules of a society reflect its values, which may be weighted differently across different societies. Furthermore, each profession creates its own ethical rules and regulations that are the basis for their profession, in which knowledge, experience and commitment to ethically acceptable goals are joined. In this article I have highlighted some value related aspects that need to be taken into account when deciding on the implementation of screening for and diagnosing FASD in adolescents and adults with special emphasis on juvenile justice and the just use of health care resources.

A diagnosis usually has moral and practical consequences as it requires further treatment or other modes of care. Increasing screening programmes and diagnostic facilities for FASD in adolescents and young adults thus leads to changes in the requirements placed on health care systems. Diagnosing FASD first in adulthood may also change the way we see the long-term effects of prenatal alcohol exposure in society. It also underlines the fact that the health and social care has failed to diagnose these persons during their childhood when interventions to prevent secondary difficulties should have been instituted. Thus the society has to be prepared not only to build resources to meet the needs of juvenile court but also to prevent these needs.

It is important to understand that the psychosocial and legal consequences of a "yes/no FASD diagnosis" may be unpredictable. Misdiagnosing either way may lead to unintended consequences. The prevailing moral values in society affect the willingness to provide the necessary resources for diagnostic procedures and identified needs of support and individual therapy. The ethical implications of implementing or refraining from the implementation of screening and diagnostic programmes for adults with FASD need to be discussed in a balanced way so that policy makers have a wider view of all possible consequences of their decision.

Worldwide, there are still many obstacles to be overcome before FASD can be reliably diagnosed at any age and the individually required therapeutic facilities will be provided. Many of the ethical implications are common to various nations but some value-laden issues are likely to be country specific. Is FASD a welcome diagnosis? Not in all countries. FASD is a heterogeneous disorder but the name emphasizes its etiological nature. The diagnosis does not give specific information about the neurocognitive and behavioural profiles but these have to be assessed individually. Nor is FASD a clearly dichotomous disorder (present or not), and the secondary effects of postnatal life may have a larger effect on one's psychosocial outcome. How can we be sure that the behavioural difficulties encountered in adolescence are due to prenatal alcohol exposure and not due to a cascade of adverse life events? This is the case especially with persons who are not identified to have major developmental difficulties in childhood but face psychosocial and adaptation difficulties when life requires them to take responsibility for independent action in daily life and to understand the moral rules of the society they live in.

The diagnostic information may be of different value to different stakeholders: the person him/herself, other family members, personnel within the legal system or the social and health care sectors or insurance system. A moral issue is also to whom the diagnostic information should be communicated—and for what purpose. How can we foresee the possible benefits and risks of an FASD diagnosis to the individual and who will weigh them? An FASD diagnosis given in adulthood can change one's image of oneself. Even more so if the diagnosis is required by authorities and not by the person him/herself. It may also have a long lasting effect on the child-parent relationship and the well-being of the family, yet it can also lead to better acceptance of one's problem.

It is important to remember that about 90 % of fertile-aged women do drink. The amount of fetuses exposed to alcohol in utero at any point is largely unknown but in many western countries it may be as high as 50 % in early pregnancy (Ahlvik et al. 2006). Only a minority of these fetuses are yet permanently harmed. Those with FASD have a life course that differs largely from normal children. A majority of children with diagnosed FASD are taken into care or are supported by the child welfare services. Of those taken into custody a large majority does not reach a stable life but are prone to have changes in their living environment repeatedly (Fagerlund et al. 2012; Koponen and Kalland 2013). After this kind of life course it is often not possible to say what is due to prenatal alcohol exposure and what is due to adverse life events.

The prevalence of FASD in adults is largely unknown. As diagnostic facilities are established, it is necessary to offer equal access to diagnostic and therapeutic possibilities for those in equal need and not only for those who are identified by the court system. Diagnosing FASD also requires that the current psychosocial wellbeing and unmet medical needs of the mother are evaluated.

In conclusion, diagnosing FASD in adults raises several questions of ethical concern. I have high-lighted only those of highest priority and which are relevant worldwide. Finally, the moral values prevailing in society affect the willingness to provide the necessary resources for prevention, diagnosis, individual care and rehabilitation.

References

Ahlvik, A., T. Haldorsen, B. Groholt, and R. Lindemann. 2006. Alcohol consumption before and during pregnancy comparing concurrent and retrospective reports. *Alcoholism, Clinical and Experimental Research* 30: 510–515.

Astley, S.J. 2006. Comparison of the 4-digit diagnostic code and the Hoyme diagnostic guidelines for Fetal alcohol spectrum disorders. *Pediatrics* 118: 1532–1544.

Autti-Rämö, I., M. Mäkelä, H. Sintonen, H. Koskinen, L. Laajalahti, R. Halila, H. Kääriäinen, R. Lapatto, K. Näntö-Salonen, P. Pulkki, M. Renlund, M. Salo, and T. Tyni. 2005. Expanding screening for rare metabolic disease in the newborn: An analysis of costs, effect and ethical consequences for decision making in Finland. *Acta Paediatrica* 94: 1126–1136.

Chudley, A.E., J. Conry, J.L. Cook, C. Loock, T. Rosales, and N. LeBlanc. 2005. Fetal alcohol spectrum disorders: Canadian guidelines for diagnosis. *CMAJ* 172(5 Suppl): S1–S21.

Fagerlund, Å., I. Autti-Rämö, M. Kalland, P. Santtila, H.E. Hoyme, S.N. Mattson, and M. Korkman. 2012. Adaptive Behaviour in Children and Adolescents with Foetal Alcohol Spectrum Disorders: A Comparison with Specific Learning Disability and Typical Development. *Eur Child Adol Psychiatr.* 12: 221–231.

Hoyme, H.E., P.A. May, W.O. Kalberg, P. Kodituwakku, J.P. Gossage, P.M. Trujillo, D.C. Buckely, J.H. Miller, A.S. Aragon, N. Khaole, D. Viljoen, K.L. Jones, and L.K. Robinson. 2005. A Practical clinical approach to diagnosis of fetal alcohol spectrum disorders; clarification of the 1996 institute of medicine criteria. *Pediatrics* 115: 39–47.

Koponen, A., M. Kalland, I. Autti Rämö, and S. Suominen. 2013. Socio-emotional development of children with foetal alcohol spectrum disorders in long-term foster family care: a qualitative study. *Nordic Social Work Research* 3: 38–58.

Kahila, H., M. Gissler, T. Sarkola, I. Autti-Rämö, and E. Halmesmäki. 2010. Maternal welfare, morbidity and mortality 6–15 years after a pregnancy complicated by alcohol and substance abuse: a register-based case-control follow-up study of 524 women. *Alcohol and Drug Dependency* 111: 215–221.

May, P, J.P. Cossage, W.O. Kalberg, L.K. Robinson, D. Buckley, M. Manning, and E.G. Hoyme. 2009. Prevalence and epidemiologi characteristics of FASD from various research methods with an emphasis on recent in-school studies. *Dev Disabil Res Rev* 2009;15: 176–192.

Mäkelä, P., H. Mustonen, and C. Tigerstedt (toim.). 2010. Suomi juo. Suomalaisten alkoholinkäyttö ja sen muutokset 1968–2008. (Finland drinks. Changes is alcohol consumption 1968–2008). Terveyden ja hyvinvoinnin laitos. Yliopistopaino. Helsinki.

Saarni, S., B. Hofmann, K. Lampe, D. Luehman, M. Mäkelä, M. Velasco-Garrido, and I. Autti-Rämö. 2008. Ethical analysis to improve decision-making on health technologies. *Bulletin of the World Health Organization* 86: 617–623.

The Danish Council of Ethics. 2001. Screening—a report. 1–118. Kööpenhamina. http://www.etiskraad.dk.

Wilson, J.M.G., and G. Jungner. 1968. *Principles and Practice of Screening for Disease*. Geneve: WHO.

Part II
Clinical Perspectives on Criminal Behaviour in FASD

Neurocognitive Difficulties Underlying High Risk and Criminal Behaviour in FASD: Clinical Implications

Katherine Wyper and Jacqueline Pei

Abstract Individuals with FASD are often characterized by impaired neurocognitive functioning, which may increase their likelihood of engaging in maladaptive, high risk, and criminal behaviour. Additionally, this group is frequently reported to experience lifelong environmental adversity, and the combination of these factors complicates efforts to intervene and support positive outcomes. As a result, individuals with FASD are over-represented in the criminal justice system, which is poorly suited for affected offenders. In this chapter, a new conceptualization of FASD and criminality is proposed, which draws on a comprehensive model of functioning to better understand the needs of offenders with FASD and ultimately reduce risk of offending behaviour.

Fetal Alcohol Spectrum Disorder (FASD)

FASD refers to a range of physical, neurological, cognitive, socio-emotional, and behavioural impairments that can result when a developing embryo or fetus is exposed to alcohol during pregnancy (Chudley et al. 2005). A leading cause of developmental and cognitive disability in the Western world, FASD has a prevalence rate of approximately 1 in 100 live births in Canada (Stade et al. 2009) and estimates as high as 2–5 % have been reported in the United States and some western European countries (May et al. 2009). Consequently, the economic impact of FASD is staggering, with an annual cost of $5.3 billion in Canada (Stade et al. 2009). One of the contributors to these costs is the high number of individuals who

K. Wyper (✉)
Department of Educational Psychology at the University of Alberta, Edmonton, Canada
e-mail: kwyper@gmail.com

J. Pei
Department of Pediatrics at the University of Alberta, Edmonton, Canada
e-mail: jpei@ualberta.ca

M. Nelson and M. Trussler (eds.), *Fetal Alcohol Spectrum Disorders in Adults: Ethical and Legal Perspectives*, International Library of Ethics, Law, and the New Medicine 63, DOI 10.1007/978-3-319-20866-4_7

come into contact with the justice system. An examination of the links between risk factors and vulnerabilities in FASD, offender characteristics, and the current state of the justice system reveals that more consideration must be given to how we can respond to these individuals in the most effective way. More effective interventions for offenders with FASD have the potential not only to reduce the high cost of the disorder, but also increase the likelihood of success in the lives of affected individuals.

Identifying FASD

As an umbrella term, FASD includes several specific diagnoses with varying degrees of impact on an individual's neurodevelopment, including: Alcohol-Related Neurodevelopmental Disorder (ARND), partial Fetal Alcohol Syndrome (pFAS), and Fetal Alcohol Syndrome (FAS). In Canada, most diagnosticians currently rely on the Canadian Guidelines (Chudley et al. 2005) to navigate the complex, inter-disciplinary diagnostic process. This process entails assessment of four domains: physical growth, facial development, central nervous system functioning, and prenatal exposure to alcohol. Each of these four domains is evaluated by a team of clinicians and ranked from 1 to 4 on severity or probability of impairment, with 1 representing no impairment or risk and 4 representing severe impairment or high risk. Assessment also involves evaluation of other pre-natal risk factors such as lack of prenatal care, prenatal complications, genetic risk factors, and in utero exposure to other teratogens; and post-natal risk factors such as abuse, disrupted living arrangements, head injuries, and exposure to violence or substance abusing caregivers (Astley 2004). Combined, this information is considered to be relevant for the process of determining differential diagnosis, as well as informing intervention approaches.

Primary Disabilities and Adverse Outcomes

The direct neurological insult due to prenatal alcohol exposure can lead to a range of neurocognitive and behavioural challenges, the most common of which are impairments in intelligence, executive functioning, learning and memory, academic skills, language and communication, visuo-spatial and motor ability, attention problems, and hyperactivity (see Mattson et al. 2011 for a review). These difficulties are evident in both standardized testing (Mattson et al. 2011) as well as through more ecologically meaningful caregiver and teacher reports (e.g., Rasmussen et al. 2007). Compared with healthy controls, IQ-matched controls, and individuals with other neurodevelopmental disabilities such as attention-deficit hyperactivity disorder, individuals with FASD show relatively intact functioning on basic neurocognitive tasks, but struggle with higher-level tasks relying on complex executive functioning skills (Mattson et al. 2011). Specifically, some studies report that individuals with FASD show relatively stronger abilities in terms of basic language (McGee et al. 2009), simple visual-perceptive tasks (Kodituwakku 2009), simple processing tasks (Burden et al. 2005a), fine motor skills, and some aspects of attention and areas of academics (Vaurio et al. 2011). Conversely, higher-level neurocognitive functions such as inhibition (Burden et al. 2009), decision-making (Kully-Martens et al. 2013), working memory (Burden et al. 2005b), and cognitive flexibility (Coles et al. 1997) tend to be more impaired among individuals with FASD.

Adverse outcomes. In addition to these direct impacts of prenatal alcohol exposure, there is also a range of adverse outcomes identified for this population. The etiologies of these outcomes are more complex, and appear to result from an interaction between direct impacts and environmental factors. Streissguth et al. (1996) and other researchers (Clark et al. 2004) have identified mental health problems, inappropriate sexual behaviours, disrupted school experiences, confinement (in hospitals or correctional settings), and substance abuse as significant concerns for adolescents and adults with FASD. Trouble with the law is also frequently noted in this group (Streissguth et al. 1991), and in the last two decades there has been a rapid growth in media and research interest around the issue of FASD in the justice system. Importantly, factors have been identified that may protect an individual against these adverse outcomes, including early diagnosis (before the age of 6 years), living in a stable and nurturing home environment, and not being a victim of violence (Streissguth et al. 2004). Unfortunately, many of these protective factors are absent for individuals with FASD.

Post-Natal Risk Factors and FASD

Further complicating the clinical picture for individuals impacted by FASD is the frequency with which they have been exposed to adverse environmental factors that may impact development. In one study, it was found that over 80 % of children born with FASD were not living with their biological mothers (Streissguth et al. 2004), and other researchers have reported that family environments of older individuals with FASD are “remarkably unstable” (Streissguth et al. 1991). Importantly, placement history has been shown to significantly impact both cognitive and behavioural functioning in children with FASD (Victor et al. 2008), indicating that caregiving environment is a crucial variable affecting outcomes. Other commonly experienced postnatal risk factors among individuals with FASD include neglect, violence toward the child, exposure to parental substance abuse, parental divorce or separation, poverty, and other major trauma (Carmichael Olson et al. 2009). Given the evidence regarding the deleterious effects of both prenatal alcohol exposure *and* adverse early life experiences on neurodevelopment, individuals with FASD experience what is sometimes referred to as “double jeopardy” with respect to these two dimensions of risk.

Adolescents and Adults with FASD

Most early studies of FASD focused on the effects of prenatal alcohol exposure in children, and only later did researchers turn their attention towards the lasting impact of the disorder into adolescence and adulthood. Researchers examining the manifestations of FASD in older ages tend to report that some physical anomalies persist, as do many of the cognitive, academic, adaptive, and developmental difficulties documented in childhood (Streissguth et al. 1991). Key functional difficulties described in these later years include behavioural and emotional problems, challenges with school and employment, and independent living (Spohr et al. 2007). Additionally, substance abuse problems and trouble with the law may increase with age among adolescents and adults with FASD (Streissguth et al. 2004).

Canadian researchers have highlighted that living with a caregiver, requiring low levels of support, and being less vulnerable to manipulation may all be related to better outcomes (Clark et al. 2004). Another important finding from this study was that IQ did not significantly affect the likelihood of adverse outcomes—but adaptive functioning did—which highlights the importance of assessing daily living skills in order to meaningfully understand an individual's needs. Because many individuals with FASD are intellectually impaired but not disabled (Mattson et al. 2011; Streissguth et al. 2004), those who experience relatively intact intelligence but limited adaptive skills may fall through the cracks of agencies relying on IQ scores alone to allocate supports.

The difficulties associated with FASD can be life-long, however the ways in which they manifest functionally may change across the lifespan. Given the increased independence and responsibility typically associated with adolescence and adulthood, older individuals with FASD face challenges different than in childhood. That is, it is possible that as an individual with FASD ages, their developmental ability may lag increasingly behind their chronological age, and with progressing environmental demands and societal expectations, neurocognitive impairment and associated behavioural challenges may become more obvious. Compounding this growing ability-expectancy discrepancy, there are fewer services and evidence-based interventions for older individuals with FASD, and despite the possibility that they may achieve independence with appropriate assistance and support (Chudley et al. 2007), there is a serious lack of services designed for affected adolescents and adults (Petrenko et al. 2013).

FASD and High Risk Behaviours

High risk behaviours are generally defined as having a negative impact on the well-being of an individual, that may cause immediate injury or contribute to cumulative negative outcomes (De Guzman and Bosch 2007). There is an important distinction between socially *acceptable* risk behaviour—such as adventure seeking (e.g., bungee jumping)—and *reckless* behaviour that threatens one's safety (e.g., substance abuse) (Arnett 1992). Common reckless high risk behaviours include unsafe sexual behaviour, reckless driving, drug and alcohol abuse, violence and deviance, criminal activity, and other actions leading to unintentional or intentional injury (Centers for Disease Control and Prevention, CDCP 2013). Adolescents and young adults (under the age of 25) are particularly vulnerable to engaging in high risk behaviours because this life stage is characterized by major changes in physical, socio-emotional, and neurological development accompanied by increased independence and autonomy (Steinberg 2008). Although some degree of experimentation and risk-taking is a normal part of development, high risk behaviours pose one of the greatest threats to the health and well-being of young people and are a significant contributor to injury and death (Butler-Jones 2011; CDCP 2013; Steinberg). High risk behaviours are concerning not only due to the dangers inherent in these behaviours, but also because high risk behaviours initiated at a young age are associated with problematic behaviours later in life (e.g., substance abuse), and certain high risk behaviours can put others at risk as well (e.g., unsafe driving, criminal behaviour) (Steinberg).

Many of the adverse outcomes noted in FASD can be considered high risk behaviours; high rates of school drop-out, substance abuse, inappropriate sexual behaviour, and suicidality certainly threaten the well-being of these individuals and can lead to injury or other negative outcomes. More specifically, some researchers have shown that individuals with FASD demonstrate maladaptive and problem behaviours such as impulsivity, teasing/bullying, dishonesty (e.g., lying, stealing, cheating), cruelty, destruction of property, physical aggression, and self-injury (LaDue et al. 1993; Nash et al. 2006; Rasmussen et al. 2008). The high rates of high risk behaviours in FASD populations could suggest a pattern of impairment that underlies their propensity to engage such behaviour.

Neurocognitive Functioning in Offender Populations

Many of the high risk behaviours described above are delinquent or illegal, and as such, interactions with the justice system are frequent for individuals engaging in these behaviours. Therefore, consideration of the neurocognitive characteristics of offender populations as a whole can provide some insight into the underlying mechanisms implicated in their behaviour, and in turn provide greater insight into potential intervention strategies to address these factors.

Executive Functioning and Self-regulation

Executive functioning refers to a set of higher-order processes that oversee thought and action under conscious control (Zelazo and Müller 2002) and guide adaptive responses to novel situations (Hughes 2011). Executive functioning undergoes rapid changes during childhood and continues to develop into adolescence and beyond (Zelazo and Carlson 2012). The neurological underpinnings of executive functioning were first established through neuropsychological studies of individuals with brain damage, and identified the frontal lobe as a key brain area associated with executive functioning (e.g., Shallice and Burgess 1991). The lifelong relevance of executive functioning is demonstrated in research showing that some early-developing executive functioning processes are predictive of later outcomes such as physical health, substance abuse, personal finances, and criminal behaviour (Moffitt et al. 2011). One way to conceptualize executive functioning is in terms of "hot" and "cold" processes. "Hot" executive functions are those involved in situations with high levels of emotion and motivation, whereas "cold" executive functions are used in emotionally-neutral situations (Zelazo and Carlson 2012). In terms of high risk behaviours, hot executive functioning processes may be particularly relevant, evinced by the finding that most adolescent risk-taking occurs during situations of emotional arousal (Steinberg 2004).

Self-regulation is another broad concept encompassing a number of sub-functions related to high risk behaviours. In describing the practical implications of self-regulation, Baumeister et al. (2007) assert that, "most of the social and personal problems that afflict people in modern western society have some element of self-regulatory failure at their root" (p. 2). Research on self-regulation originates from multiple schools of thought, but for the purpose of this review it is defined as the "self altering its own responses or inner states," (Baumeister et al. 2007, p. 5) involving "active responding and decision-making" (Murtagh and Todd 2004, p. 22), which overlaps considerably with executive functioning. Self-regulation includes emotional, behavioural, and cognitive regulation, and is used interchangeably with self-control in this review.

Executive functioning and self-regulation impairments in offenders. Delinquent and criminal behaviour has been related to numerous impairments falling under the umbrellas of executive functioning and self-regulation. Multiple studies have shown that impairments in both hot and cold executive functioning are related to violent, aggressive (Hancock et al. 2010; Hoaken et al. 2003), and antisocial behaviour (De Brito et al. 2013; Ogilvie et al. 2011). These studies parallel brain imaging research that reveals reduced structure and function of the brain areas related to executive functioning and self-regulation (i.e., prefrontal cortex) in antisocial and violent individuals (Yang and Raine 2009).

Inhibition/impulse control. Inhibition can be broadly defined as voluntary control over cognition or behaviour, and in particular as the ability to resist a prepotent or impulsive response, with neurological roots in the frontal lobe. Impaired inhibition is considered by some researchers to result in impulsivity (Aron 2007), thus these two concepts will be used synonymously for the purpose of this review. Poor inhibition and impulsivity are some of the most widely cited deficits among individuals who engage in delinquent or criminal behaviours. Early longitudinal work has implicated teacher-rated impulsivity measured in kindergarten to self-reported delinquent and antisocial behaviours in adolescents (Tremblay et al. 1994). The same group of researchers later found that inhibition may serve as a protective factor against self-reported delinquency in adolescents (Kerr et al. 1997). More recent research also suggests that impulsivity may moderate the relationship between protective factors (i.e., family warmth and parental knowledge) and delinquency (Chen and Jacobsen 2013). In adult populations, impaired behavioural inhibition has been found in offenders with antisocial personality disorder compared with healthy controls (Dolan 2012). Recent neurophysiology research, using transcranial magnetic stimulation and electroencephalography, has implicated the prefrontal cortex in the inhibition deficits among some groups of offenders (Hoppenbrouwers et al. 2013). Importantly, the relationship between low inhibition/high impusivity, and criminal behaviour may be specific to the type of offence, as Hancock et al. (2010) found that lower performance on these measures was related to a higher number and greater severity of violent but not non-violent offences.

Decision-making. The term "decision-making" is broad, but in general this process requires choosing one option over another or several other alternatives (Seguin et al. 2007). It involves both "cold" cognitive functions and "hot" emotional processes, both of which influence the outcome of the decision. Cold processes include weighing risks and benefits (incorporating aspects of reward, punishment, and future consequences), and holding alternatives in mind while comparing them, whereas hot processes involve the affective response to various alternatives (based on emotional or visceral reactions to situations, often under time pressures) (Seguin et al. 2007), all of which may have neurological roots in the prefrontal lobe (Bechara 2005; Bechara et al. 2000; Syngelaki et al. 2009).

Decision-making impairments have been documented in young offender populations as well as youth with conduct disorder, who have been shown to make more risky choices than controls (Fairchild et al. 2009; Syngelaki et al. 2009). The cause of this impairment may be explained as an imbalance in reward-punishment sensitivities (Fairchild et al.). Similar impairments have been noted in adult populations, with antisocial offenders both with and without comorbid psychopathy showing deficits in adaptive decision-making (De Brito et al. 2013). Though some researchers suggest that experimental measures of decision-making may not accurately reflect real-world situations (Steinberg 2004), recent work found that laboratory-measured decision-making was indeed predictive of recidivism in offenders at 3- and 6-month follow-up, supporting the ecological validity of this construct (Beszterczey et al. 2013).

Working memory. Baddeley's (1992) seminal research on working memory defines the construct as "temporary storage and manipulation of the information necessary for complex cognitive tasks," (p. 556) and this function is also closely associated with the prefrontal region of the brain (Owen et al. 1990). Baddeley underscores the influence of working memory on many important tasks including language processing and reasoning, and forensic research also commonly cites working memory as an area of deficit among offenders. For instance, several studies have demonstrated that juvenile offenders display weaker spatial working memory than controls (Cauffman et al. 2005; Syngelaki et al. 2009), and impairments in verbal working memory have also been noted in adult offenders both with and without antisocial personality disorder and psychopathy (De Brito et al. 2013; Hoppenbrouwers et al. 2013).

Cognitive flexibility and set-shifting. Cognitive flexibility is generally known as the ability to shift attention, adapt one's thoughts and responses, and generate novel ideas, and it is often used interchangeably with set-shifting. Deficits in these neurocognitive functions have been noted in both violent and non-violent adult offenders (Bergvall et al. 2001; Tuominen et al. 2014), adult offenders with antisocial personality disorder (Dolan 2012), and in young offenders (Syngelaki et al. 2009). Weaker performance on measures of cognitive flexibility and concept formation has also been related to a higher number and greater severity of violent crimes in adult offenders (Hancock et al. 2010).

Verbal ability. Broadly defined, verbal ability is an individual's capacity to understand language, and is localized neurologically in the left hemisphere of the brain. Verbal ability is heavily influenced by learning environment (Romi and Marom 2007), and impairment in this domain is a strong risk factor for antisocial and delinquent behaviour (Munoz et al. 2008). For example, Romi and Manom (2007) compared intellectual ability between delinquent youth and controls and found that delinquent youth performed significantly worse than controls in verbal domains, but not performance domains of the Wechsler Intelligence Scale for Children–Revised. Similarly, Manninen et al. (2013) reported that low verbal intellectual ability was highly predictive of later offending (and violent crimes in particular) among youth with conduct disorder. Lower verbal skills have also been related to physical aggression in adolescents (Barker et al. 2011), and longitudinal work suggests that language impairments in childhood lead to higher parent-reported delinquency as well as self-reported rates of criminal arrests and convictions at 19 years old; however this relationship was only found in males (Brownlie et al. 2004).

Although these findings provide a compelling case for the link between verbal impairment and delinquency/crime, Hancock and colleagues (2010) found that offenders with greater expressive verbal skills were more likely than offenders with weaker skills to commit severe violent offences. To explain these findings, Hancock et al. hypothesized that because violent offences are often predicated by verbal exchanges, individuals with greater expressive language may be more likely to escalate the situation.

Attention. The concept of attention has been defined from a number of perspectives, but one that is particularly relevant clinically is that proposed by Sohlberg and Mateer (1987): "the capacity to focus on particular stimuli over time and to manipulate flexibly the information" (p. 117). Impairments in attention and concentration have been commonly noted in delinquent youth (Manninen et al. 2013), and these problems may be worse for youth living in residential institutions compared with non-institutionalized delinquent youth and non-delinquent controls

(Romi and Marom 2007). Indeed, clinically significant attention difficulties are frequently noted in delinquent and offender groups, who display high rates of attention-deficit hyperactivity disorder (Belcher 2014; Young and Thome 2011). Moreover, offenders with attention deficits have been shown to have a higher number of convictions than those without (Tuominen et al. 2014).

The Biopsychosocial Perspective

Although it is clear from the above review that brain development has a marked impact on cognition and subsequent behaviour, it is important to note that these processes do not exist in isolation. That is, although neurological vulnerability may increase the likelihood that an individual could engage in high risk behaviours, a variety of factors will influence behavioural outcomes, and the interaction between these factors is significant. For instance, impaired brain function may interact with other adverse environmental factors such as poverty, trauma, and substance abuse to lead to heightened levels aggression (Fishbein and Tarter 2009). That is, impaired brain function may increase the likelihood that an individual experiences environmental adversity while at the same time rendering that individual more vulnerable to the negative impacts of these experiences. Moreover, the combined influence of biological and social risk factors leading to antisocial and criminal behaviours is exponentially larger than the influence of either risk factor alone (Raine 2002). Although the current chapter focuses primarily on one individual factor—namely neurocognitive functioning—behaviour in general, and delinquency and crime in particular, is perhaps best understood from a biopsychosocial perspective, whereby equal consideration is given to all of the biological, psychological, and social variables that may impact an individual throughout their life, as well as the reciprocal interaction between these variables.

Bringing It Together: The FASD-Affected Offender

Prenatal exposure to substances has been shown to increase vulnerability to criminal behaviour later in life (Heffron et al. 2011) and one mechanism through which prenatal alcohol exposure may lead to these problem behaviours is neurocognitive impairment. As described above, individuals with prenatal alcohol exposure and FASD have repeatedly been shown to display broad deficits in executive functioning (Connor et al. 2000; Mattson et al. 1999; Rasmussen et al. 2007) and self-regulation (Kodituwakku et al. 1995). In fact, executive functioning is one of the most frequently documented impairments among individuals with FASD (Mattson et al. 1999), and especially higher-level executive functioning tasks (Mattson et al. 2011). In particular, individuals with heavy prenatal alcohol exposure (with or without an FASD diagnosis) have displayed difficulties in planning, cognitive flexibility, inhibition, concept formation, and reasoning (Mattson et al. 1999). Executive functioning deficits as reported by parents and teachers of children with

FASD may account for a significant proportion of the social difficulties they experience, and behavioural regulation and metacognition ("thinking about thinking") are noted as key concerns (Schonfeld et al. 2006). Affective or "hot" executive functioning impairments may be especially relevant in terms of high risk behaviours in this population, as children with FASD have been shown to be impaired on decision-making and risk-taking tasks, and may not learn from negative consequences to make more positive choices (Kully-Martens et al. 2013).

Given the correlation between neurocognitive ability and high risk behaviours as described above, it is little wonder that individuals with FASD are especially vulnerable to engaging in such behaviours. Deficits that often characterize FASD may be a "recipe for disaster" for engaging in criminal behaviours and becoming involved with the justice system (Byrne 2002). These deficits include poor adaptive skills, slow information-processing, inflexibility and perseveration, limited understanding of consequences and ability to learn from past mistakes, receptive language impairments, distractibility and inattention, immaturity, social ineptness, vulnerability to manipulation/suggestibility, dishonesty, weaker moral judgment, and poor socio-moral values (Byrne; Malbin 2004; Mela and Luther 2013; Page 2001; Schonfeld et al. 2005), many of which have roots in neurocognitive functioning.

Other Complicating Factors

Beyond the specific neurocognitive factors that may increase their vulnerability to engaging in high risk and criminal behaviour, offenders with FASD have also been shown to experience other co-morbid conditions. For example, substance-related disorders, attention-deficit hyperactivity disorder, conduct disorder, antisocial personality disorder, opposition defiant disorder (Vidovic 2012), schizophrenia, anxiety, and other medical conditions (Mela et al. 2013) may complicate the situation further in terms of predicting vulnerability to offend as well as providing interventions to reduce the likelihood of reoffending. Additionally, given the adverse circumstances that often characterize the lives of individuals with FASD, from a biopsychosocial perspective it is important to note other risk factors beyond prenatal alcohol exposure that may be significant predictors of high risk behaviours. For instance, individual factors such as substance use, perception of life stress, and family factors such as low caregiver supervision and warmth, and verbally aggressive family conflict have been shown to correlate with delinquency in youth with prenatal alcohol exposure (Lynch et al. 2003).

Researchers exploring the perspectives of family members of individuals with FASD (Radford-Paz 2013) have echoed the complexity of the situation, suggesting that neurocognitive and social impairments, as well as overwhelming environmental demands contribute to criminal behaviour in adults with FASD. Equally importantly, family members identified supervision and structure, education and employment, positive peers, and financial and social support as protective factors against criminality (Radford-Paz).

FASD and the Justice System

Given the neurocognitive and environmental vulnerabilities experienced by individuals with FASD, it is no wonder that 60 % of affected individuals in some studies have reported experiencing trouble with the law (Streissguth et al. 2004). Indeed, both youth (Popova et al. 2011) and adults (MacPherson et al. 2011) with FASD are over-represented in the Canadian justice system with prevalence estimates of 10–23 % (Fast et al. 1999; MacPherson et al. 2011) vastly exceeding the rate of 1 % for FASD in the general population. One longitudinal study found that 35 % of adolescents and adults surveyed reported ever being incarcerated for a crime (Streissguth et al. 2004). It is also believed that many offenders have been prenatally exposed to alcohol but remain undiagnosed (Fast and Conry 2009) due to the "invisibility" of the disorder, where affected individuals may show no physical signs and display strong expressive vocabularies, and service professionals may lack awareness around FASD (Malbin 2004; Page 2001). Individuals with FASD who enter the justice system may become vulnerable targets for victimization, may end up learning more criminal acts without understanding why they were incarcerated in the first place, and may be unable to conform to the custodial environment (Byrne 2002; Conry and Fast 2000).

In the juvenile system specifically, youth with FASD have been shown to come into contact with the system earlier, with a higher number of charges (though less serious) than non-FASD young offenders, and have a greater risk of re-offending (McLachlan 2012). Adult offenders with FASD have been shown to have numerous convictions as juveniles, a higher number of convictions as adults, a greater likelihood of failing under community supervision, have disciplinary problems during incarceration, and to be re-incarcerated within 6 months than offenders without FASD (MacPherson et al. 2011). Once involved in the system, the neurocognitive, adaptive, and social difficulties associated with FASD may influence an affected individual's ability to navigate the legal process, as recent work found high rates of psycholegal impairment among young offenders with FASD related to understanding arrest, interrogation, and trial processes (McLachlan et al. 2013). Individuals with FASD may also be more inclined to give false confessions or false testimonies (Roach and Bailey 2009).

Fundamentally, the Canadian criminal justice system is not set up for individuals with low levels of functioning. Traditional justice practices assume that an offender can understand the link between inappropriate behaviour and negative consequences (Malbin 2004), and even in the case of offenders with mental disorders, it is assumed that with treatment, the disorder may be rectified (Roach and Bailey 2009). The system fails to accommodate for FASD (Roach and Bailey), and for offenders with the disorder, this system falls short:

The inability of the present system to effectively provide meaningful rehabilitation or deterrence to individuals suffering from FAS/ARND combines with the increased vulnerability to further victimization within penal facilities clearly sends a message that it is not appropriate for the Canadian judicial system to continue to minimize the very real impact FAS/ARND has on individuals. (Chartrand and Forbes-Chilibeck 2003, p. 61).

Given these issues, there is a dire need for increased FASD education and training for justice personnel (Byrne 2002; Cox et al. 2000), access to timely assessment and diagnosis of FASD, enhanced methods for screening, identifying, and treating offenders suspected of having FASD (Burd et al. 2004), and for a review of the "yes or no" determination of criminal responsibility when dealing with FASD-affected offenders (Mela and Luther 2013). Although some judges have begun to consider the impact of FASD in their sentencing decisions, formal assessments for the disorder are rarely undertaken (Vidovic 2012) and this consideration holds more weight for juvenile as opposed to adult offenders, who are deemed less capable of rehabilitation (Verdun-Jones and Butler 2013).

Recent developments in the justice system. A new perspective is required in all dimensions of the justice system to enhance the appropriateness of this system for offenders with FASD. Researchers have suggested that this new perspective should involve a greater understanding of the cognitive impacts of FASD (Moore and Green 2004), as well as a shift from punitive to therapeutic approaches (Mela et al. 2013). There has been a surge in media and research attention regarding justice issues and FASD over the last decade, and a growing awareness of FASD in Canadian criminal law (Roach and Bailey 2009). In fact, this is such an important topic that in 2013, two formal gatherings occurred in Canada; one meeting "FASD and the Law: A Conversation About Current Research and Practices" was held in Vancouver, British Columbia and the "Consensus Development Conference on Legal Issues of FASD" took place in Edmonton, Alberta. The conversation continued, with "FASD at the Frontline," an event held in 2014 in Regina, Saskatchewan to highlight justice-related FASD research, and a second "FASD and the Law Day 2015", again held in Vancouver. There have also been efforts to develop screening tools that enable parole officers to identify youth who may be affected by FASD (Conry and Asante 2010). As well, two recent Canadian Bar Association (CBA) resolutions were passed because the "…normative assumptions and the sentencing principles such as specific and general deterrence are not valid for those with FASD" (CBA 2010) and there is a need to "improve access to justice" and accommodate the disability of FASD (CBA 2013). The Edmonton Consensus Conference resulted in a 60-page document exploring the implications of FASD for the justice system, including how to better identify offenders with FASD; how the criminal justice system, family court, and family/child welfare system can better respond to and meet the needs of these offenders; how guardianship, trusteeship, and social support can be enhanced for these individuals; and what legal measures are currently in effect to prevent the disorder. Despite these efforts, there is a concerning lack of research published to date that focuses on justice interventions for offenders with FASD.

Current Approaches to Offender Rehabilitation

There is a wide range of prison-based offender treatment programs, broadly categorized into criminogenic, non-criminogenic, employment, educational, recreational, and administrative (Wilkinson 2012). Across programs, a number of factors influence treatment outcomes, including the nature of intervention, length of program, treatment setting, qualifications of staff, and offender characteristics (Brazao et al. 2013). Interestingly, although much has been written on theories and models of offender rehabilitation, less *practical* work has been done to establish effective methods of service delivery (Day et al. 2013).

The majority of treatment programs for youth and adult offenders are based on the cognitive-behavioural model (Brazao et al. 2013) and locate the cause of criminal behaviour *within the individual* (Day 2011). Specifically, cognitive-behavioural approaches focus on changing maladaptive thoughts that lead to problem behaviours (Day), and most use educational approaches to build different skill sets (Brazao et al. 2013). These interventions are typically led by a therapist and delivered in small groups, and require a strong alliance and collaboration between therapist and offender, as well as consistent offender engagement (Day et al. 2013). Cognitive-behavioural interventions have been shown to reduce various types of criminal behaviour among offenders (Wilson et al. 2005), especially those interventions that emphasize self-awareness around what leads an offender to engage in criminal behaviour (Wilson et al.).

Regardless of the approach, interventions thought to be "best practice" involve assessing and targeting individual risk and needs, increasing motivation, improving skills though practice, offering positive reinforcement, seeking continuous environmental support, evaluating outcomes, and proving feedback to offenders (Wilkinson 2012). Moreover, effective programs should be grounded in theory, target a variety of factors for change, build healthy relationships with clear supports and boundaries while being sensitive to offender diversity, be implemented with integrity, and based in the community (Brazao et al. 2013).

Meeting the Needs of Offenders with FASD

The needs of offenders with FASD do not always align with traditional justice approaches. Justice interventions may be most appropriate and effective in reducing recidivism when treatment is psychologically informed and guided by the offender's risk to society, profile of needs, and responsivity to treatment (Andrews et al. 1990). As described above, most contemporary offender rehabilitation programs rely heavily on cognitive approaches, and assume that offenders have a strong degree of control over their own behaviour, particularly in terms of personal attitudes and abilities such as problem-solving, self-monitoring and evaluation, and resisting temptation (Day 2011). Unfortunately, as described above, individuals with FASD tend to show marked deficits in these areas, struggling with attention, executive functioning, and adaptive functioning (MacPherson et al. 2011).

Some preliminary work has been undertaken to explore how the justice process may be improved for offenders with FASD. A 2007 Roundtable on the economic impact of FASD in Canada suggested that matching release requirements with the unique strengths and weaknesses of offenders with FASD would reduce recidivism, as well as the costs associated with it (Public Health Agency of

Canada 2007). Community collaborations dedicated to providing appropriate case plans and supports for individuals with FASD may also help to reduce recidivism (Olsen and Debolt 2002). These initiatives should focus on individual needs and strengths, and also incorporate external factors and emphasize environmental modifications and supports. Notably, some researchers assert that, "the most effective sentence for people with FASD may be one that aims to change their living or social situation, rather than their behavior" (Fast and Conry 2004, p. 164).

Although these efforts are promising, specific research on evidence-based justice interventions for FASD-affected offenders is very limited. Brown et al. (2012) recently published one article on potentially promising approaches for working with incarcerated youth with FASD. These approaches include implementing programs that are highly structured, consistent, predictable, individualized, and involve behavioural reinforcement; targeting adaptive and social functioning; and using behaviour therapy rather than cognitive therapy to replace maladaptive behaviours. They also suggested that treatment approaches should avoid peer group interventions, which may be inappropriate due to FASD-related social difficulties. Beyond the treatment program itself, the authors further suggested removing youth from chaotic home environments, helping youth to translate new skills, building long-term individualized support and supervision in plans for the future, and engaging a mentor to assist in transition back to the community. McLachlan et al. (2013) also recently published a report based on a survey of substance abuse service providers, which explored promising practices for addictions treatment in young offenders with FASD. This report listed practices similar to Brown et al.'s (2012) recommendations, as well as additional strategies to compensate for FASD-related neurocognitive deficits, including: less use of insight-based approaches, verbal strategies, and approaches that rely on attention/concentration/memory; more use of skill-building, extra support for youth to understand treatment content, and multi-modal learning integrating senses and physical learning; using concrete rather than abstract concepts; and teaching memory strategies. Finally, Mela and colleagues (2013) recently proposed that interventions for adult offenders with FASD should be life-long, extending beyond the justice system into transitional programs and the broader community, and that these systems of intervention may be most effective when coordinated by a mentor. The notion of mentoring aligns with Sterling Clarren's idea of the "external brain" (Kellerman 2003), which asserts that individuals with FASD may achieve the most success with the help of a responsible person to navigate the responsibilities of daily life.

Conclusions

FASD poses a significant social and economic challenge in our communities. Individuals with FASD are a unique population, confronted with neurocognitive vulnerabilities that increase the likelihood of engaging in maladaptive, high risk, and criminal behaviours. Examination of the general offender population reveals core characteristic neurocognitive deficits, many of which are often identified in FASD. Additionally, individuals with FASD tend to encounter adverse environmental experiences throughout their lives, which may increase their level of risk by amplifying the vulnerability of neurocognitive impairment. As a consequence of this two-fold risk, individuals with FASD are disproportionately represented in the justice system. Unfortunately, a review of the current justice system practices reveals that this system is a poor fit for the FASD population, which may, yet again, amplify the likelihood of negative outcomes.

From a strengths-based perspective, we know that when individuals with FASD are provided with appropriate and sufficient support, they experience greater success. To continue this early progress, and build momentum in the justice context, conceptualization of FASD and criminality needs to shift to a biopsychosocial model of understanding whereby neurocognitive functioning, psychological/mental health, and social/environmental experiences are considered. For example, an intervention approach that incorporates an understanding of potential self-regulatory limitations, depression, and the impact of childhood abuse may be more effective in addressing *all* risk factors and equipping an individual with FASD with comprehensive supports and strategies to best manage. Training initiatives, collaboration, conversation, examination of outcomes, and a willingness to think outside of the box are all crucial components of accomplishing this comprehensive approach to treatment and reducing recidivism.

References

Andrews, D.A., I. Zinger, R.D. Hoge, J. Bonta, P. Gendreau, and F.T. Cullen. 1990. Does correctional treatment work? A clinically relevant and psychologically informed meta-analysis. *Criminology* 28: 369–404.

Arnett, J. 1992. Reckless behavior in adolescence: A developmental perspective. *Developmental Review* 12: 339–373.

Aron, A.R. 2007. The neural basis of inhibition in cognitive control. *The Neuroscientist* 13: 214–228.

Astley, S.J. 2004. *Diagnostic guide for fetal alcohol spectrum disorders: The 4-digit diagnostic code*, 3rd ed. Seattle, WA: University of Washington Publication Services.

Baddeley, A. 1992. Working memory. *Science* 255: 556–559.

Barker, E.D., R.E. Tremblay, Pol A.C. van Lier, F. Vitaro, D.S. Nagin, J. Assaad, and J.R. Séguin. 2011. The neurocognition of conduct disorder behaviors: Specificity to physical aggression and theft after controlling for ADHD symptoms. *Aggressive Behavior* 37: 63–72.

Baumeister, R.F., B.J. Schmeichel, and K.D. Vohs. 2007. Self-regulation and the executive function: The self as controlling agent. In *Social psychology: Handbook of basic principles*, 2nd ed, ed. A.W. Kruglanski, and E.T. Higgins. New York: Guilford.

Bechara, A. 2005. Decision making, impulse control and loss of willpower to resist drugs: A neurocognitive perspective. *Nature Neuroscience* 8: 1458–1463.

Bechara, A., D. Tranel, and H. Damasio. 2000. Characterization of the decision-making deficit of patients with ventromedial prefrontal cortex lesions. *Brain* 123: 2189–2202.

Belcher, J.R. 2014. Attention deficit hyperactivity disorder in offenders and the need for early intervention. *International Journal of Offender Therapy and Comparative Criminology* 58: 27–40.

Bergvall, A., H. Wessely, A. Forsman, and S. Hansen. 2001. A deficit in attentional set-shifting of violent offenders. *Psychological Medicine* 31: 1095–1105.

Beszterczey, S., P.G. Nestor, A. Shirai, and S. Harding. 2013. Neuropsychology of decision making and psychopathy in high-risk sex-offenders. *Neuropsychology* 27: 491–497.

Brazao, N., C. da Motta, and D. Rijo. 2013. From multimodal programs to a new cognitive-interpersonal approach in the rehabilitation of offenders. *Aggression and Violent Behavior* 18: 636–643.

Brown, N.N., P.D. Connor, and R.S. Adler. 2012. Conduct-disordered adolescents with fetal alcohol spectrum disorder: Intervention in secure treatment settings. *Criminal Justice and Behavior* 39: 770–793.

Brownlie, E.B., J.H. Beitchman, M. Escobar, A. Young, L. Atkinson, C. Johnson, and L. Douglas. 2004. Early language impairment and young adult delinquent and aggressive behavior. *Journal of Abnormal Child Psychology* 32: 453–467.

Burd, L., R. Selfridge, M. Klug, and S. Bakko. 2004. Fetal alcohol syndrome in the United States corrections system. *Addiction Biology* 9: 169–176.

Burden, M.J., S.W. Jacobson, and J.L. Jacobson. 2005a. Relation of prenatal alcohol exposure to cognitive processing speed and efficiency in childhood. *Alcoholism, Clinical and Experimental Research* 29: 1473–1483.

Burden, M.J., S.W. Jacobson, R.J. Sokol, and J.L. Jacobson. 2005b. Effects of prenatal alcohol exposure on attention and working memory at 7.5 years of age. *Alcoholism, Clinical and Experimental Research* 29: 443–452.

Burden, M.J., C. Andrew, D. Saint-Amour, E.M. Meintjes, C.D. Molteno, H.E. Hoyme. et al. (2009). The effects of fetal alcohol syndrome on response execution and inhibition: an event-related potential study. *Alcoholism: Clinical and Experimental Research* 33:1994–2004

Butler-Jones, D. 2011. *The Chief Public Health Officer's report on the state of public health in Canada, 2011: Youth and young adults—life in transition*. Public Health Canada. Retrieved from http://publichealth.gc.ca/CPHOreport.

Byrne, C. 2002. The criminalization of Fetal Alcohol Syndrome (FAS). Unpublished report. http://www.americanbar.org/content/dam/aba/migrated/child/PublicDocuments/cfas.pdf.

Canadian Bar Association. 2010. *Resolution 10-02-A: Fetal alcohol spectrum disorder in the criminal justice system*. http://www.cba.org/cba/resolutions/pdf/10-02-A.pdf.

Canadian Bar Association. 2013. *Resolution 13-12-A: Accommodating the disability of FASD to improve justice services*. http://www.cba.org/cba/resolutions/pdf/13-12-A.pdf.

Carmichael Olson, H., R. Oti, J. Gelo, and S. Beck. 2009. "Family matters:" Fetal alcohol spectrum disorders and the family. *Developmental Disabilities Research Reviews* 15: 235–249.

Cauffman, E., Steinberg, L., and Piquero, A.R. 2005. Psychological, neuropsychological and physiological correlates of serious antisocial behavior in adolescence: The role of self-control. *Criminology: An Interdisciplinary Journal, 43*, 133–176.

Centers for Disease Control and Prevention. 2013. *Youth risk behavior surveillance system*. http://www.cdc.gov/HealthyYouth/yrbs/index.htm.

Chartrand, L.N., and E.M. Forbes-Chilibeck. 2003. The sentencing of offenders with fetal Alcohol Syndrome. *Health Law Journal* 11: 35–70.

Chen, P., and K.C. Jacobson. 2013. Impulsivity moderates promotive environmental influences on adolescent delinquency: A comparison across family, school, and neighborhood contexts. *Journal of Abnormal Child Psychology* 41: 1133–1143.

Chudley, A.E., J. Conry, J.L. Cook, C. Loock, T. Rosales, and N. LeBlanc. 2005. Fetal alcohol spectrum disorder: Canadian guidelines for diagnosis. *Canadian Medical Association Journal* 172: S1–S21.

Chudley, A.E., A.R. Kilgour, M. Cranston, and H. Edwards. 2007. Challenges of diagnosis in fetal alcohol syndrome and fetal alcohol spectrum disorder in the adult. *American Journal of Medical Genetics* 145C: 261–272.

Clark, E., J. Lutke, P. Minnes, and K. Ouellette-Kuntz. 2004. Secondary disabilities among adults with Fetal alcohol spectrum disorder in British Columbia. *Journal of FAS International* 2: 1–12.

Coles, C.D., K.A. Platzman, C.L. Raskind-Hood, R.T. Brown, A. Falek, and I.E. Smith. 1997. A comparison of children affected by prenatal alcohol exposure and attention deficit, hyperactivity disorder. *Alcoholism, Clinical and Experimental Research* 21: 150–161.

Conry, J., and K.O. Asante. 2010. *Youth probation officers' guide to FASD screening and referral*. Maple Ridge, BC: Asante Center. http://www.asantecentre.org/_Library/docs/Youth_Probation_Officers_Guide_to_FASD_Screening_and_Referral_Printer-Friendly_Format_.pdf.

Conry, J., and D.K. Fast. 2000. *Fetal alcohol spectrum disorder and the criminal justice system*. Maple Ridge, BC: British Columbia Fetal Alcohol Syndrome Resource Society Vancouver & Law Foundation of British Columbia.

Connor, P.D., P.D. Sampson, F.L. Bookstein, H.M. Barr, and A.P. Streissguth. 2000. Direct and indirect effects of prenatal alcohol damage on executive function. *Developmental Neuropsychology* 18: 331–354.

Cox, L.V., D. Clairmont, and S. Cox. 2000. Knowledge and attitudes of criminal justice professionals in relation to fetal alcohol spectrum disorder. *Canadian Journal of Clinical Pharmacology* 15: e306–e313.

Day, A. 2011. Offender rehabilitation: Current problems and ethically informed approaches to intervention. *Ethics and Social Welfare* 5: 348–360.

Day, A., C. Kozar, and L. Davey. 2013. Treatment approaches and offending behavior programs: Some critical issues. *Aggression and Violent Behavior* 18: 630–635.

De Brito, S.A., Viding, E., Kumari, V., Blackwood, N., Hodgins, S. 2013. Cool and hot executive function impairments in violent offenders with antisocial personality disorder with and without psychopathy. *Plos One, 8*, Art e65566.

De Guzman, M.R., and Bosch, K.R. 2007. *High risk behaviors among youth*. University of Nebraska-Lincoln. Retrieved from http://www.ianrpubs.unl.edu/pages/publicationD.jsp?publicationId=786.

Dolan, M. 2012. The neuropsychology of prefrontal function in antisocial personality disordered offenders with varying degrees of psychopathy. *Psychological Medicine* 42: 1715–1725.

Fairchild, G., S.H.M. van Goozen, S.J. Stollery, M.R.F. Aitken, J. Savage, S.C. Moore, and I.M. Goodyer. 2009. Decision making and executive function in male adolescents with early-onset or adolescence-onset conduct disorder and control subjects. *Biological Psychiatry* 66: 162–168.

Fast, D.K., and J. Conry. 2004. The challenge of Fetal Alcohol Syndrome in the criminal legal system. *Addiction Biology* 9: 161–166.

Fast, D.K., and J. Conry. 2009. Fetal alcohol spectrum disorders and the criminal justice system. *Developmental Disabilities Research Reviews* 15: 250–257.

Fast, D.K., J. Conry, and C.A. Loock. 1999. Identifying fetal alcohol spectrum disorder among youth in the criminal justice system. *Developmental and Behavioral Pediatrics* 30: 370–372.

Fishbein, D., and R. Tarter. 2009. Infusing neuroscience into the study and prevention of drug misuse and co-occurring aggressive behaviour. *Substance Use and Misuse* 44: 1204–1235.

Hancock, M., J.L. Tapscott, and P.N.S. Hoaken. 2010. Role of executive dysfunction in predicting frequency and severity of violence. *Aggressive Behavior* 36: 338–349.

Heffron, W.M., S. Babalonis, M. Staton-Tindall, L. Molenaar, K.M. Lommel, A.N. Kazura, and C.A. Martin. 2011. Evidence-based practice in criminal justice settings—a biological/genetic perspective: The addicted brain. In *Handbook of evidence-based substance abuse treatment in criminal justice settings*, ed. J.M. Ramos, T.P. Gullotta, and J. Gregrich, 23–44. Columbia: Springer Science+Business Media.

Hoaken, P.N.S., V.K. Shaughnessy, and R.O. Pihl. 2003. Executive cognitive functioning and aggression: Is it an issue of impulsivity? *Aggressive Behavior* 29: 15–30.

Hoppenbrouwers, S.S., D.R. De Jesus, T. Stirpe, P.B. Fitzgerald, A.N. Voineskos, D.J.L.G. Schutter, and Z.J. Daskalakis. 2013. Inhibitory deficits in the dorsolateral prefrontal cortex in psychopathic offenders. *Cortex* 49: 1377–1385.

Hughes, C. 2011. Changes and challenges in 20 years of research into the development of executive functions. *Infant and Child Development* 20: 251–271.

Kellerman, T. 2003. *External brain*. FASD Community Resource Center. http://come-over.to/FAS/externalbrain.htm.

Kerr, M., R.E. Tremblay, L. Pagani, and F. Vitaro. 1997. Boys' behavioral inhibition and the risk of later delinquency. *Archives of General Psychiatry* 54: 809–816.

Kodituwakku, P.W. 2009. Neurocognitive profile in children with fetal alcohol spectrum disorders. *Developmental Disabilities Research Reviews* 15: 218–224.

Kodituwakku, P.W., N.S. Handmaker, S.K. Cutler, E.K. Weathersby, and S.D. Handmaker. 1995. Specific impairments in self-regulation in children exposed to alcohol prenatally. *Alcoholism, Clinical and Experimental Research* 19: 1558–1564.

Kully-Martens, K., S. Treit, J. Pei, and C. Rasmussen. 2013. Affective decision-making on the iowa gambling task in children and adolescents with Fetal alcohol spectrum disorders. *Journal of the International Neuropsychological Society* 19: 137–144.

LaDue, R.A., Streissguth, A.P., and Randels, S.P. 1993. Clinical considerations pertaining to adolescents and adults with fetal alcohol syndrome. In T. B. Sonderegger (Ed.), *Perinatal substance abuse: Research findings and clinical implications* (pp. 104-31.). Baltimore, MD: The Johns Hopkins University Press.

Lynch, M.E., C.D. Coles, T. Corley, and A. Falek. 2003. Examining delinquency in adolescents differentially prenatally exposed to alcohol: The role of proximal and distal risk factors. *Journal of Studies on Alcohol* 64: 678–686.

MacPherson, P.H., A.E. Chudley, and B.A. Grant. 2011. *Fetal alcohol spectrum disorder (FASD) in a correctional population: Prevalence, screening and characteristics.* Ottawa, ON: Correctional Service of Canada.

Malbin, D.V. 2004. Fetal alcohol spectrum disorder (FASD) and the role of family court judges in improving outcomes for children and families. *Juvenile and Family Court Journal* 55: 53–63.

Manninen, M., M. Lindgren, M. Huttunen, H. Ebeling, I. Moilanen, H. Kalska, and S. Therman. 2013. Low verbal ability predicts later violence in adolescent boys with serious conduct problems. *Nordic Journal of Psychiatry* 67: 289–297.

Mattson, S.N., N. Crocker, and T. Nguyen. 2011. Fetal alcohol spectrum disorders: Neuropsychological and behavioral features. *Neuropsychology Review* 21: 81–101.

Mattson, S.N., A.M. Goodman, C. Caine, D.C. Delis, and E.P. Riley. 1999. Executive functioning in children with heavy prenatal alcohol exposure. *Alcoholism, Clinical and Experimental Research* 23: 1808–1815.

May, P.A., P. Gossage, W.O. Kalberg, L.K. Robinson, D. Buckley, M. Manning, and E. Hoyme. 2009. Prevalence and epidemiologic characteristics of FASD from various research methods with an emphasis on recent in-school studies. *Developmental Disabilities Research Reviews* 15: 176–192.

McGee, C.L., O.A. Bjorkquist, E.P. Riley, and S.N. Mattson. 2009. Impaired language performance in young children with heavy prenatal alcohol exposure. *Neurotoxicology and Teratology* 31: 71–75.

McLachlan, K. 2012. *An examination of the psycholegal abilities, risks, and needs of adolescents and young adults with fetal alcohol spectrum disorder (FASD) in the criminal justice system.* (Unpublished doctoral dissertation). Burnaby, BC: Simon Fraser University.

McLachlan, K., R. Roesch, J.L. Viljoen, and K.S. Douglas. 2014. Evaluating the psycholegal abilities of young offenders with fetal alcohol spectrum disorder. *Law and Human Behavior* 38: 10–22.

McLachlan, K., K. Wyper, and A. Pooley. 2013. *Promising practices in substance abuse treatment for justice-involved youth with FASD.* Maple Ridge, BC: Asante Centre.

Mela, M., and G. Luther. 2013. Fetal alcohol spectrum disorder: Can diminished responsibility diminish criminal behaviour? *International Journal of Law and Psychiatry* 36: 46–54.

Mela, M., Tait, C., Levine, G., and Nicolaichuk, T. (2013). *Managing mentally disordered offenders with FASD in the community.* http://www.motherisk.org/documents/FAS13/Mela.pdf.

Moffitt, T.E., L. Arseneault, D. Belsky, N. Dickson, R.J. Hancox, H. Harrington, and A. Caspi. 2011. A gradient of childhood self-control predicts health, wealth, and public safety. *Proceedings of the National Academy of Sciences of the United States of America* 108: 2693–2698.

Moore, T., and M. Green. 2004. Fetal alcohol spectrum disorder (FASD): A need for a closer examination by the criminal justice system. *Criminal Reports* 19: 99–108.

Munoz, L.C., P.J. Frick, E.R. Kimonis, and K.J. Aucoin. 2008. Verbal ability and delinquency: Testing the moderating role of psychopathic traits. *Journal of Child Psychology and Psychiatry* 49: 414–421.

Murtagh, A.M., and S.A. Todd. 2004. Self-regulation: A challenge to the strength model. *Journal of Articles in Support of the Null Hypothesis* 3: 19–51.

Nash, K., J. Rovet, R. Greenbaum, E. Fantus, I. Nulman, and G. Koren. 2006. Identifying the behavioural phenotype in fetal alcohol spectrum disorder: Sensitivity, specificity and screening potential. *Archives of Women's Mental Health* 9: 181–186.

Ogilvie, J.M., A.L. Stewart, R.C.K. Chan, and D.H.K. Shum. 2011. Neuropsychological measures of executive function and antisocial behavior: A meta-analysis. *Criminology* 49: 1063–1107.

Olsen, J., and Debolt, D. (2002). *Community justice project: Case management of fetal alcohol spectrum disorder*. Retrieved from http://www.cvfasd.org/pdf/commjustice.pdf.

Owen, A.M., J.J. Downes, B.J. Sahakian, C.E. Polkey, and T.W. Robbins. 1990. Planning and spatial working memory following frontal lobe lesions in man. *Neuropsychologia* 28: 1021–1034.

Page, K. 2001. Fetal Alcohol Spectrum: The hidden epidemic in our courts. *Juvenile and Family Court Journal* 52: 21–31.

Petrenko, C.L.M., N. Tahir, E.C. Mahoney, and N.P. Chin. 2013. Prevention of secondary conditions in fetal alcohol spectrum disorders: Identification of systems-level barriers. *Maternal and Child Health Journal*. Epub ahead of print.

Popova, S., S. Lange, D. Bekmuradov, A. Mihic, and J. Rehm. 2011. Fetal alcohol spectrum disorder prevalence estimates in correctional systems: A systematic literature review. *Canadian Journal of Public Health* 102: 336–340.

Public Health Agency of Canada. 2007. *Report: National roundtable on the development of a Canadian model for calculating the economic impact of FASD*. http://www.phac-aspc.gc.ca/hp-ps/dca-dea/prog-ini/fasd-etcaf/publications/ei-me07-eng.php.

Radford-Paz, E. 2013. *Risk and protective factors for criminality among adults with FASD*. (Published Master's thesis). Laurentian University, Sudbury, ON. http://zone.biblio.laurentian.ca/dspace/handle/10219/2128.

Raine, A. 2002. The role of prefrontal deficits, low autonomic arousal, and early health factors in the development of antisocial and aggressive behaviour in children. *Journal of Child Psychology and Psychiatry* 43: 417–434.

Rasmussen, C., R. McAuley, and G. Andrew. 2007. Parental ratings of children with fetal alcohol spectrum disorder on the behaviour rating inventory of executive function (BRIEF). *Journal of FAS International* 5: 1–8.

Rasmussen, C., V. Talwar, C. Loomes, and G. Andrew. 2008. Brief report: Lie-telling in children with fetal alcohol spectrum disorder. *Journal of Pediatric Psychology* 33: 220–226.

Roach, K., and A. Bailey. 2009. The relevance of fetal alcohol spectrum disorder in Canadian criminal law from investigation to sentencing. *U.B.C. Law Review* 41: 68.

Romi, S., and D. Marom. 2007. Differences in intelligence between nondelinquent and dropout delinquent adolescents. *Adolescence* 42: 325–336.

Schonfeld, A.M., S.N. Mattson, and E.P. Riley. 2005. Moral maturity and delinquency after prenatal alcohol exposure. *Journal of Studies on Alcohol* 66: 545–554.

Schonfeld, A., B. Paley, F. Frankel, and M.J. O'Connor. 2006. Executive functioning predicts social skills following prenatal alcohol exposure. *Child Neuropsychology* 12: 439–452.

Seguin, J.R., L. Arseneault, and R.E. Tremblay. 2007. The contribution of "cool" and "hot" components of decision-making in adolescence: Implications for developmental psychopathology. *Cognitive Development* 22: 530–543.

Shallice, T., and P.W. Burgess. 1991. Deficits in strategy application following frontal lobe damage in man. *Brain* 114: 727–741.

Sohlberg, M.M., and C.A. Mateer. 1997. Effectiveness of an attention-training program. *Journal of Clinical and Experimental Neuropsychology* 9: 117–130.

Spohr, H., J. Willms, and H. Steinhausen. 2007. Fetal alcohol spectrum disorders in young adulthood. *Journal of Pediatrics* 150: 175–179.

Stade, B., A. Ali, D. Bennett, D. Campbell, M. Johnston, C. Lens, and G. Koren. 2009. The burden of prenatal exposure to alcohol: Revised measurement of cost. *Canadian Journal of Clinical Pharmacology* 16: e91–e102.

Stade, B.C., W.J. Ungar, B. Stevens, J. Beyene, and G. Koren. 2004. The burden of prenatal exposure to alcohol: Measurement of costs. *Pediatric Research* 55: 71A–72A.

Steinberg, L. 2004. Risk taking in adolescence: What changes, and why? *Annals of the New York Academy of Sciences* 1021: 51–58.

Steinberg, L. 2008. A social neuroscience perspective on adolescent risk-taking. *Developmental Review* 28: 78–106.

Streissguth, A.P., F.L. Bookstein, H.M. Barr, P.D. Sampson, K. O'Malley, and J.K. Young. 2004. Risk factors for adverse life outcomes in fetal alcohol syndrome and fetal alcohol effects. *Journal of Developmental and Behavioural Pediatrics* 25: 228–238.

Streissguth, A.P., J.M. Aase, S.K. Clarren, S.P. Randels, R.A. Ladue, and D.F. Smith. 1991. Fetal alcohol syndrome in adolescents and adults. *Journal of the American Medical Association* 265: 1961–1967.

Syngelaki, E.M., S.C. Moore, J.C. Savage, G. Fairchild, and S.H.M. Van Goozen. 2009. Executive functioning and risky decision making in young male offenders. *Criminal Justice and Behavior* 36: 1213–1227.

Tremblay, R.E., R.O. Pihl, F. Vitaro, and P.L. Dobkin. 1994. Predicting early onset of male antisocial behavior from preschool behavior. *Archives of General Psychiatry* 51: 732–739.

Tuominen, T., T. Korhonen, H. Hämäläinen, S. Temonen, H. Salo, J. Katajisto, and H. Lauerma. 2014. Neurocognitive disorders in sentenced male offenders: Implications for rehabilitation. *Criminal Behaviour and Mental Health* 24: 36–48.

Vaurio, L., E.P. Riley, and S.N. Mattson. 2011. Neuropsychological comparison of children with heavy prenatal alcohol exposure and an IQ-matched comparison group. *Journal of the International Neuropsychological Society* 17: 463–473.

Verdun-Jones, S.N., and A. Butler. 2013. Sentencing neurocognitively impaired offenders in Canada. *Canadian Journal of Criminology and Criminal Justice* 55: 495–512.

Vidovic, P. J. 2012. *Neurocognitive impairments and the criminal justice system: A case analysis of the impact of diagnoses of FASD and ADHD on the sentencing of offenders in the courts of three Canadian provinces*. (Published doctoral dissertation). Burnaby, BC: Simon Fraser University. http://summit.sfu.ca/item/12312.

Victor, A., J.R. Wozniak, and P. Chang. 2008. Environmental correlates of cognition and behavior in children with fetal alcohol spectrum disorders. *Journal of Human Behavior in the Social Environment* 18: 288–300.

Wilkinson, R.A. 2012. *Breaking the cycle—Tasmanian Corrections plan (2010-2020): Best practice in offender rehabilitation*. Tasmania Department of Justice. http://www.justice.tas.gov.au/correctiveservices/breaking_the_cycle.

Wilson, D.B., L.A. Bouffard, and D.L. Mackenzie. 2005. A quantitative review of structured, group-oriented, cognitive-behavioral programs for offenders. *Criminal Justice and Behavior* 32: 172–204.

Yang, Y., and A. Raine. 2009. Prefrontal structural and functional brain imaging findings in antisocial, violent, and psychopathic individuals: A meta-analysis. *Psychiatry Research: Neuroimaging* 174: 81–88.

Young, S., and J. Thome. 2011. ADHD and offenders. *World Journal of Biological Psychiatry* 12: 124–128.

Zelazo, P., and S. Carlson. 2012. Hot and cool executive function in childhood and adolescence: Development and plasticity. *Child Development Perspectives* 6: 354–360.

Zelazo, P.D., and U. Müller. 2002. Executive function in typical and atypical development. In *Blackwell handbook of childhood cognitive development*, ed. U. Goswami, 445–469. Malden, MA: Blackwell Publishers Ltd.

Medico-Legal Interventions in Management of Offenders with Fetal Alcohol Spectrum Disorders (FASD)

Mansfield Mela

Evidence is accumulating as to the effective psychotherapeutic and pharmacological interventions applied to those with Fetal Alcohol Spectrum Disorder (FASD). Such interventions have not been studied in the criminal justice system. Yet results of studies among youth with FASD and those involved with the criminal justice system should be transferable to adults. Similarly, when dealing with an offender with FASD, changes in or modifications to the legal framework and approaches are suggested as potential additional interventions to be considered. Education of and increasing FASD knowledge among key officials in the criminal justice system are pivotal for understanding the disability and how to intervene in offenders with FASD. To improve outcomes therefore, offenders should be promptly identified and existing forensic mental health teams should be trained in both FASD assessment and intensive case management. Interventions with scientific evidence should inform the management of offenders as they pass through the different stages of the criminal justice system. Appropriate expectations, positive relational approach, and the provision of support, structure and supervision are critical especially when offenders are in the community. A collaborative team effort is needed to reduce the consequences of the disability of FASD and reduce victimization of both the offender and the members of the society.

Introduction to FASD and Criminality

Fetal Alcohol Spectrum Disorder (FASD) is the leading cause of intellectual developmental disorder (Dykens 2000; Gillberg 2003). The disorder arises from prenatal alcohol exposure (PAE) to the fetus. This is suspected, in predisposed individuals, as the cause of neuropathological neuronal loss mediated through apoptosis.

M. Mela (✉)
University of Saskatchewan, Saskatoon, Canada
e-mail: mansfieldmela@gmail.com

M. Nelson and M. Trussler (eds.), *Fetal Alcohol Spectrum Disorders in Adults: Ethical and Legal Perspectives*, International Library of Ethics, Law, and the New Medicine 63, DOI 10.1007/978-3-319-20866-4_8

This typically results in poor adaptive functioning. The latter becomes obvious through secondary disabilities when intellectual demands are exerted on the adult and adolescent. Increased criminal activities are some of the prominent secondary disabilities found among those who are prenatally exposed to alcohol (Cleary and Freeman 1992; Ressler et al. 1988; Streissguth et al. 1999).

Fetal Alcohol Spectrum Disorder and the Criminal Justice System

Primary disabilities such as neurocognitive abnormalities of poor memory, impulsivity, poor judgment, inability to link action and consequences, suggestibility, gullibility, and secondary disabilities such as high rates of mental disorders, alcohol and substance use disorders as well as inappropriate social skills are suspected as contributing to the increased rates of offending and reoffending among those with FASD. Such offenders with FASD are frequently repeat offenders and revolve through the Criminal Justice System. The direct and indirect costs especially from prosecution, incarceration and loss of income are high.

The overrepresentation of FASD among offenders has medical, social and economic implications. Given the regularity with which FASD is associated with criminal involvement, dealing with offenders with FASD and mental disorder through the criminal justice system has been unsatisfactory (Chartrand 2003). Consequently, the Canadian Bar Association (CBA) and the American Bar Association (ABA) simultaneously adopted specific resolutions, calling on all levels of government to provide treatment, supports, and rehabilitation in place of incarceration. The intent of this recommendation is to promote the modification of the traditional approaches to the dispensation of justice. As those with FASD appear to be verbally competent in the face of poor comprehensive abilities, the disorder has earned the title of the "invisible disorder". This is not surprising when only about a fifth of those with the disorder have any distinguishing facial features (Streissguth et al. 1999; Astley 2010).

Mentally Disordered Offenders (MDOs) with FASD in the Criminal Justice System

Statistics

Taking a lifelong perspective among those exposed to alcohol prenatally, a substantial number will develop a variety of mental disorders and commit criminal offences (Streissguth et al. 1999). It goes without saying that many victims traversing the Criminal Justice system also have FASD even if not formally diagnosed. Current evidence linking PAE to criminality is based on a positive family history of alcohol problems in 60 % of 36 multiple murderers studied just about two decades after FASD was characterized (Ressler et al. 1988). One of the approaches used to examine the relationship between offending and FASD is to follow up diagnosed individuals and estimate how many among them commit offences. Using this approach, criminal involvement, termed "trouble with the law" was reported among approximately 60 % of an American cohort of 415 individuals diagnosed with FASD (Streissguth et al. 1999). The cohort has now been followed up for over 30 years. Similar findings have been replicated in a Canadian study and in another cohort comprised of 1400 individuals diagnosed with FASD (Astley 2010). Follow up of a similar cohort of FASD diagnosed individuals in British Columbia, Canada, produced a 45 % involvement with the law suggesting a close relationship between having FASD and criminal activities.

Another approach used to explore the potential mechanism by which FASD is linked to the propensity to criminal and antisocial activities is to study the rates of FASD in different offender populations. In a specific project for determining the rate of diagnosis of FASD among probationers, a rate of 92 % FASD was found among 47 probationers. The population was biased by being a select population for whom the diagnosis of FASD was highly suspected. This study method has been used among youth and adults admitted for psychiatric assessment and coming into the federal penitentiary in Canada. The calculated rate of FASD diagnosis in that adult study was 10–28 % depending on the level of certainty in confirming maternal use of alcohol during pregnancy. It is safe to assume this rate to be an underestimate of the rate in the adult offender population. This is primarily because the study excluded all those thirty (30) years old and over as well as those who were already in custody at the time of the study. Other studies reported on rates of FASD among offenders and youth offenders under community supervision.

The average rate of FASD was about 10 % among offenders (MacPherson and Chudley 2008; Rojas and Gretton 2007). In an active case study, the rate of FASD found among youth offenders referred for psychiatric assessment (23.3 %) was also supportive of the association between PAE and subsequent offending (Fast et al. 1999). The unpublished incidence study mentioned above adopted a similar active case finding method. The study adhered to the Canadian guidelines for FASD diagnosis (MacPherson and Chudley 2008). Table 1 below describes the studies examining the rate of FASD among offender populations. These studies are a reasonable source for public planning but not sufficient to categorical answer if PAE predisposes to criminal activities. Cohort studies with their ability to control

Table 1 Rates of FASD in different offender populations

FASD rates among offenders		
Authors	Study	Rates (%)
Rojas and Gretton (2007)	File review 230 youths (13–18)	10.9
Murphy and Chittenden '07	Survey 137 youth (14–19)	11.7
Fast et al. (1999)	Diagnostic 287 youth (12–18)	22.3
McPherson, Chudley and Grant	Diagnostic 91 adult (19–30)	9.9 (18)

confounders hold the key for determining the impact of PAE, separate from other criminogenic factors, in criminality. Studies of that nature with criminal activities as outcomes are not yet available.

In the absence of these powerful studies, some meta-analytic studies summarizing the findings in the existing literature are emerging and adding support for the relevance of PAE as a risk factor in the etiology of criminality. In a systematic review of studies on incarcerated youths, a systematic literature review concluded that youth with FASD are 19 times more likely to be incarcerated than youth without FASD (Popova et al. 2011). While these results are associations rather than causal findings, the correlation between FASD and subsequent offending is evident. Such correlations are bolstered by another research finding. In the Canadian context, FASD is present throughout and encountered in the Criminal Justice system. FASD is however overrepresented among Indigenous peoples (Rojas and Gretton 2007). This likely contributes to the disproportionate overrepresentation of indigenous people in the criminal justice system.

FASD and Type of Offences

Mentally disordered offenders and especially those with FASD are suspected as contributing a lot to the number of offences prosecuted. This stems from the repeated nature of the offences and other factors such as lack of identification, knowledge of how best to support them and the absence of the relevant support especially in the community. The repeated nature of the offences is gleaned from results of some studies. As well, certain types of offences thrive in the context of the specific impaired cognitive functions linked with FASD such as impulsivity, poor memory and executive function. The aforementioned are essentially the primary disabilities described among those with prenatal alcohol exposure and consequent dysfunctions in FASD (Chartrand 2003). Those affected with FASD contribute to that surge in offence statistics, as they often appear in the criminal justice system due to repeat offences, including breaches and failures commonly committed by 'unsophisticated criminals' (Roach and Bailey 2009).

The types of offences committed by those with FASD have been studied from the first follow up result of the FASD group that were monitored for 15–25 years (Streissguth et al. 1999). In that study, half of those described as being in trouble with the law committed theft, burglary, assault, murder, domestic violence, and child molestation; all of these were categorised as offences against the person (Streissguth et al. 1999). The most common first criminal behaviour was shoplifting. Additional evidence for more minor types of offending comes from the Correctional Service of Canada finding that, among offenders with FASD with an average of at least 15 convictions, theft and breaches of conditions were the most prominent offences (MacPherson and Chudley 2008). Mentally disordered offenders with FASD are frequent repeat offenders and continually revolve through the criminal justice system. It has been estimated that for each substantive offence, 0.7 system-generated offences such as breaches of conditions are prosecuted and adjudicated. Not only are the prosecutions expensive, the offences are largely preventable as they are caused by cognitive risk factors such as lack of attention to time and date, poor memory in remembering conditions, and poor executive functioning. Prosecutions not only increase criminal statistics, but can also force these offenders to face lamentable injustices.

Experiences Within in the Criminal Justice System and the Need for Alternative Approaches

The 1992 death sentence and execution of Robert Harris in California epitomizes a form of injustice associated with the criminalization of disability, such as FASD. Specifically, in spite of a clear diagnosis of FAS and accompanying cognitive deficits, the courts made no exemption, reasoning that excusing the crime would open the 'flood gates' for others to use FASD to avoid punishment (Cleary and Freeman 1992). Variations on this unfortunate ruling continue in courts around the world. Another experience of the criminal justice system with FASD is the frustration of those who deal with the offenders with FASD. When Justice Trueman in the province of British Columbia, Canada became frustrated in the case of a repeat offender diagnosed with FASD, she wrote in her judgement (Chartrand 2003):

> *The cognitively challenged are before our courts in unknown numbers. We prosecute them again and again and again. We sentence them again and again and again. We imprison them again and again and again. They commit crimes again and again and again. We wonder why they do not change. The wonder of it all is that we do not change.*

Justice Trueman's intention was clearly to communicate that a different approach to offenders and mentally disordered offenders (MDOs) with FASD was advisable and should be separate from traditional sentencing practices. Such an attempt to change the system and think outside the box clearly depends on thoughtful alternatives to sentencing. The record of adopting rehabilitative principles and practices instead of sentencing offenders with FASD is regrettably dismal. In a review of 40 FASD offenders sentenced by Canadian courts, researchers described many instances of misunderstanding, a lack of alternatives and the disadvantages of incarceration (Chartrand 2003). This examination led to the recommendation that alternative approaches should be used. For example, one alternative that was invoked by a judge in the province of Saskatchewan, Canada was to order the province to provide an accused with FASD with the necessary community resources and services to meet his needs. Unfortunately, the appeal court in Saskatchewan disagreed with the judge and the status quo of applying traditional sentencing principles to FASD offenders was maintained (Roach and Bailey 2009). Thus, when offenders with FASD repeatedly present before the courts, incarceration becomes the default system, which is problematic because there is evidence that offenders with FASD do not do well in penal institutions (Luther and Mela 2006; Fast and Conry 2009). However, there are viable alternatives. For instance, one potential approach would be to create a mental health court and utilize different verdicts of diminished responsibility to divert offenders to the services they need to promote law-abiding behaviour (Mela and Luther 2013). Of course, this type of approach is contingent upon legislative action.

Alternative approaches that accommodate an offender with FASD should focus on various points of contact with the criminal justice system. Therapeutic jurisprudence is a set of principles using the law to provide alternatives for those with disabilities. These principles are based on promoting the psychological and physical wellbeing of persons subject to legal proceedings. At various junctions of the system, the offender with FASD should be considered and handled based on the disabilities present. There are variations of how this can be practicalized. An offender may be considered a high risk to offend on account of pre-crime factors such as poor education, impulsivity and poor socioeconomic status (see Fig. 1). The disadvantages of poor education, childhood adversity and lower socioeconomic factors are well recognised factors that increase criminality in many offenders. Preventative measures of educational upgrading, employability and trauma related intervention assume significance at this stage. In this chapter, many of the stages and junctions will be discussed in greater detail to point out how the law can be used to intervene.

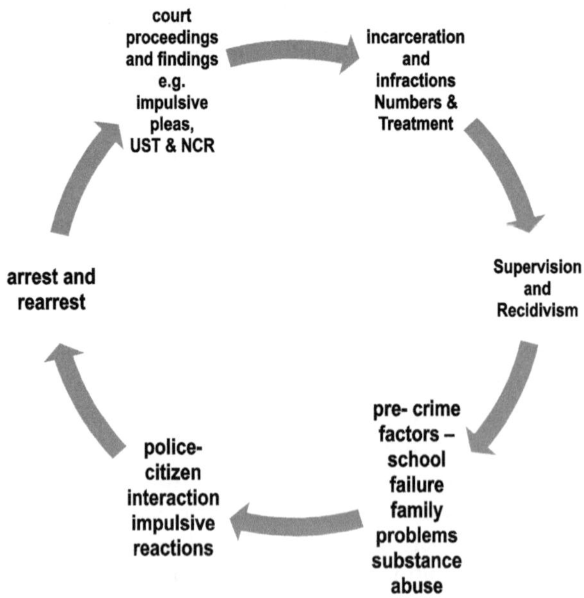

Fig. 1 Relevant points in the criminal justice system that create both aggravating and mediating interface with FASD accused

Issues Impacting Interventions with Offenders with FASD

Based on available evidence, it is safe to conclude that a strong link exists between PAE and increased rates of disability, mental disorder, and criminality (Fast et al. 1999; Popova et al. 2011; Streissguth and O'Malley 2000). This fact should inform strategies for better identification of FASD and mental disorders among offenders. Those strategies in turn should provide effective and relevant interventions. As a result, a better standard of care could be achieved and would provide highly necessary evidence for effective treatment. Establishing effective interventions is especially challenging due to the multiplicity of comorbid presentations of those with FASD and related conditions.

There is substantial variation and multiplicity in the compositions and comorbid presentations of those with FASD and other related conditions (Streissguth and O'Malley 2000). In fact, with such a high rate of comorbid mental conditions and the co-occurrence of other disabilities, having FASD with no additional conditions is the exception and not the norm. That is, many affected by FASD will have additional mental disorders as well as a history of offending. The term '*triply troubled*' has been coined to describe pregnant adolescents who used opiates and engaged in criminal behaviour (Crome and Kumar 2007). This term can also easily be applied to the complex linkages between FASD, mental disorders, and engagement in criminal activity. The result of the apparent complexity of issues surrounding the comorbidity of FASD and mental disorders is that individuals with these conditions crisscross various stages of the criminal justice system. The exact mechanism by which the offenders with FASD and additional conditions that perpetuate offending are not clearly studied yet. To adequately intervene conceptually and practically, research should focus on better understanding of these mechanisms.

Pre-crime Risk Factors and Initial Encounters with the Criminal Justice System

There are a number of relevant pre-crime factors that can be indicators of later criminality, such as low educational achievement, poor self-esteem, and a lack of self-identity commonly reported in those with FASD. Combined with other risk factors such as family stresses and substance use problems, the risk for encountering the criminal justice becomes elevated (Streissguth et al. 2004). Furthermore, primary disabilities such as a lack of risk awareness and a tendency toward impulsivity are likely to inflame any interactions a person with FASD has with and during encounters with law enforcement agents. It has been suggested, though not formally studied, that such interactions may contribute to the increased rate of criminal activity among those with FASD. Specifically, poor social skills and sensory sensitivity, and intolerance to touch can play a role in resisting arrest or a propensity for violent behaviours during arrest (Fig. 1). The resulting struggle could lead to a range of outcomes that shift outcomes from simple arrest to charges of resisting arrest, assault on a peace officer, or even fatal results.

Another key issue is that, among individuals with FASD, cognitive disorders are frequently concealed by seemingly normal or even the appearance of superior communication skills (that are accompanied by a lack of understanding). Unfortunately, this is a substantial disadvantage to offenders who are judged as competent, even though they face tremendous challenges with the criminal justice system based on the true lack of cognitive abilities. Once arrested, assessments should be requested in order to determine the accused triablility (competence) and culpability (criminal responsibility). Evidence is accumulating that highlights the need for assessment as part of criminal justice proceedings. For instance, a major financial settlement was provided to an individual who falsely confessed to six sexual offences. In 1997 Mr. Simon Marshall was arrested in connection with sex crimes. Those sexual assaults were said to have occurred in the Sainte-Foy area of Quebec, Canada. Suggestibility, vulnerability and poor consequential thinking are suspected to have played a great role in confessing to those crimes. He was later acquitted by DNA evidence and compensated by the government. These psychological abnormalities suspected as the mechanism for such confession are equally recognised as potential crime aggravating factors in FASD. However, many individuals with FASD continue to pass through the criminal justice system without a diagnostic and functional assessment.

An assessment is usually requested so that the clinical picture can, when possible, provide an explanation for both the cause of offending and the preventative intervention in the accused with FASD. Some jurisdictions have defences and verdicts that accommodate the functional disability suspected in a person with FASD when it is associated with the offence in question. When such options are missing or are inadequate in taking account of FASD disability, traditional punishment is invoked to the disadvantage of the person. In fact, only one of 82 court cases involving individuals with FASD has produced a discussion about the Not Criminally Responsible (NCR) designation and there have, to date, been no cases where this designation has been applied (Mela and Luther 2013). As the finding of not criminally responsible is essential a threshold one separating those culpable and responsible and those not, a 'grey zone' approach with diminished capacity as an alternative verdict should be a positive outcome.

Why Incarceration Is not an Appropriate Intervention

Some have justified incarceration of individuals with FASD, based on the assumption that it will provide much needed structure for the offender. However, victimization and involvement as pawns for the more sophisticated gang members in prison are a few of the negative impact of incarceration. Thus, once in jail, primary disabilities such as unawareness of risk, desire to fit in, and willingness to be used make them vulnerable as prey for the more experienced and antisocial offenders (Fig. 1). For example, they are often young and naïve enough to be physically and/or sexually assaulted. Moreover, they frequently end up serving as 'errand boys' for prison trade in contrabands, bearing the brunt of the consequences alone due to the 'con code' that is established in these environments. The result is an increased number of disciplinary actions taken against them and consequently elongation of their sentence.

If an offender with FASD is able to return to the community, it is likely that he or she will be accompanied by a significant list of disadvantages including: a lack of social and survival skills, being more criminalized and institutionalized, and a wider range of criminal associates and peers (Boland et al. 1998). This more 'streetwise' individual is then more likely to engage in more criminal activity and lack a sense of connectedness with society due to prior removal from mainstream society for a significant amount of time. The consequence of these factors is a propensity toward recidivism and the facilitation of a more versatile offending pattern. This leads to a cycle of crime and is a primary source of increased crime statistics among this group. Furthermore, cognitive deficits related to FASD often lead to a cumulative default behaviour pattern of offending mediated through the failure to attend appointments because of a criminalized and oppositional approach to life. There are however times when the offender may manifest genuine amnestic consequences of FASD. Research suggests that those with FASD have difficulty learning and remembering. As a result, these individuals will often commit more offences, breaches of court orders, and then become reincarcerated. In addition, the situation is not improved by FASD-related deficits such as poor planning and challenges with consequential thinking, which make compliance to conditions even more unlikely. As a whole, apart from increasing criminal statistics, the cost of crime is increased substantially in offences perpetuated by individuals with FASD.

Interventions for Offenders with FASD

Interventions among offenders generally can be conceptualized in several ways; these include philosophical approach, specialty specific methods, and the goals and desires of the person and the society at large. In the realm of interventions for those with FASD, terms such as cure, alleviate, improve, care, minimize, and maintain health are suggested so as to intervene appropriately and feasibly. Some of the strategies used do not fit the traditional treatment where specific changes are expected in treatment. To apply any approaches to FASD will require the different contexts and various disabilities involved. Moreover, these approaches may replace the comfortable use of terms such as 'healing' or 'treatment,' which may have the connotation of the complete extermination or at least the reversal of the condition. The reality is that until final pathophysiological mechanisms and specific interventions are developed to address some or all of the deficits or issues faced by offenders with FASD, it is necessary to rely on principles of care and maintenance in creating and implementing appropriate interventions and management. With knowledge of the interface between the features of FASD and the criminal justice system, interventions that are medico-legally informed can be considered and applied.

There is a range of general interventions that are effective for offenders with FASD. Conceptualized correctly, the right interventions should include the risk need responsivity principles, the biopsychosocial intervention model and the legal 'hook' approach. Understanding the interaction between individuals' criminal tendencies and potential with environmental opportunity for crime creates seamless interweaving approaches worthy of mention. Even though there is more information and research in children and adolescents with FASD, some lessons are transferable to interventions with an adult offender with FASD. Interventions in children and adolescents can be assumed to have some value in adults. This is because the profile of disabilities resulting from PAE are on a continuum from children to adults.

Clinicians and experts, Conry and Fast, cautioned those who work with FASD offenders suggesting they must understand the implications of FASD in the individuals to prevent further victimization (Fast and Conry 2009). The importance of appropriate expectations cannot be overemphasized. Realistic expectations allay the anxiety of the FASD affected and protect against frustration in the medical, psychiatric or legal agents dealing with the accused. It allows a reasonable adaptation to the deficits inherent in FASD. For instance, expecting a sleep disordered FASD accused to attend court in the early part of the morning without assistance is bound to result in failure to attend. Without much imagination, one can estimate the level of frustration by the court officials.

In the community where an offender with FASD is expected to keep some conditions under supervision, challenges can arise in communication with the officer supervising. Deficits in comprehension, a well-recognised problem in those affected by FASD is likely to contribute to misunderstanding which will negatively affect both the accused and the relationship he/she has to the supervisor. Judges should request an accused with FASD to recount what was told them and what they understood by the directions and instructions given them. Verbalization of such information supports understanding and likely prevents more charges that could result from lack of understanding. FASD accused are likely to suggest they understood something even when they have not in order not to look stupid in front of everyone.

As well, family therapy has a high level of efficacy when it comes to re-entry into the community. Specifically, it provides an additional support system that the offender can rely on. However, this type of support may not always be available, so any other supports (be it people or community services) can be particularly valuable. Finally, professional aftercare and intensive case management is crucial to effective transitions into the community. This can range from working with various healthcare practitioners to group home operators and caseworkers. Overall, the biopsychosocial model of intervention has been found to be the most successful, and includes a combination of: diagnostic interventions, psychosocial interventions, and pharmacological interventions. In addition, mentorship programs have been found to be particularly effective (these are discussed in greater detail in—Collaborative psychosocial interventions for offenders with FASD).

Diagnostic Interventions

Diagnostic processes are not usually seen as specific interventions. In FASD, however, diagnosing someone before the age of six was found to be preventive to the bad outcome in a follow up among those diagnosed with FASD (Streissguth et al. 1999). The reasons for this are that a diagnosis facilitates a better understanding of behaviour and challenges, leverages social support, changes expectations, and thus provides more positive adjustment and can lessen the adverse consequences of brain deficits. The diagnostic process itself is crucial in understanding symptoms and it forms the basis for planning evaluation and treatment of those diagnosed. Assessments ordered by Judges in the criminal justice system should therefore fulfil similar goals. However, as already mentioned above, this request for assessment does not frequently occur. In Fig. 1, the potential for diversion exist in the courts but this is unlikely unless there are accurate and up to date assessment reports. Having a diagnostic assessment identifies strengths and weaknesses helpful in guiding relevant and effective interventions. In adults, therefore, efforts at replicating the findings of better outcome on account of completing the diagnostic process is highly necessary (Streissguth et al. 1999). Diagnosis as a form of intervention has the capacity to be an effective intervention for individuals with FASD.

FASD Diagnosis and the DSM 5

FASD has long been a brain disorder that the DSM has struggled to accommodate (Mela 2006). Unfortunately, FASD languished in a sort of 'no man's land' in the field of psychiatry after being excised from the DSM IIIR in 1988. However, prenatal alcohol exposure and FASD have regained their place in the DSM 5, at least to some degree. This change has had mixed reviews. The term FASD itself is discouraged as a diagnostic term. The various entities that can be diagnosed (e.g., static encephalopathy) all fall within the spectrum of disorders called FASD. Overall, it has been described as a continuum of complex neuropsychiatric, cognitive, behavioural, social, language, communication, and other multisensory deficits (Streissguth and O'Malley 2000). Modifications of the DSM 5 have not incorporated a specific category pertaining to FASD as a spectrum, but the changes in the intellectual disability (Intellectual Developmental Disorder) section of the DSM should accommodate most of those whose neurocognitive abnormalities and disabilities arise on account of being prenatally exposed to alcohol.

According to the DSM 5, a person affected by FASD can be diagnosed even if he or she does not have a low IQ score. One of the ways to do so is to invoke the Intellectual Developmental Disorder criteria and use specifications for severity and etiology to qualify the disorder. Removing the multi-axial system and less dependence on IQ as a particular qualifier can facilitate diagnosis. Second, it is also possible to invoke the section of *not otherwise specified* under the Neurodevelopmental Disorder classification. A third approach is to use the criteria for Neurobehavioural Disorder that is associated with prenatal alcohol exposure (ND-PAE). Although these proposed criteria require further study, this research criterion recognizes "more than minimal exposure to alcohol exposure during gestation" and allows confirmation to be obtained from "maternal self report…medical or other records, or clinical observation."

The recognition by the DSM of the influence and effects of prenatal alcohol exposure is exemplified by the frequent use of the phrase *associated with prenatal alcohol exposure* in a number of diagnoses. This applies to neurodevelopmental disorder, attention deficit hyperactivity disorder and developmental coordination disorder. By so doing the DSM recognised PAE as a significant risk and prognostic factor.

The benefit of these changes is that the level of evidence required for confirmation of maternal alcohol use is less rigorous that in previous guidelines for diagnosis (Astley and Clarren 2000; Chudley et al. 2005). The significance of this change is that previous diagnostic systems required formal maternal confirmation of alcohol exposure during pregnancy, which could be very challenging to obtain. This confirmation is especially difficult when the offender is estranged from his or her family. These new criteria open the door to diagnosing individuals who may otherwise have not been able to obtain a diagnosis due to rigid criteria. Since early and accurate diagnosis is associated with more positive outcomes, this may represent a step forward in the FASD diagnostic process and efficacy of interventions. Furthermore, in the case of offenders with FASD, it may facilitate easier diagnosis and the accompanying supports and services that accompany it. Increased diagnosis may also promote an understanding of the underlying cognitive, executive function, and other primary disabilities that contribute to a lack of success in the traditional criminal justice system and provide a stronger case for legal alternatives.

Collaborative Psychosocial Interventions for Offenders with FASD

Interventions for offenders with FASD (and frequently comorbid mental disorders) must be lifelong—there is no beginning or ending point. They must be multifaceted, and incorporate assistance and support beginning at the institutional level and expanding through the transition into the community. Mentorship programs, both within institutions and in the community, are an effective intervention for offenders with FASD.

The incorporation of a mentor into all types of interventions ensures that there is a continual source of healthy social and emotional support that may not otherwise exist within the offender's life. This is part of the biopsychosocial model, where effective transitions to the community must, at the very least, begin with intensive case management and assertive treatment, followed by continued management and soft support (e.g., by mentors or family). In a typical rehabilitation cascading level of independence, the person of interest progresses from semi-independence to finally full independence. The multiple intervention approaches that are used to intervene with individuals with FASD have varying degrees of supporting evidence, including randomized control trials, a high level study for proving effectiveness of a treatment.

Pharmacological Interventions

Although at first glance, pharmacological interventions can be seen as strictly the domain of psychiatrists, they have a higher level of efficacy when others such as the mentor and family physician are involved. In particular, this can lead to higher levels of compliance and monitoring. There have been a number of studies assessing the impact of various pharmacological interventions, each with their pros and cons. First, methylphenidate has been utilized specifically to decrease hyperactivity. However, it does not increase attention or memory to the same extent (Dommet et al. 2008). Second, stimulants have been effectively used as a means of decreasing hyperactivity and impulsivity associated with FASD (Doig et al. 2008). Third, there has been a preferential response to dexamphetamine (O'Malley et al. 2000). However, many of the medications prescribed for treatment of individuals with FASD have adverse side effects such as weight loss and headaches that could potentially decrease the level of compliance with pharmaceutical regimes. Currently, atomoxetine and lisdexamfetamine appear to be potential options for treatment, but data about their efficacy is still being established (Peadon et al. 2012).

In summary, there are a number of different options available for pharmaceutical interventions for offenders with FASD. Being that those are results from double blind placebo controlled studies, the evidence is reasonably robust. However, it is important to weigh the benefits against potentially adverse side effects to ensure compliance. On a positive note, new evidence for the efficacy of existing and new interventions continues to emerge. Pathophysiologically informed interventions include the use of specific sleep aid medications to alleviate hypersomnia and insomnia. The prevalence of restless leg syndrome can be reduced by occupational therapy assessment and intervention as well as neuropathic pain medications. Pharmacological interventions include treatment of psychosis, mood disorders, anxiety as well as impulsivity as specific pathophysiologically informed treatment targets.

Legal Interventions

Education of judges and those involved in the courts and criminal justice system was suggested as one of the significant improvements needed in the field of FASD. After setting some educational targets, the authors of a study reviewed and evaluated the level of education in the criminal justice system and opined that better education resulted in better knowledge about FASD to the benefit of the FASD affected (Moore and Green 2004). Using the law for therapeutic purposes is termed therapeutic jurisprudence. There is a high level of frustration with the propensity of the FASD offender to continually circulate through the criminal justice system and the associated challenges that presents. Consequently, the legal principles adopted in contending with offenders with FASD are different from traditional advocacy and litigation methods for the system. For instance, it has been reported by many legal observers that often offenders with FASD are like strangers in the system who suffer many ills such as poor access to diagnosis (Streissguth et al. 1999; Roach and Bailey 2009). This is problematic because, as noted in Diagnostic Interventions, diagnosis can be the crucial element of intervention and rehabilitation. Why does this occur? The answer is, primarily due to unawareness and a lack of education, as well as a low availability of diagnostic services. For instance, in Canada only about 2000 diagnoses are completed yearly. When compared to the 3000 new cases expected each year, not to mention the innumerable existing undiagnosed cases, this represents a serious issue.

Increasing the Efficacy of Legal Interventions Through Changes to the Criminal Justice System

There are a number of actions that can be taken to facilitate better interventions in the context of the legal system. To begin, it is essential to educate existing forensic mental health teams starting with the forensic psychiatrists. Educational and practical diagnostic training could feasibly be provided to interested teams of the field of forensic mental health. In addition, these psychiatrists could be educated about the implications of FASD in regard to the legal system by packaging them with modules for continuous medical education points (CME). Earlier, the importance of diagnosis as a means of intervention was discussed and the specifics of the potential for the DSM 5 was also addressed. Another method to improve legal interventions is to make psychiatrists aware of the new avenues for FASD diagnosis under the DSM 5. Every expert conducting an assessment for the court needs to consider the possibility of FASD. In the current dispensation under the DSM-5, no diagnostic group can be excluded when completing an assessment. Likewise diagnosis is less cumbersome as it requires only two of the four regular domains for diagnosis.

Usually the alcohol exposure, brain deficiency, growth and facial abnormalities are examined. In DSM-5 only the first two are required.

Existing forensic mental health teams have the skill set and capacity to be the default diagnostic and intervention teams for those with FASD who are part of the criminal justice system. The intensive and active case management model adopted by the forensic mental health teams likely explain the great success they have had in terms of the accused that are found not criminally responsible, called NCR accused in Canada. The recidivism rate of those with mental disorder who pass through the interdisciplinary supervisory (NCR Review) board compared to those through the Parole Board of Canada is about a third of the latter. This superior approach in managing persons with mental disorder is essential in helping and can be emulated by the criminal justice system to handle the deficits associated with FASD when those affected have offended. Mental health teams charged with the task of caring for FASD accused after adjudication are likely to serve the society well by the comprehensive collaboratively focused approaches for care.

Another factor to consider is that legislative action may be needed to create a verdict of diminished capacity involving FASD and other neurodevelopmental disorders (especially those that are 'invisible'). Many countries signed up to the 1998 Rome statute of the International Criminal Court. This provides for an additional arm of the insanity defence. Such an arm provides for consideration of any inability to conform behaviour with the law. Its threshold is in the grey zone provided by the black and white compartmentalization of the strict insanity law. This is consistent with the principles of proportionality depending on the blameworthiness of the offender so diminished by cognitive deficits. Those falling under this verdict, in Canada for instance, could then be diverted to the NCR Review board for dispositions. Moreover, post sentence diversion should be instituted for FASD offenders completing a custodial sentence and found to pose a high risk based on past repeat offending. At the start of their sentence, all offenders should be screened with one of the FASD screening tools. Those screening positive should be sent to prison-based diagnostic teams. At the end of their sentence, offenders with FASD can be diverted to the forensic mental system, the logical team equipped with the ability to manage the medico-legal risks of such offenders.

Finally, consideration should be given to the modification of the court process of ordering assessments to include FASD as a mental condition. In such a case, the courts will take judicial notice of FASD and consider it as a potential factor in understanding the offender. This is where legislation can define what the disorder is and what the criminal justice system should do in recognising and accommodating those with the disorder. A call made clearly by the North American legal community through their Bar associations. As such, costing and relevant outcomes should be studied over time to review social justice, safety, economic, and societal outcomes.

Psychosocial Interventions

Lessons can be gleaned for adult FASD offenders from various strategies from effective studies in adolescents and from other psychological approaches of intuitive validity. In a relational manner, provision of support structure and supervision of the FASD offender in the community are the corner stone of intervention. Boulding and Brooks cited Bruce Perry with respect to the six important factors to be applied in dealing with adult FASD affected. These can be extended to the offenders as they are instructive of the strength based approach needed to eliminate blame and shame, and foster socialization and prosocial living. They suggested a relational approach that sees the individual first as a person and not as an offender (Boulding and Brooks 2010). The goals and approach is to focus on the present and delivery should be repetitive to reinforce understanding. This approach is consistent and linked with rewards that build successes using strength –based tasks. Rhythmic consistencies by paying attention to the rhythms of life are encouraged including day and night, meals and exercise schedules. Above all, respect for the individual is seen as a crucial step for involving the offender (Boulding and Brooks 2010).

Brown et al. (2012) have reviewed and reported on strategies useful in adolescents. These include providing structure, contending with unexpected situations, avoiding 'contagion' effect from peers and behavioural therapy focuses on replacing maladaptive behaviours with positive ones (Brown et al. 2012). These could be transferable to the adult population and further studied in line to determine effectiveness. There is much more to be studied in the adult offender with FASD and attending to mental conditions serve to alleviate some of the pressures faced by the individuals.

Conclusion

The current level of knowledge and research is that FASD is over represented in the criminal justice system. While the mechanism by which this occurs is not exactly known, there are possible childhood, mental disorder and criminogenic risk factors that make this likely. Knowledge of the disorder among professionals and practitioners in the system is essential to understand the offender with FASD and to accommodate the deficits and dysfunction using proxies of approaches used in other similar settings. The courts are placed rightly to request assessments that identify the deficits and strengths of accused. The results of such assessments should inform interventions that would likely reduce criminality and victimization. The role of FASD in the criminal justice system affects offenders, witnesses and victims alike. Planning for these accommodations among those affected in addition implementing other recognised and researchable methods of intervention are proposed as the important strategies for helping those in the justice system.

Acknowledgments Stacey McHenry is recognised for her assistance with the initial draft.

References

Astley, S.J. 2010. Profile of the first 1,400 patients receiving diagnostic evaluations for Fetal Alcohol Spectrum Disorder at the Washington State Fetal Alcohol Syndrome Diagnostic and Prevention Network. *The Canadian journal of Clinical Pharmacology* 17(1): e132–e164.

Astley, S.J., and S.K. Clarren. 2000. Diagnosing the full spectrum of fetal alcohol-exposed individuals: Introducing the 4-digit diagnostic code. *Alcohol and Alcoholism* 35(4): 400–410.

Boland, F.J., R. Burril, M. Duwyn, and J. Karp. 1998. *Fetal alcohol syndrome: Implications for correctional service*. Ottawa: Canada Correctional Service.

Boulding, D., and S. Brooks. 2010. Trying differently: A relationship centered approach to representing clients with cognitive challenges. *International Journal of Law and Psychiatry* 33: 448–462.

Brown, N.N., P.D. Connor, and S.A. Adler. 2012. Conduct-Disordered adolescents with Fetal Alcohol Spectrum Disorder Intervention in secure treatment settings. *Criminal Justice and Behavior* 39(6): 770–793.

Chartrand, L.N., and E. Forbes-Chilibeck. (2003). The sentencing of offenders with fetal alcohol syndrome. *Health Law Journal* 11: 35–70.

Chudley, A.E., J. Conry, J.L. Cook, C. Loock, T. Rosales, and N. LeBlanc. 2005. Fetal Alcohol Spectrum Disorder: Canadian guidelines for diagnosis. *Canadian Medical Association Journal* 172(5 suppl): S1–S21.

Cleary, S.F., and K. Freeman. 1992. Exploring the link between fetal alcohol syndrome and sociopathic behaviour. *Free Inquiry in Creative Sociology* 22(2): 5.

Crome, I.B., and M.T. Kumar. 2007. Epidemiology of drugs and alcohol use in young women. *Seminars in Fetal and Neonatal Medicine* 12(1): 98–105.

Doig, J., J.D. McLennan, and W.B. Gibbard. 2008. Medication effects on symptoms of Attention-Deficit/Hyperactivity disorder in children with Fetal Alcohol Spectrum Disorder. *Journal of Child and Adolescent Psychopharmacology* 18(4): 365–371.

Dommet, E.J., E.L. Henderson, M.S. Westwell, and S.A. Greenfield. 2008. Methylphenidate amplifies long-term plasticity in hippocampus via noradrenergic mechanisms. *Learning and Memory* 15(8): 580–586.

Dykens, E.M. 2000. Psychopathology in children with intellectual disability. *Journal of Child Psychology and Psychiatry* 41(4): 407–417.

Fast, D.K., and J. Conry. 2009. Fetal Alcohol Spectrum Disorders and the criminal justice system. *Developmental Disabilities Research Reviews* 15(3): 250–257.

Fast, D.K., J. Conry, and C.A. Loock. 1999. Identifying fetal alcohol syndrome among youth in the criminal justice system. *Journal of Developmental and Behavioral Pediatrics* 20(5): 370–372.

Gillberg, C. 2003. Learning disability. *The Lancet* 362(9386): 811–821.

Luther, G.E., and M.A. Mela. 2006. Ten top issues in law and psychiatry. *Saskatchewan Law Review* 69: 40.

MacPherson, P., and A.E. Chudley. 2008. Fetal Alcohol Spectrum Disorder (FASD): Screening and estimating incidence in an adult correctional population. In: 2nd international conference on Fetal Alcohol Spectrum Disorder: Research, policy, and practice around the world, Vancouver.

Mela, M.A. 2006. Accommodating the Fetal Alcohol Spectrum Disorders in the diagnostic and statistical manual of mental disorders (DSM V). *JFAS Int e* 23: 10.

Mela, M., and G. Luther. 2013. Fetal Alcohol Spectrum Disorder: Can diminished responsibility diminish criminal behaviour? *International Journal of Law and Psychiatry* 36(1): 46–54.

Moore, T., and M. Green. 2004. Fetal Alcohol Spectrum Disorder (FASD): A need for closer examination by the criminal justice system. *Criminal Reports* 19(6): 99–108.

O'Malley, K.D., B. Koplin, and V.A. Dohner. 2000. Psychostimulant clinical response in fetal alcohol syndrome. *Canadian Journal of Psychiatry* 45: 90–91.

Peadon, E., D. Thomas, and E.J. Elliot. (2012). Pharmacological interventions for ADHD symptoms in children with Fetal Alcohol Spectrum Disorders. Cochrane Database of Systematic Reviews.

Popova, S., S. Lange, D. Bekmuradov, A. Mihic, and J. Rehm. 2011. Fetal Alcohol Spectrum Disorder prevalence estimates in correctional systems: A systematic literature review. *Canadian Journal of Public Health* 102(5): 336–340.

Ressler, R., A.P. Burgess, and D.J. Douglas. 1988. *Sexual homicides: Patterns and motives*. New York: Lexington Books.

Roach, K., and A. Bailey. 2009. The relevance of Fetal Alcohol Spectrum Disorder and the criminal law from investigation to sentencing. *University of British Columbia Law Review* 42: 1–68.

Rojas, E.Y., and H.M. Gretton. 2007. Background, offence characteristics, and criminal outcomes of Aboriginal youth who sexually offend: A closer look at Aboriginal youth intervention needs. *Sex Abuse* 19(3): 257–283.

Streissguth, A.P., and K. O'Malley. 2000. Neuropsychiatric implications and long-term consequences of Fetal Alcohol Spectrum Disorders. *Seminars in Clinical Neuropsychiatry* 5: 177–190.

Streissguth, A.P., H. Barr, J. Kogan, and F. Bookstein. 1999. Primary and secondary disabilities in fetal alcohol syndrome. In *The challenge of fetal alcohol syndrome: Overcoming secondary disabilities*, ed. A.P. Streissguth, and J. Kanter. Seattle: University of Washington Press.

Streissguth, A., F. Bookstein, H. Barr, P. Sampson, K. O'Malley, and J. Young. 2004. Risk factors for adverse life outcomes in fetal alcohol syndrome and fetal alcohol effects. *Developmental and Behavioral Pediatrics* 25: 226–238.

Oral Language and Communication Factors to Consider When Supporting People with FASD Involved with the Legal System

Linda Hand, Megan Pickering, Sally Kedge and Clare McCann

The stories that parents tell about their children with Fetal Alcohol Spectrum Disorder overwhelmingly reflect how important communication skills are for managing life. Combine language difficulties with cognitive issues, and the chances that the person will get into trouble with the law are high—too high for us to be comfortable with what all too often happens with these young people.

An excellent example is this account given by a parent (abridged) (for the full version, please see Yee, nd).

> Some grocery store personnel had rudely evicted (my son) and his friends (at 3.00 a.m.) when they had been doing nothing wrong. He went back three weeks later and purposely right in front of them shoplifted some sandwiches. He wanted these guys to see him stealing. It was his idea of "revenge." Three store employees chased him out of the store, across the parking lot (where he dropped the sandwiches), across a street, and into an alley. My son turned to face them. He held up his 3" blade pocket knife and said, "Leave me alone! What do you want? I've given you back your stuff!" (After some physical violence, during which he did not resist) my son was arrested and charged with Armed Robbery.

In this account, her son showed a combination of not understanding consequences, not reading situations accurately, and not understanding the concepts of language involved. It is often the case that such problems are not seen as 'language' difficulties, but the concepts involved are embedded in language—concepts like 'stealing' and 'revenge' and 'consequences'.

L. Hand (✉) · S. Kedge · C. McCann
Talking Trouble Aotearoa New Zealand, The University of Auckland,
Auckland, New Zealand
e-mail: l.hand@auckland.ac.nz

S. Kedge
e-mail: sallykedge@gmail.com

C. McCann
e-mail: c.mccann@auckland.ac.nz

M. Pickering
Hawkes Bay District Health Board, Waipukurau, New Zealand
e-mail: Megan.Pickering@hawkesbaydhb.govt.nz

M. Nelson and M. Trussler (eds.), *Fetal Alcohol Spectrum Disorders in Adults: Ethical and Legal Perspectives*, International Library of Ethics, Law, and the New Medicine 63, DOI 10.1007/978-3-319-20866-4_9

The Nature of Language Difficulties in FASD

Researchers have consistently reported that language difficulties are a major aspect of FASD, from the days when Fetal Alcohol Syndrome only was considered (e.g. Becker et al. 1990) through to the models that incorporate all the possibilities of the Fetal Alcohol Spectrum (Mattson and Riley 1998; Streissguth et al. 1996; and Coggins et al. 2007).

However, this can be confusing. Many children and young people with FASD may appear very verbally fluent, and have normal speech and grammar.[1] The difficulties are not so much at this level as in higher level language skills, where skills are needed to comprehend and express longer stretches of oral language (such as 'explanations' and stories), or to be flexible enough with words to manage multiple or figurative meanings, to draw inferences or indirect messages correctly, or to incorporate social schemas so as to manage the politeness systems of language (Brown and Levinson 1987). These are all skills that most of us can take for granted. So much are they taken for granted that the average person is rarely aware that they are using such skills on a constant basis, and equally importantly, not aware that they automatically expect the same skills of others.

A model of the neuropsychological domains often used in discussing FASD is that seen in Fig. 1 (e.g., Chudley et al. 2005; Mattson et al. 2011). This is not a model of language per se, and yet the shaded domains show where language skills either dominate (e.g., "communication") or are heavily involved in these neuropsychological domains (academic achievement such as literacy, for example, is underpinned by oral language skills) (Kamhi and Catts 2005).

As a model, Fig. 1 is rather static, and fails to capture the dynamic and interactive nature of human communication. A better model for our purposes is that found in Coggins et al. (2007) (see Fig. 2). They termed this model "a framework of social communication", which it is, and highlighting the social nature of language enables us to focus on the aspects we saw in the opening vignette from Dianne Yee. They concern the fact that what one person in the interaction had in their mind as to what was going on was different to what the others had in theirs. Both, however, assumed that the others had the same expectations and interpretations that they themselves did. Hence language is about, as Fig. 2 says "influencing people and interpreting events", with certain associated behaviours and decision-making and planning, which all centrally depend on mutually held understandings.

So what happens when this mutual understanding does not hold? Mostly what happens is that the representative(s) of the majority (typical person, normal language user) assumes the other actually does understand the same things they do, and therefore inaccurately attributes their behaviours as deliberate violations of these mutual expectations. These actions are assumed to arise by choice or from

[1]Although deficits have been reported in these skills for some, e.g. Abel 1990, cited in Mattson and Riley (1998) p. 289.

Cognition	**Academic achievement**	**Adaptive behaviour, social communication, social skills**
Executive functioning and abstract reasoning	**Hard & soft neurological signs – sensory and motor**	**Communication**
Attention deficit /hyperactivity	**Memory**	**Brain structure**

Fig. 1 Neuropsychological domains

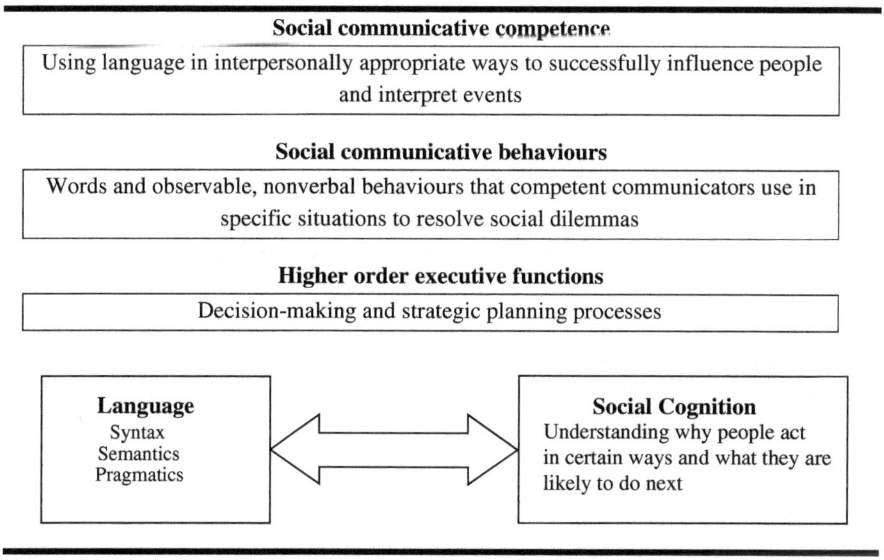

Fig. 2 A model of higher level language (from Coggins et al. 2007, p. 118)

defects of character or behaviour. The other person is felt to be "aggressive", "uncooperative", "arrogant", and/or "criminal", and hence deserved everything that happened to them. This is very evident in the parent experience we saw at the beginning of the chapter.

This understanding of language and language skills as central to interactions and of the role of mutual understanding and its failures, helps us see why the consequences of language difficulties can be so overwhelming. The website "Fetal Alcohol Spectrum Disorder and Justice" (http://www.fasdjustice.ca/) presents a number of vignettes which illustrate this (see Boxes): At best the behaviour of the person concerned seems inconsistent, out to fool others or to lie.

Because this is so puzzling to interpret, even experts in young people and/or in the law can find themselves assuming the worst possible interpretation.

B. came to his court date for an assault charge and suddenly decided to plead guilty. His lawyer explained that he had a good chance of being found not guilty at trial, and advised him, strongly, against pleading. B insisted. When asked by the judge if he had anything to say before sentencing, B gave a long, rambling explanation of why he was innocent. (He couldn't handle the suspense of waiting any longer and didn't understand the court process.)

C. was a defence witness. On the stand, the defence lawyer gently tried to ask her questions that would help C. present helpful facts. Although the questions were basic, C. found them difficult and reacted with hostility to the lawyer. (She knew the lawyer was defending her friend, but did not understand his role or his questions).

C. also asked the judge to stop people in the court from speaking to each other. (The people were her family members, there to help. However, she had difficulty concentrating on the questions because of all of the ambient noise.)

The Consequences of Language Difficulties in FASD

Conry and Fast (2000) said "People with FASD even when not mentally handicapped, have great difficulty understanding and processing language; in this regard they may be unintentionally abused by the system" (pp. 34–35). The boxes above have prepared us for this conclusion.

Language is involved in every stage of the legal process. An initial encounter with the police or other front-line staff and a young person can often go either way, depending on how the young person communicates. Are they able to 'talk their way out of it'? Present a credible case for themselves? Recognise what it is that the front-line person is likely to want to hear, in terms of explanation and appropriate recognition of the rights of others? If they are, there may be few, if any, consequences beyond that point. Should it go further, are they able to explain their story to a lawyer or advocate so that person can present their best case, and also to understand and follow the recommendations and expectations of that lawyer or advocate? Are they able to understand the processes and language of the court, and the meaning of judges' rulings, of orders or sentences, and what they now need to do? If placed in a rehabilitation or other programme, are they able to participate in the (usually) verbal activities central to them (many of these programmes are centred around the 'talking therapies'), and to comprehend what they are expected to achieve from it?

There is ample evidence that a high proportion of young people in the justice systems of the UK, Australia and the USA, have oral language difficulties.

The figure most commonly cited is 50–60 %, compared to some 7–12 % of their peers (see Snow and Powell 2012, for a summary). An unknown number of these young people are on the spectrum of Fetal Alcohol Disorders. Hughes et al. (2012) showed very high numbers of neurodevelopmental disabilities in their populations of young people who offend in the UK, of which the most highly represented is communication difficulties (cited as 60–90 %). FASD is listed as a lower figure—11–12 %—but they recognise that diagnosis is unlikely to be consistently made with this group. Some of the higher figure young people would also be in the FASD group if properly assessed.

So significant is this that Conry and Fast (2000 p. 25) considered that trouble with the law is a screening indicator for FASD, as shown in their diagram below.

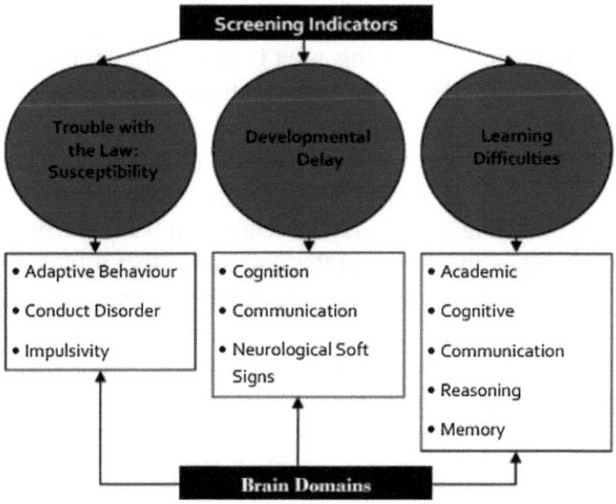

If 'trouble with the law' is a screening indicator, then it is clear that the problems concerned are being detected far too late. Hughes et al. (2012) discuss the idea that neurodisability itself (such as FASD) places a young person at risk for offending, confounded with other known risk factors such as truancy, poor socialisation, inconsistent nurturing and school failure. Some children with FASDs may have a history of dysfunctional caregiving and are at risk of co-occurring disabilities such as head injury and depression.

What Can Be Done?

The trajectory so often seen, as illustrated above, is not inevitable. What it has highlighted is that there is a need for a great deal more understanding of the problems of language difficulties within a wide range of professions and in the community. This understanding has at least two purposes; to minimise unnecessary consequences for the young person, arising from misinterpretation, and secondly to allow these disciplines and persons to begin to make adaptations; to their interaction patterns, to instruments of assessment, and to interventions, as well as to day-to-day interactions. Early intervention is required to ensure that children's language needs are provided for throughout their childhood to ensure that they are equipped with the language and social communication skills they need to participate in all the interactions they may have in life, whether in education, health or legal contexts.

Recognition and consistent diagnosis of FASD is still not well established except in isolated pockets. Canada has perhaps the best record to date. But ways of assessing and working with language difficulties have a longer history, and there are new interventions emerging which are proving effective in the youth justice sector (e.g. Gregory and Bryan 2009; Department of Health (UK) 2011; Lanz 2009). Additionally, work such as that of Cross (2011) has many points to contribute.

Another action which would greatly enhance the outcomes for young people and the systems they are within is to employ more speech language therapists (SLTs). The contribution they can make in this area is considerable, but it has consistently surprised those new to the idea. Lord Ramsbotham, Chief Inspector of Prisons for England and Wales 1995–2001, said in 2006 "I have to admit that in all the years I have been looking at prisons and the treatment of offenders, I have never found anything so capable of doing so much for so many people at so little cost as the work that speech and language therapists carry out." Similarly the Governor of Polmont Youth Offenders Institution in the UK, stated that if by a mischance he had to get rid of all of his staff, the last one out of the gate would his SLT. Justice is a new area of intervention for SLT, but it is one clearly within the scope of practice.

There are many ways for professionals, or anyone interacting with the young person, to adjust their own language. Comprehensive professional development courses are desirable, such as those offered by the Asante Centre (Bodaly 2011). The website "Fetal Alcohol Spectrum Disorder and Justice" (http://www.fasdjustice.ca/) also offers lawyers some cogent advice, such as;

- Keep your questions short to avoid asking too much at once. Complicated and wordy questions can be especially difficult to process and respond to, by a person with FASD.
- Listen actively to the person's responses. If he or she is not responsive, try to reword your questions accordingly.
- Be patient.

An excellent source of information and advice on best communication practices for legal practitioners who work with vulnerable children and young people is the Advocate's Gateway's website. This can be found at http://www.theadvocatesgateway.org. Their UK based project provides essential training and support for legal professionals who work with vulnerable individuals and they provide from a range of 'toolkits' and videos demonstrating excellent interaction strategies for use in court. Their team members have had pivotal roles in establishing intermediary services in the UK for vulnerable individuals to help ensure that they fully understand what is said to them and can participate during legal situations e.g. in court, during police interviews. These intermediary services need to be established in jurisdictions around the world as many vulnerable young people are currently missing out on the opportunity to properly participate in the legal processes they are involved with.

The UK Department of Health (2011) handbook offers practical advice for communicating with those offenders with learning disabilities, such as in the box below. Much of this would seem applicable for the FASD population:

- Remember to always speak clearly and slowly, using plain language and avoid jargon.
- Explain any technical words to the person.
- Ensure they understand things before moving on.
- Use the person's name when addressing them at the start of a question or comment.
- Always explain to the person concerned exactly why they are in a new situation, what they should expect and when.
- Prepare the person for each stage of the communication, for example, "David, I will now ask you some simple questions" or "David, I will now explain what we are going to do next."
- Emphasise important words.
- Avoid using acronyms, abbreviations, sarcasm and metaphors.
- Be patient and calm whilst communicating, do not rush the person you are talking to. They are likely to need longer to process the questions and think about their answers.
- Use concrete terms not abstract references, for example, "At breakfast time" rather than "Early on."
- Break large chunks of information into smaller chunks, with short breaks.
- It may help to use visual aids when asking questions. For example, you could use photos or drawings to illustrate a point.
- Ask questions in the chronological order in which they happened. You could use a calendar and spoken prompts. For example, "Was it the day after you went to the doctors?"
- Keep the questions precise and not vague, for example, 'Where were you yesterday afternoon?' rather than 'Where have you been recently?'
- Avoid double-negative and unclear questions. For example, instead of "You were not in the shop, were you?" you could ask "Were you in the shop?"
- Be aware that repeating questions may suggest to the person that they have given the wrong answer the first time.
- The person must not feel pressured into a response. Many people are more suggestible or they may be eager to give what they think is the 'desired' answer.
- It is important to check that the person understands what has been said. You can ask them to repeat the question or message in their own words, or they could be questioned further.

Conclusion

There is much work needed on the topic of language, FASD and the legal system. A greater understanding of the profiles of language and communication difficulties associated with FASD is required. How these profiles interact with language and communication difficulties experienced by other groups with associated difficulties needs to be established. Further development of efficient methods of assessment of language for populations of children and young people affected by alcohol is needed to inform the design of effective practical interventions that can be robustly evaluated. Evidence based intervention approaches that directly develop language skills in children and young people with FASD, starting as soon as possible, are required.

Alongside this direct work on building language skills in those with FASD, we also need to raise awareness amongst practitioners and the community of the centrality and importance of language across all stages of life. This will help the teams around each person with FASD to ensure that they are carefully considering how they communicate within their own interventions to ensure that language difficulties are not going to be preventing their work from being effective. Equipping those affected by FASD with strong language and communication skills, together with ensuring that clinicians and those in the legal professions are aware of the implications of language difficulties (and how they can help make sure that these are catered for) will have a powerful preventative and protective impact on this highly vulnerable population.

References

Bodaly, K. 2011. *FASD and communication disability: Strategies for youth in the legal system. A curriculum for professionals working with youth in the justice system. Professional participants' manual*. British Columbia, Canada: The Asante Centre for Fetal Alcohol Syndrome.

Becker, M., Warr-Leeper, G. A., and Leeper, H. A. 1990. Fetal alcohol syndrome: a description of oral motor, articulatory, short-term memory, grammatical, and semantic abilities. *Journal of Communication Disorders* 23:97–124.

Brown, P., and S. Levinson. 1987. *Politeness; some universals in language usage*, 2nd ed. Cambridge: Cambridge University Press.

Catts, H.W., and A.G. Kamhi eds. 2005. *Language and reading disabilities*. Boston: Pearson.

Chudley, A., J. Conry, J. Cook, C. Loock, T. Rosales, and N. LeBlanc. 2005. Fetal Alcohol Spectrum Disorder: Canadian guidelines for diagnosis. *Canadian Medical Association Journal* 172(5 suppl): S1–S21.

Coggins, T.E., G. Timler, and L. Olswang. 2007. A State of double Jeopardy: Impact of prenatal alcohol exposure and adverse environments on the social communicative abilities of school-age children with Fetal Alcohol Spectrum Disorder. *Language Speech and Hearing Services in Schools* 38(2): 117–127.

Conry, J., and D.K. Fast. 2000. *Fetal Alcohol Syndrome and the criminal justice system*. Vancouver: Law Foundation of British Columbia.

Conry, J., and K. Asante. 2010. *Youth probation officers guide to RASD screening and referral*. Canada: Asante Centre for Fetal Alcohol Syndrome. http://www.asantecentre.org/_Library/docs/Youth_Probation_Officers_Guide_to_FASD_Screening_and_Referral_Booklet_Format_.pdf.

Cross, M. 2011. *Children with social, emotional and behavioural difficulties and communication problems: There is always a reason* (2nd ed). London: Jessica Kingsley.
Department of Health (UK). 2011. Positive practice positive outcomes: A handbook for professionals in the criminal justice system working with offenders with learning disabilities. Leeds. http://www.gov.uk/government/publications/positive-practice-positive-outcomes-a-handbook-for-professionals-in-the-criminal-justice-system-working-with-offenders-with-a-learning-disability.
Gregory, J., and K. Bryan. 2009. Evaluation of the leeds speech and language therapy service provision within the intensive supervision and surveillance programme provided by the leeds youth offending team leeds, The University of Surrey/National Health Service. Leeds. http://www.leedscommunityhealthcare.nhs.uk/document.php?o=2803.
Hughes, N., H. Williams, P. Chitsabesan, R. Davies, and L. Mounce. 2012. *Nobody made the connection: The prevalence of neurodisability in young people who offend*. Office of the Children's Commissioner, UK. www.childrenscommissioner.gov.uk/content/publications/content_633.
Kellerman, T. 2000–2002. Secondary disabilities in FASD. http://www.come-over.to/FAS/fasconf.htm.
Lanz, R. 2009. Speech and language therapy within the Milton Keynes youth offending team. A summary report outlining the findings of 4 month pilot project examining the speech, language and communication needs of the young people accessing the Milton Keynes youth offending team. Unpublished report.
Mattson, S.N., and E.P. Riley. 1998. A review of the neurobehavioral deficits in children with Fetal Alcohol Syndrome or prenatal exposure to alcohol. *Alcoholism, Clinical and Experimental Research* 22(2): 279–294.
Mattson, S. N., N. Crocker, and Tanya T. Nguyen. 2011. Fetal alcohol spectrum disorders: Neuropsychological and behavioral features. *Neuropsychology Review* 21(2):81–101.
Sellman, D., and J. Connor. 2009. In utero brain damage from alcohol: A preventable tragedy. *New Zealand Medical Journal* 122 (1306): 6–8.
Snow, P. and M. Powell. 2012. Youth (in)justice: Oral language competence in early life and risk for engagement in antisocial behaviour in adolescence. *Trends & Issues in Crime and Criminal Justice* 435:1–6 (Australian Institute of Criminology).
Streissguth, A.P., H.M. Barr, J. Kogan, and F.L. Bookstein. 1996. Understanding the occurrence of secondary disabilities in clients with Fetal Alcohol Syndrome (FAS) and fetal alcohol effects (FAE). Final report to the Centers for Disease Control and Prevention (CDC), August, 1996, Seattle: University of Washington, Fetal Alcohol & Drug Unit, Technical report No. 96-06.
Yee, D. (nd). Keep out! (of the criminal justice system). http://www.come-over.to/FAS/keepout.htm.

Cross, M. (2011). Children with social, emotional and behavioural difficulties and communication problems: There is always a reason (2nd ed.). London: Jessica Kingsley.
Department of Health (UK). (2011). Positive practice positive outcomes: A handbook for professionals in the criminal justice system working with offenders with learning disabilities. Leeds. http://www.gov.uk/government/publications/positive-practice-positive-outcomes-a-handbook-for-professionals-in-the-criminal-justice-system-working-with-offenders-with-learning-disabilities
[illegible]

Part III
Judicial Perspectives

Evolution of the Diagnosis of Fetal Alcohol Spectrum Disorder from DSM-IV-TR to DSM-5: The Justice System in the United States—Time for a Paradigm Shift!

Anthony P. Wartnik, Jerrod Brown and Sarah Herrick

The development of FASD throughout the DSM-IV-TR era has had significant impact in criminal cases, especially regarding death penalty litigation. This chapter will focus on individual rights afforded to indigent defendants as it relates to ineffective assistance of counsel, competency, mitigation, and death penalty cases and explore the potential impacts of the DSM-5 on the legal process.

At the outset, it is important to recognize that some may take issue with case law that does not deal directly with FASD-related cases and basic concepts of due process. It needs to be pointed out that lawyers frequently cite case law that has precedence on basic constitutional and other legal concepts that do not deal with the specific issue at hand, yet the lawyer believes because of their similarities, they are applicable. As well, now that disabilities attributable to PAE have been recognized by the DSM-5 as mental or developmental/intellectual disabilities, the legal and mental health communities, in order to effectively serve criminal defendants afflicted with a FASD, as well as judges who are duty bound to enforce the individual rights of the accused, need to become better informed about FASD.

It should be recognized that to effectuate a paradigm shift, a change of culture within legal and mental health professional communities is necessary. This is a priority challenge and the remainder of this chapter will provide a basis to encourage this change. The American Bar Association (ABA) (2012) has embraced this challenge through its Resolution 112B on FASD.

A.P. Wartnik (✉)
Seattle, Washington, USA
e-mail: TheAdjudicator@comcast.net

J. Brown · S. Herrick
St. Paul, Minnesota, USA
e-mail: jerrod01234brown@live.com

S. Herrick
e-mail: sarahherrick@msn.com

M. Nelson and M. Trussler (eds.), *Fetal Alcohol Spectrum Disorders in Adults: Ethical and Legal Perspectives*, International Library of Ethics, Law, and the New Medicine 63, DOI 10.1007/978-3-319-20866-4_10

Structure and Historical Underpinnings

It may prove beneficial to consider the historical development of FASD. The problems of combining pregnancy and alcohol were recognized in biblical times. According to the English Standard Bible (2009), specifically the book of Judges, verse 13:7 offers "behold, thou shalt conceive and bear a son: And now drink no wine or strong drink."

In his Gin War writings, Henry Fielding (1751) pondered the fate of infants conceived in gin and nourished in the womb and by the breast. In 1973, four University of Washington doctors made the connection between alcohol consumption and pregnancy based on identifiable facial features unique to children born following prenatal alcohol exposure, referring to this condition as Fetal Alcohol Syndrome (FAS) (Jones et al. 1973).

Some 23 years later, links between prenatal alcohol exposure and the development of secondary disabilities were reported in a study commissioned by the Centers for Disease Control and Prevention (CDC) (Streissguth et al. 1996). Very few appellate cases involving FAS and Fetal Alcohol Effects were decided in the late 1980s. There has been an increase in these cases since 1996.

A recent survey using the ACHA-National College Health Assessment II was conducted on 1,035 University of Alaska college students (Brems et al. 2014). It included three questions to assess knowledge of FASD risk factors and two questions regarding general FASD knowledge. These responses were compared to those given by 2,292 professionals (e.g., corrections, educators, physicians, public health nurses, social workers and substance abuse counselors) in a similar survey. The results indicated the college students were adequately informed about FASD risk factors and general FASD knowledge compared to the physicians, who were the least knowledgeable about FASD risk factors. The physicians placed in the middle of all the professions surveyed on FASD general knowledge. It is conjectured those professionals involved with the legal system may be far less informed regarding FASD compared to college students and other professionals (Brems et al. 2014).

Female college students remain at higher risk for alcohol-exposed births due to higher rates of alcohol consumption and binge drinking compared to non-college peers. An increase in alcohol consumption among female college students has been noted, as well as a sharp rise in frequent binge drinking (Wechsler et al. 2000). Although most sexually active college women use contraceptives, alcohol consumption is associated with increasingly ineffectual use of contraception. Another study of 2,012 college women showed, 13 % of risk (at risk for alcohol-exposed pregnancies) drinkers used ineffective contraception and 31 % were risk drinkers who engaged in sexual activity without consistently using condoms (Brems et al. 2014; Ingersoll et al. 2008).

Development of Law and Level of Understanding of FASD in the Legal System: Pre-DSM-5

Knowledge of the U.S. Criminal Justice System is necessary to appreciate the delays in recognizing the presence of FASD within forensic and legal settings. The U.S. Constitution provides American citizens with certain individual rights and protections. States establish criminal laws, provided such laws do not conflict with these rights and protections. Thus, criminal cases originating under state law must be tried and reviewed in state court before accessing federal court review of questions or concerns involving the Constitution. This is referred to as "Exhaustion of State Remedies" and the procedure for moving a State case to the Federal Court is known as *Habeas Corpus*. State cases cannot be removed to federal court until the exhaustion of state remedies. A *Writ of Habeas Corpus* derives from the English Common Law, meaning "bring the body before the court."

The umbrella term "FASD" was unknown in 1999. More was known about FAS than other conditions caused by prenatal alcohol exposure (PAE). FAS, compared to other FASD's, may not necessarily present as more serious or severe forms of brain damage compared to the other PAE related conditions that make up the umbrella. Misunderstandings about FASD have been noted by others:

> While researchers have made significant progress in the identification and understanding of FASD, the legal system has been slow to grasp its impact on juvenile crime. Many juvenile justice system professionals are unable to deal appropriately with individuals with FASD because they lack critical knowledge and training about the disorders. The result is a failure to recognize individuals with FASD and to ensure that they receive appropriate services and support, from arrest through reentry into society. One must first become familiar with the nature and cause of FASD (Williams 2006, pp. 4–5).

Although court decisions reflect general recognition that FASD impacts behavior relevant to the justice system, they also reflect an imperfect understanding of the diagnostic process, symptoms, and behavioral implications. It is necessary to identify fundamental concepts to those responsible for such decisions within the justice system. As an example, our Anglo-Saxon common law suggests it is a fundamental condition upon which criminal responsibility reposes that individuals have the capacity to reason right from wrong, and thus choose between right and wrong. It is these capacities—reason and choice—which give the moral justification to imposing criminal responsibility and punishment onto offenders. If a person can reason right from wrong and has the ability to choose right from wrong, then attribution of responsibility and punishment is morally justified or deserved when that person consciously chooses wrong.

Furthermore, our common law assumes that criminal offenders must be treated as rational and autonomous with the ability to make choices. The problem is those afflicted with FASD are often not rational, autonomous, choosing agents, able to reason right from wrong, or able to choose right from wrong. To put these principles in perspective, one must consider:

> Evidence concerning certain alcohol-related conditions has long been admissible during the guilt phase of a criminal proceeding to show lack of specific intent… Just as the harmful effect of alcohol on the mature brain of an adult imbiber is a matter within the common understanding, so too is the detrimental effect of this intoxicant on the delicate, evolving brain of a fetus held in utero… (Such) evidence should be admitted at the guilt phase of a trial if offered to show that defendant lacked the mental state (here premeditation) that is part of the crime (Dillbeck v. State of Florida 1994, pp. 1029–1030).

Dillbeck v. State of Florida (1994) also stands for appointing expert witnesses to assist defendants with a mental health defence "in order to level the playing field".

Despite these clearly stated fundamental principles, FASD is still misunderstood within the legal system (McLachlan et al. 2014). There is general lack of appropriate education for those who work in or with our justice system (Cox et al. 2008). Furthermore, it is only 40 years post-discovery of FAS which is reflected in the current level of knowledge. Many mental health professionals continue to have difficulty diagnosing individuals with FASD, and available support/referral services for both the professional and client are minimal. Until 2013, with the development of DSM-5, FASDs were not viewed as developmental disorders.

Historically, justice system professionals have experienced difficulties when considering the nexus between the condition and criminal behavior outside of mental retardation or developmental disabilities as noted in the previous editions of the DSM. This is particularly true, when, unlike FAS, there are no physical features identifying the impact of prenatal alcohol exposure. It is easier to consider the intellectual disability as being causally related to a person's behavior. It becomes difficult to relate invisible organically caused conditions to one's subsequent behavior in the context of the widely variable behaviors that are associated with FASD. Those who are neurologically impacted by PAE, including those who lack the facial features related to FAS, often present with developmental delays and challenges; experience learning disabilities; engage in self-injurious behaviors; are impulsive; and demonstrate disturbances in emotional, as well as social adaptive life skills (Sampson et al. 1997).

Those impacted by PAE are at increased risk to engage in criminal behavior because of poor impulse control; deficits with processing information; problems understanding behavioral cause and effects; a lack of understanding of societal norms; impairment with short-term memory; compromised personal boundaries; poor stress tolerance resulting in confusion; challenges in understanding abstract concepts; emotional regulation difficulties; and/or poor judgment (McCreight 1997).

Also, individuals impacted by FASD can be easily influenced by others and may be viewed as naïve (Williams 2006). Given this prior profile of a "non-developmental" condition, it is understandable why many in the criminal justice system often perceive defendants with FASD as willful and disobedient "bad actors."

However, these individuals are often unable to control their behaviors and/or understand what motivates their actions.

On the other hand, if this same defendant suffers from a developmental disorder, it is much easier to see the relationship between the disorder and the criminal behavior, though the complexity of the neurodevelopment disorder can still make it difficult for the uneducated to comprehend. In summary:

> When the defendant enters into the court arena, he/she must deal with a complex system of order, processes, and fact finding or trial. The defendant is also expected to comprehend the gravity of the situation and the consequences of his or her alleged actions. The defendant with FASD is ill equipped to navigate the court system. Officers of the court are equally unprepared. They approach the situation as if dealing with just another youth or adult in trouble, rather than with one whose actions result from damage inflicted before birth. The defendant is placed in juvenile detention or jail, an environment that he or she again is ill equipped to navigate. For an individual with FASD, the objective of general deterrence cannot be achieved beyond the period of detention because (1) the actions probably were not taken with the intent to commit a crime, (2) the underlying cause of the dysfunctional behavior has not been addressed, and (3) appropriate treatment geared to the special needs of a person with an FASD has not been provided (Williams 2006, pp. 8–9).

The Issues and Case Law: The DSM-IV/DSM-IV-TR Era

Competency, mitigation and death penalty issues are often encountered in FASD cases, underscored by rights to: diagnosis, expert witnesses on guilt-innocence, mitigation of sentence, and judicial preference for hands-on examination of defendants by the experts. This may include issues of diminished capacity, competency to stand trial, entry of a plea, and waiving constitutional rights. Confessions made by those with FASD, testimony by victims/witnesses with FASD, vulnerable victims with FASD, ineffective assistance of counsel, and presence of intellectual disability to disqualify imposition of the death penalty are also legal process considerations. The law is settled regarding many of these issues, but there is still confusion and misunderstanding regarding judicial application to FASD.

A. *Right to Diagnosis*

Indigent defendants are entitled to have experts diagnose and present applicable evidence at public expense. The leading case involving FASD is *Castro v. Oklahoma* (1995) which established the right to appointed experts for diagnosis and testimony. It requires a threshold showing that the issue being pursued is clear, genuine, fairly debatable or in doubt, and it is likely that, if established, the defendant's mental condition could be a significant mitigating factor (*Castro v. Oklahoma* 1995). *Castro v. Oklahoma* (1995) did not discuss diagnosis relating to issues of competency to stand trial, entering a plea and waiver of constitutional rights, guilt and innocence, effective assistance of counsel, and assessment of intellectual disability as protection against the death penalty, but clearly these rights follow from the appointment of diagnostic experts.

Ten years earlier, *Ake v. Oklahoma* (1985) decided that when a defendant makes a preliminary showing that his sanity at the time of the offence is likely to be a significant factor at trial, due process requires states to provide a psychiatrist's assessment. *Lambert v. Blodgett* (2003) involved evidence of alcohol exposure in utero, and highlighted ineffective assistance of counsel and the duty to provide experts with meaningful information, such as a thorough psychiatric assessment. It should be noted *Lambert v. Blodgett* (2003), was later reversed on procedural grounds based on Habeas Corpus (*Lambert v. Blodgett* 2004). *Silva v. Woodford* (2007) explored the duty to investigate and consider FAS for possible mitigation. *Rompilla v. Beard* (2005), involved a defendant with a third grade level of cognition despite nine years of school. This case provided evidence the birth mother consumed alcohol during the pregnancy and included expert testimony of organic brain damage and an extreme mental disturbance significantly impairing several of his cognitive functions.

In addition, *Rompilla v. Beard* (2005), considered the duty to obtain prison, school, medical and other records that might reveal significant mitigating evidence and/or which would likely be relied on by the prosecutor for purposes of proving aggravating factors. In *Lambert v. Blodgett* (2003), the Court ruled since the U.S. Supreme Court had only addressed the right to the appointment of expert witnesses in a death penalty case (*Ake v. Oklahoma* 1985), an attorney cannot be found ineffective for not developing a mitigating factor in a non-death penalty case. This outcome lacked common sense.

FASD is a mitigating factor, just as insanity (*Ake v. Oklahoma* 1985) is a defence regardless of the nature of the crime. It is senseless for FASD to serve as mitigation in only death penalty cases. In addition, if the *Lambert v. Blodgett* (2003) rationale is to be considered reasonable, then the right to appointed experts in non-death penalty cases could be subject to challenge which has not happened in the 30 years since *Ake v. Oklahoma* (1985). Regardless, absent a nexus between brain damage and criminal conduct, behavioral problems are likely to be viewed by the jury as an aggravating circumstance.

B. *Habeas Corpus, Mitigation and Ineffective Assistance of Counsel*

FASD typically surfaces when newly appointed *Habeas* counsel learns from the client or family that the person may have FASD, which was not pursued by prior counsel, and brings in experts who then make the diagnosis. The Supreme Court revisited this issue also in *Martinez v. Ryan* (2012) and *Trevino v. Thaler* (2013). *Martinez v. Ryan* (2012) did not involve FASD but involved ineffective trial and appellate counsel in a State proceeding. The Court granted an exception to the *Cullen v. Pinholster* (2011) rule.

In *Trevino v. Thaler* (2013), the structure and design of the Texas system made it virtually impossible for an ineffective assistance claim to be presented on direct review by the appellate court. On petition for *Habeas Corpus*, the federal district court appointed new counsel who, for the first time, raised the issue of ineffective trial and appellate counsel for failing to investigate and present FASD as mitigating evidence in State Court. The Supreme Court ruled in both cases that defendants had a right to proceed in federal court.

C. *Competency, Juvenile Capacity and Diminished Capacity*

Competency to stand trial is the ability to assist counsel in one's defence. Subsumed issues include: defendant's competency to enter a plea, to waive the right against self-incrimination (i.e., the *Miranda* rights to remain silent, to have an attorney present during questioning), and the right to oppose warrantless searches of property. Absent a finding of incompetency, juvenile defendants may claim incapacity to commit a crime in many jurisdictions, which is typically based on age and maturity and whether the child understands the nature and effects of his or her actions. The age factor is usually established legislatively ranging from a non-rebuttal presumption of lack of the necessary capability, to a rebuttable incapacity range, to a non-rebuttal that has the necessary capacity range.

No cases have addressed the question of capacity of a defendant whose chronological age is at the "capable" level, but whose intellectual/developmental age is in either the "incapable" or "presumed incapable" range. However, an argument should be made that incapacity should apply to defendants with this FASD profile. This argument is also useful for educating the Court as to the client's disabilities.

An accused person, either a juvenile or an adult, may claim diminished capacity. Diminished capacity renders a defendant incapable of forming the specific intent to commit the "mens rea" element of the crime. Diminished capacity reduces the accused's liability or culpability. Absent "mens rea," a defendant cannot be found guilty of that crime. However, a finding of guilt is still available for lesser offences that do not require specific intent. Diminished capacity is an affirmative defence. The defendant has the burden of proving it "beyond a reasonable doubt." This defence should be considered early in the proceeding. When diminished capacity is alleged, especially when involving suspected FASD-impacted individuals, FASD experts should be obtained to assess both conditions. However, some states do not permit a diminished capacity defence.

The competency rule lacks uniform application. Many lawyers resist bringing competency motions because they feel judges and juries do not take the issue seriously, or misapply the law, making a successful ruling highly unlikely. They do not view contested competence adjudications as a useful challenge (Philipsborn 2008). Judges have been known to find defendants competent for inappropriate reasons. Examples include, the defendant is competent because of ability to identify the colors of the American flag, sit at a chessboard and play chess (even though very young and immature children can learn basic moves) or forages for food (even though dogs learn to prowl garbage cans for food which is a survival instinct).

Competency is the ability to understand information that is relevant to making decisions, to have knowledge of being confronted with a legal decision, to think rationally about available options, and to express choices. Defendants with FASD may "sound" or "look" competent, but may function at a level much younger than their chronological age. Individuals with FASD may present with many conditions that could impact their competency. There have been multiple cases where appellate courts have recognized trial counsel's specific duty to investigate signs of mental health, including organic and other brain damage, issues thoroughly (Stetler 2014). Trial counsels' duties also include the wise choice of experts and providing experts with the appropriate background information that will enable them to render trustworthy opinions (Stetler 2014). *Bloom v. Calderon* (1997), *Caro v. Calderon* (1999), and *Skaggs v. Parker* (2000) are excellent examples.

As a result of the DSM-5 including a diagnosis of Other Specified Neurodevelopment Disorder (Associated with Prenatal Alcohol Exposure) and consideration of Neurobehavioral Disorder Associated with Prenatal Alcohol Exposure as a future diagnosis, with the U.S. Supreme Court's adoption of the DSM-5 definition of Intellectual Disability (ID), FASDs and the adaptive behavioral deficits associated with FASDs are now in the realm of mental health issues. As such, the impact of FASD on competency should be explored, as should obtaining testimony from properly trained expert witnesses.

There are certain realities involving competency that add clarity to, and explain why, there is imperfect understanding. This applies to the diagnostic process, observed symptoms, behavioral consequences, and, most importantly, any brain damage caused by PAE resulting in impaired cognitive and executive functioning which may affect a defendant's response to the legal system. This includes the ability to make rational decisions and to maintain impulse control in the courtroom. If the brain damage is significant, competency to stand trial and the other aforementioned rights may need to be considered. There is no significant distinction between competence to enter a plea, waive rights, to consent to search, or to stand trial (*Godinez v. Moran* 1993), but competency is far more than just being oriented to time and place, and having knowledge of some of the events.

This places a heavy burden on defence counsel to provide clients with effective assistance of counsel. The leading case on competency is *U.S. v. Dusky* (1960). It preceded FAS discovery by 13 years and did not deal with FASD. The Court stated "the test must be whether the defendant has sufficient present ability to consult with his lawyer with a reasonable degree of rational understanding and whether he has a rational as well as factual understanding of the proceedings against him" (*U.S. v. Dusky* 1960, p. 403).

Furthermore, unless found competent, a defendant cannot be tried for a crime. A trial of an incompetent person violates the right to substantive due process (*Pate v. Robinson* 1966). To support this statement, Justice Kennedy emphasized the primary thrust of *U.S. v. Dusky* (1960), as follows:

> Competence to stand trial is rudimentary, for upon it depends the main part of those rights deemed essential to a fair trial, including the right to effective assistance of counsel, the right to summon, to confront, and to cross-examine witnesses, and the right to testify on one's own behalf or to remain silent without penalty for doing so (*Cooper v. Oklahoma* 1996, citing *Drope v. Missouri* 1975, p.172).

A plea agreement cannot be entered because an incompetent person cannot so agree (*State v. Ortiz* 2001). The Court in *Lafferty v. Cook* (1991) concluded that

U.S. v. Dusky (1960) and *Drope v. Missouri* (1975) place a responsibility on the trial court as follows:

> It is not enough for the district judge to find defendant (is) oriented to time and place and (has) some recollection of events. The test must be whether he has sufficient present ability to consult with his lawyer with a reasonable degree of rational understanding, and whether he has a rational as well as factual understanding of the proceedings against him. (*Lafferty v. Cook* 1991, p. 1548).

There seems to be little room for lawyers and judges to misunderstand or misinterpret what is necessary to find a defendant competent to stand trial. However, a thoughtful decision by a federal trial judge suggests a reason why this occurs. In *U.S. v. Duhon* (2000), the court recognized a difference between incompetency due to mental illness and incompetency due to mental retardation, as well as the need to utilize different criteria for making those competency determinations. The judge said "erroneous diagnosis threatens a fundamental component of our criminal justice system, the basic fairness of the trial itself" (*U.S. v. Duhon* 2000, p. 670, relying on *Cooper v. Oklahoma* 1996):

> After making the 'profound' choice whether to plead guilty, the defendant who proceeds to trial will ordinarily have to decide whether to waive his 'privilege against compulsory self-incrimination' (*Boykin v. Alabama* (1969), by taking the witness stand,… he may have to decide whether to waive his 'right to trial;' and, in consultation with counsel, he may have to decide whether to waive his 'right to confront (his) accusers,' by declining to cross-examine witnesses for the prosecution. With the assistance of counsel, the defendant is also called upon to make myriad smaller decisions.. . The importance of the rights and decisions demonstrates that an erroneous determination of competence threatens a 'fundamental component of our criminal justice system,' the basic fairness of the trial itself (*Cooper v. Oklahoma* 1996, p. 352, citing *Godinez v. Moran* 1993, p. 398).

In *U.S. v. Duhon* (2000), the judge indicated the court should consider mental illness may be reversible with treatment and that mental retardation may be static and untreatable, which may make a person more susceptible to suggestion, citing the American Bar Association (ABA) (1998). He believed courts should be wary of the arbitrary use of general competency assessment techniques and standards in assessing a mentally retarded defendant's competency:

> There is general recognition that competence is based on a specific set of cognitive abilities and the functional capacity to exercise those abilities. Thus, competency scales or structured interviews can be used with persons who have mental retardation. *However, because persons with mental retardation are cognitively impaired, not mentally ill, the strongly cognitive elements of a competency evaluation need to be given special attention as to functional behavior. A defendant with mental retardation may also be limited as to functional behavior. Thus, a defendant with mental retardation might be seemingly 'restored' to competency by instructing that individual about trial elements, but he or she may not be able to make intelligent legal decisions [emphasis added]* (American Bar Association 1998, p. 169).

Without mentioning FASD, this judge appears to have a clear understanding of the diagnostic process, symptoms, and behavioral consequences of being afflicted with cognitive and functional deficits, which, as it turns out, are also primary markers of FASD.

An example of limited understanding appears in a Montana Supreme Court ruling in *The Matter of SMS, a youth* (2010). S.M.S. was charged with sexually offending his six-year-old sister. He sought suppression of his confession. He claimed "severe cognitive delay" rendered involuntary both his waiver of *Miranda* rights and his confession. Four separate psychological evaluations revealed S.M.S. possessed a very low level of intellectual functioning, needed special education classes, and suffered from FAS, PTSD and ADHD. The court stated:

> S.M.S. met the age requirements of (the statute) for providing a valid waiver without his parents or counsel present. S.M.S.'s educational level and cognitive abilities raise questions as to whether S.M.S. understood the gravity of the circumstances and his confession. Two doctors provided conflicting testimony regarding S.M.S.'s cognitive abilities. The youth court sat in the best position to evaluate the competing testimony along with the videotape of the interview to determine whether S.M.S. had confessed voluntarily (*The Matter of SMS, a youth* 2010, p. 104).

In retrospect, it is wondered if the doctors and judges understood the extent of S.M.S's problems. It is also unclear whether the Court considered the other two criteria for the admissibility of a confession, whether the confession was also knowingly and intelligently given in addition to being voluntary. The trial and appellate courts put a lot of weight on the literal language of the statute in question and the care taken by the investigating police officer. They may not have carefully considered S.M.S.'s cognitive and executive functional deficits may have led to an unknowing, unintelligent, and involuntary admission.

Furthermore, regarding videotaped interrogations, it is not uncommon for individuals with FASD to indicate verbally they understand while lacking true comprehension. It is not clear whether the Montana court had this information when viewing and evaluating the videotape. Also, the record does not indicate whether the interrogating police officer asked S.M.S. leading questions, fed information to him, or asked open-ended questions. Open-ended questions are more likely to accurately reveal one's true cognitive ability and understanding. To avoid the problems similar to those experienced in *The Matter of SMS, a youth* (2010) in preparing experts to testify, counsel should suggest that the expert address the concerns described by the judges in *U.S. v. Dusky* (1960) and *U.S. v. Duhon* (2000) in their testimony.

In Washington State, judges received FASD education through judicial conferences. This education impacted judicial competency rulings, as demonstrated by *State v. Scholler* (2009) and *State v. Thornton* (2011). These cases involved juveniles who were below grade level in reading and comprehension. Both were found to be permanently incompetent based on the testimony of FASD experts. One juvenile was diagnosed with FAS, and the other was diagnosed with an Alcohol Related Neurodevelopment Disorder (ARND). Both suffered from significant executive functioning deficits. In one case, the child's verbal skills masked these deficits to a large extent.

The Court found the juvenile did not have the capability to meaningfully assist his attorney in his own defence as a result of FAS. The other juvenile suffered from a static maturation well below chronological age due to ARND, Bipolar

Disorder, and Attention-Deficit Hyperactivity Disorder (ADHD). The trial judge in *State v. Thornton* (2011) found Thornton was skilled at covering up gaps in memory. Neither case was appealed by the State. The Court ruled:

> Overall, the results of the (psychological and neuropsychological) assessment indicated that Jason is suggestible and quite prone to changing his responses in the face of negative feedback, making him highly susceptible to a skilled cross-examination, which would defeat the integrity of this court. I'm persuaded that reliability flies out the window if you have the two dynamics going on, as in this case, the impaired executive functioning along with the suggestibility and the memory impairment, which causes confabulation… reliability is nil… essentially with the perfect storm of deficits. The more significant the memory deficits, the more effect they're going to have on competency. So those two aspects of fetal alcohol disorder make it qualitatively different from the run-of-the-mill mental retardation cases (he was diagnosed as mildly retarded)… Adaptive disabilities are also present. It is persuasive and part of his condition. The executive functions are the foundation for competent adaptive deficits in this case (*State v. Thornton* 2011, p. 15).

D. *Eighth Amendment to the U.S. Constitution, the Death Penalty and Mental Retardation*

The Constitution's Eighth Amendment prohibits "cruel and unusual" punishment. In *Atkins v. Virginia* (2002), the Supreme Court prohibited the imposition of the death penalty when a defendant is mentally retarded. In doing so, the court utilized the "evolving standards of decency that mark the progress of a maturing society" test. In 1989, the Supreme Court ruled juveniles under the age of 16 could not be executed (*Thompson v. Oklahoma* 1989). *Roper v. Simmons* (2005) extended it to age 18, recognizing the brain has not yet fully matured.

Mental retardation cases followed a similar route. According to common law, "idiots" and "imbeciles" could not be executed. Yet the Supreme Court held that execution was not automatically forbidden for persons with mental retardation (*Penry v. Lynaugh* 1989) until *Atkins v. Virginia* (2002), exempted defendants with mental retardation from the death penalty. However, unfortunately, *Atkins v. Virginia* (2002) left it to individual states to establish rules for implementing the Court's decision. Atkins had a full scale IQ score of 59, below the "about 70" mentioned in the opinion, together with executive, cognitive and adaptive behavioral deficits (*Atkins v. Virginia* 2002). The Court referred to the DSM-IV/DSM-IV-TR definition of mental retardation, which required a full scale IQ score of 70 or below; the presence of executive, cognitive, and adaptive behavioral deficits; and the onset of mental retardation prior to age 18. Atkins suffered many adaptive behavioral deficits identified in DSM-IV/DSM-IV-TR, deficits that are also commonly associated with FASD.

State definitions varied significantly, some not having IQ cut-offs and some having cut-offs ranging from 65 to 75. Some states take into consideration the Standard Error of Measurement (SEM) margin of error, while others do not. One state presumes mental retardation with a score of 70 or below. Some states do not have a full scale IQ requirement, relying instead on "overall functioning" (*Hall v. Florida* 2014). This made it difficult to pursue findings of mental retardation under *Atkins v. Virginia* (2002).

Moreover, the Supreme Court held juveniles convicted as adults for non-death penalty crimes cannot be sentenced to "life in prison without the possibility of parole" (*Graham v. Florida* 2010). Finally, in the combined cases of *Miller v. Alabama* (2012) and *Jackson v. Hobbs* (2012), involving two 14-year-olds convicted of murder, the Court held that imposition of mandatory life in prison sentences (without possibility of parole) constituted cruel and unusual punishment. Thus, the decisions in *Atkins v. Virginia* (2002), *Roper v. Simmons* (2005), *Graham v. Florida* (2010), *Miller v. Alabama* (2012), and *Jackson v. Hobbs*, all Eighth Amendment "cruel and unusual" punishment rulings, made prior to DSM-5, recognized defendants with mental retardation and juveniles should not be held to the same level of accountability as other persons for their criminal actions due to their reduced culpability and responsibility caused by their intellectual disability.

Intellectual Disability and FASD: The Era of DSM-5

A. *Intellectual Disability*

The DSM-5 advances the FASD/intellectual disability/disorder jurisprudence and it promotes FASD as a diagnosable developmental/intellectual disorder/disability, rather than just a medical condition. It is significant that the DSM-5's adaptive deficits for intellectual disability are consistent with neurodevelopment/neurobehavioral disorders related to PAE.

Following publication of the new DSM-5, the U.S. Supreme Court revisited its *Atkins v. Virginia* (2002) decision in the case of *Hall v. Florida* (2014). It was concerned with Florida's motive in enacting legislation the Florida Supreme Court interpreted to exclude many defendants from protection against the death penalty as cited in *Atkins v. Virginia* (2002). The statute required a full scale IQ score of 70 or below; the Florida Supreme Court then interpreted this to mean unless the IQ score was 70 or below, a defendant could not present evidence of adaptive behavioral deficits which were part of the statute. The U.S. Supreme Court found this interpretation ran afoul of *Atkins v. Virginia* (2002). *Hall v. Florida* (2014) adopted the DSM-5 definition and criteria for the diagnosis of intellectual disability. The DSM-5 puts full scale IQ on equal footing with the adaptive behavioral deficits.

The Court indicated intellectual disability is a "condition, not a number" (*Hall v. Florida* 2014, p. 21), citing the importance of considering deficits in addition to intelligence test results. In addition, *Hall v. Florida* (2014) referenced an "approximately" 70–75 approach, using the generally accepted plus/minus 5 points standard error of measurement (SEM) when conducting IQ testing. The Florida Supreme Court ignored the error factor. *Hall v. Florida* (2014), in looking at the SEM in combination with the DSM-5's (American Psychiatric Association 2013) list of deficits, suggested that "a person with an IQ score above 70 may have such a severe adaptive behavioral problem. .. that the person's actual functioning is comparable to that of individuals with a lower IQ score" (*Hall v. Florida* 2014, p. 21).

The Court voiced concerns that because of the differences in the laws of each state, a defendant could be subject to the death penalty in one state, yet found in another state to have an intellectual disability and not be subject to the death penalty. The Court also acknowledged it was appropriate to consider a standard adopted by medical experts as opposed to relying on a single factor. The Court closed with the following:

> Florida's law contravenes our Nation's commitment to dignity and its duty to teach human decency as the mark of a civilized world. The States are laboratories for experimentation, but those experiments may not deny the basic dignity the Constitution protects (*Hall v. Florida* 2014, p. 22).

Interestingly, the trial judge who sentenced Hall in 1978, preceding *Atkins v. Virginia* (2002) by 24 years, found Hall was retarded, but lacking a nexus between the mental retardation and the murder, sentenced Hall to death. After *Atkins v. Virginia* (2002) was issued, Hall brought a motion in State Court to set aside the death sentence because of his mental retardation. Before the State court ruled on the motion, the legislature passed the statute in question. The Florida Supreme Court, applying its "70 or below" cut-off standard, denied the motion. It was following that decision that Hall sought Certiorari to the U.S. Supreme Court.

As a matter of historical reference, the DSM-IV (American Psychiatric Association, 1994) and DSM-IV-TR (American Psychiatric Association, 2000) defined mental retardation as: having an IQ score of 70 or below; coupled with adaptive functioning that was also below culturally and age-appropriate standards that impacted at least two of the following areas: communication skills, ability to provide self-care, living skills, social/interpersonal skills, the ability to access community resources, provide and follow self-direction, functional skills related to academics, work skills, leisure activities, health, and safety. The onset of the condition needed to occur before the age of 18.

Currently, the DSM-5 changed mental retardation to intellectual disability and established the following diagnostic criteria:

A. Deficits in intellectual functions, such as reasoning, problem solving, planning, abstract thinking, judgment, academic learning, and learning from experience, confirmed by both clinical assessment and individualized, standardized intelligence testing.
B. Deficits in adaptive functioning that result in failure to meet developmental and sociocultural standards for personal independence and social responsibility. Without ongoing support, the adaptive deficits limit functioning in one or more activities of daily life, such as communication, social participation, and independent living, across multiple environments, such as home, school, work, and community.
C. Onset of intellectual and adaptive deficits during the developmental period (American Psychiatric Association 2013, p. 17)

The *Hall v. Florida* (2014) Court, in adopting the DSM-5 definition of intellectual disability, recognized the propriety of permitting exercise of professional judgment and discretion by diagnosticians in reaching diagnostic opinions. In doing so, the Court again confirmed that the human brain does not reach full maturity (e.g., the developmental period) by age 18.

B. *FASD*

The DSM-5 established, under Neurodevelopmental Disorders, a subcategory designated as "Other Specified Neurodevelopment Disorder" which can include a specifier of "Associated with Prenatal Alcohol Exposure". This category applies to presentations in which symptoms characteristic of a neurodevelopment disorder that cause impairment in social, occupational, and/or other important areas of functioning predominate, but do not meet the full criteria for any of the disorders in the neurodevelopment disorders diagnostic class. It is a diagnosis by exclusion for any specific developmental disorder and can include a range of developmental disabilities following exposure to alcohol, or other substances, in utero (American Psychiatric Association 2013).

The DSM-5's "Proposed Criteria" for "Neurobehavioral Disorder Associated with Prenatal Alcohol Exposure," which are neurodevelopment in nature (American Psychiatric Association 2013), match up closely with the adaptive functioning deficits criteria for intellectual disability. This will require education and training of diagnosticians in order to accomplish meaningful "further study." The proposed criteria provides for more than minimal exposure to alcohol during gestation, which can be confirmed from maternal self-report of alcohol use in pregnancy, clinical observation, and/or medical, as well as other records. The other criteria categories are impaired neurocognitive functioning as manifested by one or more of the following impairments: global intellectual performance (70 or below IQ), executive functioning, learning, memory, and/or visual-spatial reasoning.

As well, impaired self-regulation is considered, as manifested by problems with one or more of the following: mood regulations, behavioral regulation, attention deficit, and/or impulse control. Impairment in adaptive functioning is considered and is manifested by two or more of the following: communication deficit; social communication and interaction; daily living skills; and motor skills. The onset of the disorder occurs in childhood and the disturbance causes clinically significant distress or impairment in social, academic, occupational, and/or other important areas of functioning.

Furthermore, the disorder is not better explained by the direct physiological effects associated with postnatal use of a substance (e.g., medication, alcohol or other drug), a general medical condition, another known teratogen, a genetic condition, or environmental neglect. The Supreme Court's emphasis in the *Atkins v. Virginia* (2002) and *Hall v. Florida* (2014) on the deficits described above, provides for the ability to diagnose neurodevelopment/neurobehavioral disorders related to PAE.

Conclusion

Although this chapter provides a general overview of legal precedent related to FASD, interested individuals are encouraged to continue exploring the ramifications of PAE with involvement in the legal system. Resolutions by the 2011 Canadian (CBA) and 2012 American (ABA) Bar Associations called for education and different treatment for those with FASD. Furthermore, supporting a paradigm shift, the DSM-5's consideration of neurodevelopment/neurobehavioral disorders associated with PAE and intellectual disability should compel professional schools as well as continuing education requirements of professional societies, to prioritize education related to all of the FASD umbrella conditions and not just those recognized by the DSM-5. This is particularly needed because psychologists and other qualified mental health professionals can now offer FASD-related diagnoses.

This field is rapidly evolving. As such, judicial system professionals of all levels, medical doctors, nurses, psychologists, social workers, and educators must continue to seek ongoing training and information focused on FASD and the associated risks.

References

American Bar Association (ABA). 1998. *Commission on Mental and Physical Disability Law.*

American Bar Association (ABA). 2012. *Resolution 112B on FASD.*

American Psychiatric Association. 1994. *Diagnostic and statistical manual of mental disorders*, 4th ed. Washington, DC: Author.

American Psychiatric Association. 2000. *Diagnostic and statistical manual of mental disorders.* (4th ed.). Washington, DC: Author.

American Psychiatric Association. 2013. *Diagnostic and statistical manual of mental disorders*, 5th ed. Washington DC: Author.

Ake v. Oklahoma. 1985. 470 U.S. 568.

Atkins. v. Virginia. 2002. 536 U.S. 304.

Bloom v. Calderon. 1997. 132 F.3d 1267, 1278 (9th Cir. 1997).

Boykin v. Alabama. 1969. 395 U.S. 238, 243.

Brems, C., M.E. Johnson, J.S. Metzger, and S.L. Dewane. 2014. College students' knowledge about fetal alcohol spectrum disorder. *Canadian Society of Pharmacology and Therapeutics* 21(2): e1159–e1166.

Canadian Bar Association (CBA). 2011. Resolution 10-02-A on FASD.

Caro v. Calderon. 1999. 165 F.3d 1223, 1226 (9th Cir. 1999).

Castro v. Oklahoma. 1995. 71 F.3d 1502 (10th Cir. 1995).

Cooper v. Oklahoma. 1996 517 U.S. 348, 349.

Cox, L., D. Clairmont, and S. Cox. 2008. Knowledge and attitudes of criminal justice professionals in relation to fetal alcohol spectrum disorder. *Canadian Journal of Clinical Pharmacology* 15(2): 306–313.

Cullen v. Pinholster. 2011. 131 S.Ct. 1388.

Dillbeck v. State. 1994. 643 S.2d 1027 (Fla. 1994).

Drope v. Missouri. 1975. 420 U.S. 162: 171–172.

English Standard Bible (n.d.). 2009. Wheaton, Illinois: Good News Publishers.

Fielding, H. 1751. *Inquiry into the Causes of the Late Increase of Robbers, etc.*, Printed for A. Millar by Henry Fielding.
Godinez v. Moran. 1993. 509 U.S. 389, 398.
Graham v. Florida. 2010. 130 S.Ct. 176.
Hall v. Florida. 2014. 572 U.S.
Ingersoll, K.S., S.D. Ceparich, M.D. Nettleman, and B.A. Johnson. 2008. Risk drinking and contraception effectiveness among college women. *Psychology Health* 23: 965–981.
Jackson v. Hobbs. 2012. 567 U.S. 2455.
Jones, K.L., D.W. Smith, C.N. Ulleland, and A. Streissguth. 1973. Pattern of malformation in offspring of chronic alcoholic mothers. *Lancet* 301(7815): 1267–1271.
Lafferty v. Cook. 1991. 949 F.2d 1546, 1548 (10th Cir.).
Lambert v. Blodget. 2003. 248 F.Supp.2d 988 (E.D. WA).
Martinez v. Ryan. 2012. 566 U.S. 1.
McCreight, B. 1997. *Recognizing and managing children with fetal alcohol syndrome/fetal alcohol effect: A Guidebook*. Washington, DC: Children Welfare League of America.
McLachlan, K., R. Roesch, J.L. Viljoen, and K.S. Douglas. 2014. Evaluating the psycholegal abilities of young offenders with fetal alcohol spectrum disorder. *Law and Human Behavior* 38(1): 10–22.
Pate v. Robinson. 1966. 383 U.S. 375, 379.
Penry v. Lynaugh. 1989. 409 U.S. 302.
Philipsborn, J.T. 2008. Dealing with experts on competence to stand trial: Suggestions and approaches—Part two. *The Champion Magazine* 3: 42.
Rompilla v. Beard. 2005. 545 U.S. 374.
Roper v. Simmons. 2005. 543 U.S. 551.
Sampson, P., A. Streissguth, F. Bookstein, R. Little, S. Clarren, P. Dehane, and J. Graham. 1997. Incidence of fetal alcohol syndrome and prevalence of alcohol-related neurodevelopment disorder. *Terotology* 56(5): 317–326.
Silva v. Woodford. 2007. 279 F.3d 825 (9th Cir. 2007).
Skaggs v. Parker. 2000. 235 F.3d 261,271 (6th Cir. 2000).
State v. Ortiz. 2001. 104 Wn.2d 479.
State v. Scholler. 2009. Snohomish County Superior Court, Nos. 08-8-00426-1 and 08-9-00443-1.
State v. Thornton. 2011. Kitsap County Superior Court, No. 09—1-01284-7.
Stetler, R. 2014. Mental health evidence and the capital defence function: Prevailing norms. *University of Missouri Kansas City Law Review* 82: 2.
Streissguth, A., Barr, H.M., Kogan, J. and Bookstein, F.L., 1996. Understanding the occurrence of secondary disabilities in clients with fetal alcohol effects (FAE). (Center for Disease Control and Prevention Grant #R04/CCR008515 m). University of Washington School of Medicine, Department of Psychiatry and Behavioral Sciences, Fetal Alcohol and Drug Unit.
The Matter of S.M.S. 2010. a Youth, 355 Mont. 102, 104.
Thompson v. Oklahoma. 1989. 487 U.S. 815.
Trevino v. Thaler. 2013. 133 S.Ct. 1911.
U. S. v. Duhon. 2000. 104 F.Supp.2d 663 (W.D. La. 2000).
U.S. v. Dusky. 1960. 362 U.S. 402.
Washington Statute RCW 9A.04.050 (1975, 2011).
Wechsler, H., M. Kao, H. Lee, and G.W. Dowdall. 2000. Environmental correlate of underage use and related problems of college students. *Journal of Preventive Medicine* 19: 24–29.
William, S.J. 2006. Is there justice in the juvenile justice system? Examining the role of fetal alcohol spectrum disorders. *Justice Policy Journal,* 3(1): 1–15.

Additional Resource List

Although the following legal cases are also included in the Reference list, they provide compelling information on the development of constitutional law and other legal aspects that may not involve FASD, but could be relevant to issues impacting the developing law on FASD.

RCWA 9A.04.050, Revised Code of Washington.

Regina v. J (T). 1999. No. 57, The Yukon Territorial Court.

Woodard v. Alabama. 2013. 571 U.S. 405.

The following FASD resources are provided to assist in expanding one's interest and knowledge related to the complexities of FASD:.

Brown, N.N., P.D. Connor, and R.S. Adler. 2012. Conduct-disordered adolescents with Fetal Alcohol Spectrum Disorder intervention in secure treatment settings. *Criminal Justice and Behavior* 39(6): 770–793.

Brown, N., G. Gudjonsson, and P. Connor. 2011. Suggestibility and Fetal Alcohol Spectrum Disorders: I'll tell you anything you want to hear. *The Journal of Psychiatry & Law* 1: 39–73.

Brown, N.N., Wartnik, A.P., Connor, P.D., and Adler, R.S. 2010. A proposed model standard for forensic assessment of Fetal Alcohol Spectrum Disorders. *Journal of Psychiatry & Law*, 38(4): 475–506.

Burd, L., D.K. Fast, J. Conry, and A.D. Williams. 2010. Fetal alcohol spectrum disorder as a marker for increased risk involvement with corrections. *The Journal of Psychiatry and Law* 38: 559–583.

Burd, L., Rachael, H., Selfridge, B.S., Klug, M.G., and Juelson, T. 2003. Fetal alcohol spectrum disorder in the Canadian corrections system. *The Journal of FAS International*, e14: 110.

Burd, L., R. Selfridge, M. Klug, and S. Bakko. 2004. Fetal alcohol syndrome in the United States corrections system. *Addiction Biology* 9(2): 169–176.

Chartrand, L.N., and E.M. Forbes-Chilibeck. 2003. The Sentencing of offenders with fetal alcohol syndrome. *Health Law Journal* 11: 35–70.

Douds, A., H. Stevens, and W. Sumner. 2013. Sword or shield? A systematic review of the roles FASD evidence plays in judicial proceedings. *Criminal Justice Policy Review* 24(4): 492–509.

Douglas, H. 2010. The sentencing response to defendants with foetal alcohol spectrum disorder. *Criminal Law Journal* 34(4): 221–239.

Fast, D., and J. Conry. 2004. The challenge of fetal alcohol syndrome in the criminal legal system. *Addiction Biology* 9(2): 161–166.

Fast, D.K., and J. Conry. 2009. Fetal alcohol spectrum disorders and the criminal justice system. *Developmental Disabilities Research Reviews* 15(3): 250–257.

Fast, D.K., J. Conry, and C.A. Loock. 1999. Identifying fetal alcohol syndrome among youth in the criminal justice system. *Journal of Developmental and Behavioral Pediatrics* 20(5): 370–372.

Malbin, D.V. 2004. Fetal Alcohol Spectrum Disorder (FASD) and the role of family court judges in improving outcomes for children and families. *Juvenile and Family Court Journal* 55(2): 53–63.

Moore, T.E., and M. Green. 2004. Fetal alcohol spectrum disorder (FASD): A need for closer examination by the criminal justice system. *Criminal Reports* 19(1): 99–108.

Popova, S., S. Lange, D. Bekmuradov, A. Mihic, and J. Rehm. 2011. Fetal alcohol spectrum disorder prevalence estimates in correctional systems: a systematic literature review. *Canadian Journal of Public Health* 5: 336–340.

Thiel, K., N. Baladerian, K. Boyce, O. Cantos, L. Davis, K. Kelly, K. Tavenner, K. Mitchell, and J. Stream. 2010. Fetal alcohol spectrum disorders and victimization: Implications for families, educators, social services, law enforcement, and the judicial system. *Journal of Psychiatry & Law* 39: 121–157.

Smart Justice and FASD in Alaska: From Prevention to Sentence Mitigation

Teri Tibbett and Michael I. Jeffery

"Smart Justice" means weighing—in every criminal case—the likely effectiveness of the actions we take.
-Then-Alaska Chief Justice Walter L. Carpeneti (2012).

A lot of people with FASD just need a constant reminder. If we can do that instead of putting them in jail, society is a lot better off and certainly the people with FASD are.
-Alaska Senator Kevin Meyer (2014).

Introduction

On September 19, 2012 Alaska became the first state in the United States to pass a statute that explicitly allows its judges to reduce a felony sentence for an offender diagnosed with a fetal alcohol spectrum disorder (FASD) condition. The statute promised more humane treatment of such offenders and significant cost savings to the state, but this promise has not yet been realized. This chapter explores the developing acceptance of "Smart Justice" in Alaska, some best practices for providing effective services to lower the risk of criminal behavior by persons affected by an FASD condition, and, for those individuals who do get involved in the criminal justice system, analyzes Alaska's FASD mitigating factor statute, including an outline of reasons for its slow implementation. It is hoped that implementation of positive supports for people affected by FASD will reduce the need to even consider application of a statutory mitigating factor for them.

T. Tibbett (✉)
Juneau, Alaska, USA
e-mail: teri@tibbett.com

M.I. Jeffery
Barrow, Alaska, USA
e-mail: mijeffery@yahoo.com

M. Nelson and M. Trussler (eds.), *Fetal Alcohol Spectrum Disorders in Adults: Ethical and Legal Perspectives*, International Library of Ethics, Law, and the New Medicine 63, DOI 10.1007/978-3-319-20866-4_11

Alaska's Move Toward Smart Justice

In 2012, the Honorable Walter L. Carpeneti, Chief Justice of the Alaska Supreme Court at the time, advocated for implementation of Smart Justice in Alaska. He first explained the concept:

> 'Smart justice' means weighing—in every criminal case—the likely effectiveness of the actions we take. Further, it means considering the costs of these actions—to our resources, to public safety, and to the collective human potential of our citizens. In practice, it means making criminal justice decisions that reserve our most costly response to crime—prison time—for those cases where other less-costly alternatives will not effectively protect the public or rehabilitate the perpetrator (Carpeneti 2012).

With this encouragement, and after experiencing the financial drain of building an expensive new prison for approximately 250 million dollars (US) between 2004 and 2012, with annual maintenance and staffing costs of approximately 50 million dollars (US) per year, Alaskan lawmakers began looking at ways the state could be "smarter" in the area of criminal justice spending (Alaska Department of Corrections 2011).

A 2009 study by the Institute of Social and Economic Research (ISER) at the University of Alaska Anchorage suggested that for certain non-violent offenders, offering specialized treatment programs inside correctional facilities, combined with adequate transition and case management services on the outside, results in lower state costs, less recidivism, and a safer general public (Institute of Social and Economic Research (ISER) 2009). These conclusions mirror national research on the value of alternatives to incarceration that present "a considerable potential in cost savings, improved outcomes for offenders, and improved public safety" (Office of National Drug Control Policy 2011).

The ISER study indicated that the State of Alaska could both reduce the number of Alaskans in prison or jail and save considerable money over the next twenty years by "spending more" up front and "strategically expanding" certain programs, including prison-based programs (such as education/vocational programs, and treatment for substance abuse and for sex offenders), transition programs for inmates with mental health disorders, alternatives to incarceration (such as mental health courts, drug and alcohol courts, electronic monitoring, and residential substance abuse treatment), programs for juvenile offenders (such as aggression replacement training, family therapy, and residential treatment), and prevention programs (such as Head Start for low-income families). "Education and substance abuse treatment programs—in prison, after prison, and instead of prison—save the state two to five times what they cost and reach the most people" (ISER 2009).

According to the Alaska Department of Corrections (DOC), the costs of incarceration and various forms of supervision in Alaska for the 2014 Fiscal Year (i.e. FY 2014) per day, per person were:

Jails (hard beds) $158.67
Community residential centers (soft beds) $85.67

Electronic monitoring $21.02
Probation and parole supervision $7.32 (Gutierrez 2015)

Supervising offenders in the community is far less costly than jail beds and can provide a supportive reentry that fosters rehabilitation and prevents recidivism. Daily cost data for promising methods for community supervision of persons with mental health issues including cognitive impairments is not available, yet programs such as "Forensic Assertive Community Treatment Teams" or "Forensic Intensive Case Managers" have been identified in an Alaska study to be effective with this type of probationer (Hornby Zeller Associates 2014).

Alaska's Criminal Justice Working Group (CJWG)[1] has had an important role in the development of Smart Justice practices in Alaska. In 2010, the Alaska Prisoner Reentry Task Force, a committee within the CJWG, issued a *Five-year Prisoner Reentry Strategic Plan, 2011–2016* (Alaska Department of Corrections 2011). The plan offers a comprehensive vision and includes practical recommendations for improving successful reentry for Alaska's returning citizens. Some of the recommendations include substance abuse treatment programs, expanded electronic monitoring, improved access to housing and employment, and improved responses to inmates' behavioral health needs.

Chief Justice Walter L. Carpeneti advocated for even more action to support Smart Justice in his State of the Judiciary address to the Alaska State Legislature in 2012, when he asked legislators to support collaborative efforts to reduce recidivism across the three branches of government (legislative, judicial, and executive), and state and local agencies. He called for Smart Justice practices like an expansion of therapeutic courts "and other problem-solving courts for misdemeanor offenders" for "individuals who pose no substantial risk" to public safety:

> Such courts address the problems at the root of criminal behavior, have solid track records across the state, and often succeed at reducing or eliminating prison terms. Expanding therapeutic courts for misdemeanor offenders would not only reserve prison beds and the prison budget for more serious felony offenders, but might well ensure that misdemeanor offenders do not become more serious felony offenders themselves (Carpeneti 2012).

The Chief Justice expressed the importance of considering the cost of judicial actions to the state's resources and on public safety, and asked legislators to include the judiciary in tailoring prison or treatment sentences to offenders:

[1]The Alaska Criminal Justice Working Group (CJWG) is comprised of representatives from the Alaska Court System, the Departments of Law, Corrections, Public Safety, Health and Social Services, and Education and Early Development, the Anchorage Police Chief, and the heads of the Public Defender Agency, Office of Public Advocacy, Division of Behavioral Health, Division of Juvenile Justice, and the Alaska Mental Health Trust Authority. It is chaired by a justice of the Alaska Supreme Court and the Lieutenant Governor and coordinated and staffed by the Alaska Judicial Council through funding by the Alaska State Legislature. The CJWG has focused on two main aspects of the criminal justice system: crime prevention and reduction of recidivism, and efficiencies in the system. Website: http://justice.uaa.alaska.edu/forum/26/1spring2009/c_cjwg.html.

> If we are ever to turn the tide of prison recidivism, we must make room in the sentencing process for 'smart justice' principles to take hold. Chief Justice Ray Price of the Missouri Supreme Court perhaps said it best in his own 2010 State of the Judiciary address: 'There is a better way. We need to move from anger-based sentencing that ignores cost and effectiveness to evidence-based sentencing that focuses on results …' (Carpeneti 2012).

Since the 2009 ISER report, Alaskan lawmakers have also been looking at what other states have done and have invited experts from those states to present their ideas for reducing recidivism and redirecting criminal justice dollars in Alaska.

Alaska has also introduced innovations of its own. As described in a separate section below, in 2012, the Alaska State Legislature unanimously passed Senate Bill 151, a statute making FASD a mitigating factor in Alaska, introduced by Senator Kevin Meyer.

In 2014, the Alaska Legislature passed an omnibus crime bill that addressed many elements of Smart Justice reforms. Some of the provisions of the bill included substance use assessments for all offenders incarcerated for 30-days or longer, a broadening of the existing authority for credit for time served in residential treatment, adding Post Traumatic Stress Disorder (PTSD) and Traumatic Brain Injury (TBI) as mitigating factors for sentencing, and establishing a "24/7" sobriety monitoring program. Additionally, Senator Meyer worked with the bill's sponsor to add language that would require DOC to provide screening for FASD and other brain-based disorders in corrections facilities, effective January 1, 2016. Governor Sean Parnell signed this bill into law on July 16, 2014.[2]

Also in 2014, Alaska Senator Pete Kelly established a workgroup to address FASD in Alaska. The workgroup, named *Empowering Hope,* made recommendations that include: supporting a public awareness campaign, expanding residential substance abuse treatment for pregnant women who experience alcoholism or drug addiction, implementing screening for FASD in prisons, developing a community network of "Natural Helpers" (persons within a community who support pro-social values), and promoting programs that reduce the time between conception and knowledge of the pregnancy (Kelly 2014).

Senator Kelly also introduced and saw pass two resolutions in 2014 addressing FASD. Senate Concurrent Resolution 13 asked the Governor to establish and support programs that address FASD, including rapid screening of sentenced offenders by the Department of Corrections, expanding residential substance abuse treatment for pregnant women, and supporting a public relations campaign. Senate Concurrent Resolution 14 encouraged the development of citizen networks to create positive community and social norms for prevention of FASD. It also encouraged the Governor to support the use of regional best practices models to address health and social services challenges with effective and cost-efficient methods.

Alaskan lawmakers are recognizing that with appropriate supports, people with FASD can live successfully in the community as contributing citizens, and that support services provide jobs for Alaskan caseworkers, clinicians, assisted living

[2]Ch.83, SLA 2014 (eff. 07/14/2014); FASD screening codified at AS 33.30.011(4)(c).

providers, mental health and substance abuse counselors, psychologists and psychiatrists, and others. There is a clear desire to see additional support for FASD-related issues as part of Alaska's Smart Justice effort.

Services that Can Effectively Keep Adults with FASD Out of the Justice System

Support services reduce the likelihood that a person with FASD will come in contact with the correctional system. Housing assistance, employment and training support, education, counseling and case management, all contribute to an individual's ability to maintain a stable and productive life (ISER 2009).

Rick Allen, Director of the Alaska Office of Public Advocacy, said that the number one need for people with FASD is housing:

> Somebody has to have basic stability in their lives, and that means a dry, warm place to sleep, first, and then they have to have food in their belly. And then, if their time can be occupied with positive things, you can keep people out of trouble … That's the whole principal behind mental health court … People who have housing first and some sort of productive thing to do, whether its low-level employment or volunteerism, or whatever, that's a huge factor in reducing recidivism and keeping people moving in a positive direction.

He said that with such support people living "with pages of misdemeanant convictions" suddenly stopped getting in trouble and started reestablishing relationships with family members. "I've seen really positive changes, so I know that it's possible," he said (Allen 2014).

Julia Erickson, former case manager and probation officer for the Juneau Alaska mental health court, named the Coordinated Resources Project, worked with individuals experiencing mental health and substance abuse disorders, including people with FASD. She assisted clients in obtaining housing, food, mental health care, employment, and other supports that help them remain stable in the community.

Erickson, a Master's level mental health clinician and Chemical Dependency Counselor I, noted that people with FASD were regularly represented in her caseload of about thirty individuals. While some had an FASD diagnosis, others showed indications of prenatal alcohol exposure, including difficult behaviors, impulsivity, difficulty concentrating, memory deficits, susceptibility to victimization, and difficulty understanding the connection between actions and the consequences.

She believes that too few Juneau agencies provide FASD-informed services, or are able to take a person with FASD into their caseload. Erickson noted that not many case workers are trained in interventions for people with extreme behaviors, who have difficulty complying with program expectations, such as keeping appointments, taking medications, or complying with the conditions of their probation/parole.

Echoing Mr. Allen, Erickson said housing was the primary need for her clients with FASD. Erickson recommended an FASD-informed assisted living environment with caregivers who understand and can work with difficult behaviors.

Second, she recommended access to diagnostic services from a qualified multi-disciplinary team that offers support after the diagnosis, so that individuals and their families can receive appropriate follow-up, case management, and other supports.

Third, she recognized specialized caseworkers are needed to assist individuals in maintaining stability to reduce the occurrence of crises. For example, her experience shows that a crisis about a pet can be just as debilitating for her clients as more traditional problems. Caseworkers help clients fill out applications/paperwork, make and keep appointments, manage money, give reminders about hygiene and appropriate social behavior.

She said that help qualifying for benefits is needed because an FASD diagnosis often fails to translate to accessible services. In Alaska, criteria for eligibility for Intellectual/Developmental Disability (IDD) services usually require that a person have an IQ below 70. Not all people with FASD experience low intellectual functioning, requiring a more nuanced application for disability benefits.

Finally, Erickson cited the need to look at delivery of behavioral health services for people with FASD:

> We need resources, but we also need flexibility … The problems associated with mental illness have been around for a long time. Social workers have known how to work with the extreme behaviors, as far as medication management, social structuring, and counseling, etc. An individual with a mental illness can be expected to recover in some ways. But we haven't really focused on people with FASD in the same way. We have this thinking that '*this* diagnosis gets *this* treatment,' but with FASD, providers don't always know what to do with the individuals they are working with. We need a variety of resources and opportunities, as well as a lot of flexibility for individuals with FASD (Erickson 2014).

Deb Evensen, MA, and director of Fetal Alcohol Consultation and Training Services (FACTS) in Alaska, works with adults and children with FASD and the school districts serving them, as well as conducting training events in and outside of Alaska. She believes that people living with FASD are "not hopeless" and can live stable and productive lives when given the appropriate supports:

> With FASD, we can't change the brain, but we can change the outcome for the person. It's not true that people living with FASD are hopeless, or that they can't learn, because they can learn, when the right supports are put in place. What's been the problem is that we've been using strategies that were developed for people who don't have FASD—people with mental illness or more typical brains. We use those techniques and they don't work with the kinds of brain differences caused by prenatal alcohol exposure. It can't be a weekly counseling session or consequences or insight therapy. That's where we lose them. That's why these people fall through the cracks—because we're not using the right strategies. We need to listen to them, figure out where they're stuck, and then provide support to those parts. It's always based on what, instead of why (Evensen 2014).

Support and structure in daily life are key for helping people with FASD live successfully in the community and stay out of trouble. Teaching *what to do* instead of *what not to do*, or *why not to do it* is what is important, Evensen said.

"We don't teach them to 'get it,' we teach them what to do in their lives and we provide structured support, and always a number that they can call and someone who checks in on them who doesn't always wait for them to make the call. We would do the same thing if it was a person with another kind of disability" (Evensen 2014).

Efforts to Build Supports for Individuals Affected by an FASD Condition

The movement to address fetal alcohol spectrum disorders is growing across communities, states, provinces, and national borders. A growing body of evidence-based practices is showing promise for effectively addressing FASD prevention and intervention in the criminal justice system.

The Substance Abuse and Mental Health Services Administration's FASD Center for Excellence and the National Organization On Fetal Alcohol Syndrome offer research, toolkits, webinars and other resources for supporting people with FASD who are involved in the justice system.[3]

In Alaska, many stakeholders are identifying and making steps toward addressing FASD in a variety of settings. With funding allocated in the 2015 Fiscal Year (FY15) from the Alaska State Legislature, the Alaska Mental Health Trust Authority launched a media campaign designed to raise awareness about not drinking alcohol during pregnancy. The Alaska Legislature allocated FY15 funding to the Institute of Circumpolar Health to study the efficacy of pregnancy tests in bars to raise awareness about the risks involved with drinking alcohol during pregnancy.

Also, as discussed above, the Legislature passed legislation in 2014 that directed the Alaska Department of Corrections (DOC) to establish by January 1, 2016 a mechanism for screening for FASD and other brain-based disabilities within their facilities. The Legislature also directed the department to establish a culturally-appropriate program that serves people with FASD.

A report produced for DOC in 2014 by Hornby Zeller Associates, Inc., *Trust Beneficiaries in Alaska's Department of Corrections*, recommends Modified Therapeutic Communities (MTCs) as an appropriate setting for people with FASD while incarcerated. MTCs are specialized units that keep people with mental illness apart from the general population, using interventions that may include

[3]Their respective websites are Substance Abuse and Mental Health Services Administration (SAMHSA), http://www.fasdcenter/samhsa.gov and National Organitzation on Fetal Alcohol Syndrome (NOFAS), http://www.nofas.org.

psycho-education, cognitive behavioral treatment, conflict resolution, medication management, etc.

> Within correctional facilities, many researchers recommend the use of integrated treatment models to meet both the mental health and substance abuse needs in a consistent and comprehensive manner. MTCs are an example of a treatment model that can provide integrated treatment, in addition to aligning with recommendations that mentally ill offenders be housed in therapeutic environments while incarcerated.

The report also suggests that participants in reentry MTCs experienced "significantly less recidivism" than participants who received traditional supervision, and lists people with FASD and traumatic brain injuries (TBI) as groups that may benefit from these specialized units.

> These populations require both intensive and structured supervision, and Anthony Wartnik, a retired judge, provides three suggestions to best serve those with FASD who are being released into their community: (1) live in a group home or facility with a structured regiment such as when to get up, when to eat, and so forth; (2) get a very structured job. Part-time is okay. Employment provides people with FASD something regularized that they need to do every day; (3) create a daily schedule with the individual that will be overseen by a parent, advocate or sponsor.

Other recommendations for people with FASD listed in the same report include: "Working out an agreement for money management; securing safe, affordable housing or a subsidized residential placement; providing in-home support to help the individual live as independently as possible; teaching and modeling parenting skills if the individual has children; referring the individual for specialized vocational training and/or job placements; ensuring medical care; arranging for a case manager to help individuals with FASD and their families access necessary services; organizing drug and alcohol treatment for the individual, if needed; serving as an advocate to ensure recommendations are implemented; acknowledging the individual's limitations, strengths and skills; accepting the FASD-impacted individual's world" (Hornby Zeller Associates 2014).

The Alaska Mental Health Trust Authority (Trust) has supported efforts to gather recommendations from stakeholders statewide for strategies to address FASD. The Trust has provided funding for conferences, training and education, and advocacy for housing and employment, community services, treatment, and other supports for people with FASD.

The Alaska Criminal Justice Working Group (CJWG) has focused its efforts on two main aspects of the criminal justice system: (1) crime prevention and reduction of recidivism; and (2) efficiencies in the system. The CJWG provides a forum at the highest level to support Smart Justice initiatives and to examine why a Smart Justice initiative like the FASD mitigating factor is not being considered in a meaningful way in Alaska's courtrooms. The CJWG is able to make credible recommendations for policies and funding requests that reduce these barriers.

William J. Edwards, Deputy Public Defender, Mental Health Branch, Los Angeles County Office of the Public Defender recommends lawmakers adopt a similar statute to that in Minnesota, which includes fetal alcohol spectrum disorder as a developmental disability. Adopting the Minnesota language could be a first step to support individuals with FASD in qualifying for appropriate community support services (case management, housing and employment assistance, behavioral health support, assisted living, etc.) and reduce the likelihood of their ending up in the criminal justice system.

Edwards states that even when his client's IQ is above 70, the client affected by an FASD condition will usually show low adaptive functioning. This condition affects a person's ability to manage normal life tasks like cooking meals, maintaining basic hygiene, navigating public transportation, keeping a job, or managing a bank account. People with low adaptive functioning can experience brain processing difficulties and behavior problems, resulting in a higher risk for incarceration, adjudication, homelessness, victimization, and recidivism. Edwards explained in an interview that for such persons, the key to effective intervention and to reducing recidivism is life-long support, not increased jail time:

> Even though you have a client with an IQ of 100, you still may have very, very low adaptive behavior skills, childlike qualities [like] impulse control problems and inability to learn from their prior mistakes. Simply putting them in jail or prison is not going to deter future criminal conduct … The way to reduce recidivism, in my opinion, is to get the support services that [people with FASD] need to live in the community, and realize that a lengthy sentence is not going to deter future criminal conduct … I think the real big picture is for judges to understand that [for] a person with 108 IQ and low adaptive [functioning], there is a need for life-long guidance and support. This is not a person who we can simply give medication to and say 'okay, he or she is going to get better.' It's brain damage; it's permanent (Edwards 2014).

In 2013, a round table of stakeholders sponsored by the Alaska FASD Partnership discussed strategies for FASD prevention and intervention, screening/diagnosis, training/education, and data/research. Experts from a variety of backgrounds and disciplines offered recommendations and informed the Trust about evidence-based strategies that have shown success in Alaska and other places. These recommendations included the following:

(1) **Parent-Child Assistance Program (PCAP).** Support funding to implement a PCAP program in Alaska that is similar to the highly successful program in the state of Washington. The program would provide support services to both pregnant and post-partum women with alcohol addiction/dependency. The model utilizes intensive case management provided by trained peer mentors (which is reimbursable for those receiving benefits from the federal Medicaid program) and does not require "bricks and mortar" because it is a home-visitation program. It is recommended that PCAP be implemented with fidelity to the Washington program, and that it be implemented in several regions in Alaska with culturally appropriate considerations.

(2) **Screen, Brief Intervention, and Referral to Service (SBIRT).** Support SBIRT and early identification and intervention for substance abuse in the following ways: (1) disseminate SBIRT among Alaskan health professionals and health professional training programs; (2) reimburse/incentivize the implementation of SBIRT during all prenatal and well-woman exams in Alaska for women ages 14–44.

(3) **CHOICES.** Integrate the CHOICES program in primary care and behavioral health settings throughout Alaska to identify risk and offer referral to services early and support training for CHOICES in regional locations. (Note: CHOICES is an evidence-based intervention program for women, through the U.S. Centers For Disease Control (CDC), designed to increase motivation and commitment to stop drinking alcohol and/or use contraception effectively to avoid alcohol-related pregnancies. It involves two to four counseling sessions, plus a contraception counseling session.)

(4) **FASD and trauma-informed practices.** Support trauma-informed training and practices in the Division of Juvenile Justice (DJJ), Department of Corrections (DOC), Office of Children's Services (OCS), and Department of Education and Early Development (DEED). Support community services that serve people with FASD. Support training and ongoing support for medical and behavioral health professionals, foster care and adoptive parents, parents, teachers, and people who work in OCS, DJJ, DOC, so they are providing FASD-informed practices.

(5) **Establish protocols for FASD screening and referral to diagnosis in DOC, DJJ, OCS.** Work with DOC, DJJ, OCS to establish protocols for FASD screening and referral to diagnosis. Research practices in other states and Canada.

(6) **IDD-capable substance abuse treatment.** Support outpatient substance abuse treatment that is Intellectual/Developmental Disability (IDD)-capable immediately to women of childbearing age who have an IDD and are currently, or at-risk of, becoming pregnant. Treatment should include residential treatment that is family-friendly and trauma-informed. Consider collaborations with tribal organizations that have existing resources that can be grown if they are willing to expand.

(7) **FASD public awareness campaign.** Support funding to implement a comprehensive statewide public awareness campaign that educates citizens about "no safe amount of alcohol during pregnancy" and asks the question "is it worth the risk?" Include in the campaign that drinkers who are not alcoholics are also at risk, even if they occasionally binge drink or drink moderately during pregnancy.

(8) **Expand capacity to diagnose in Alaska.** (1) Provide existing diagnostic teams with adequate resources to increase number of diagnoses per year; (2) develop screening and diagnostic capabilities in un-served or underserved communities through mechanisms such as travelling diagnostic teams to support local professionals; and, 3) increase outreach to referral sources.

(9) **Expand *FASD Into Action* and *Families Moving Forward* training.** Support funding for *FASD Into Action* and *Families Moving Forward* training that reaches across professional disciplines (such as corrections, courts, juvenile justice, child welfare and education/early development), is culturally appropriate, and includes a strategy to provide affordable training for parents.

(10) **Evaluate Alaska's FASD diagnostic data.** The evaluation should include how much service Alaska has provided, with demographics; evaluate the quality of services and fidelity to the 4-digit FASD diagnostic model developed by the University of Washington Fetal Alcohol and Drug Unit, customer satisfaction, and access to timely services, to discover how the diagnosis has helped support long-term outcomes.

FASD-informed supports require substantial funding and training and hiring staff that are informed about FASD, and having such supports in Alaska's vast rural areas is a particular challenge. Unfortunately, if supports are not available, many adults with an FASD condition will continue to become involved with the criminal justice system, including felony crimes that trigger the presumptive sentencing laws now in effect in Alaska. As stated by Canadian Judge Mary Ellen Turpel-LaFond at a 2008 Whitehorse, Yukon Territory conference concerning FASD and the justice system, there is a need to understand persons with an FASD condition "and keep them out of the justice system and to ensure that the justice system is not used as a substitute for appropriate social services and supports for some of the most vulnerable citizens" (Fraser 2008).

Background for the Need for a Statutory FASD Mitigating Factor in Alaska Sentencing Law

The intent underlying Alaska's sentencing structure, that people will modify their behaviors based on the criminal justice system's response to their crimes, is not usually met when the current presumptive sentencing ranges are applied to individuals with FASD. The disabilities on the FASD spectrum most often manifest as deficits in executive function, which can result in impaired adaptive behavior, memory difficulties, inability to plan, and failure to recognize the consequences of actions (Streissguth 1996). Advocates believe it is important for the courts to take these deficits into account during sentencing. Neither the offender, nor society, benefits from holding individuals with FASD to community standards they cannot attain (Jeffery 2010).

Pre-2012 sentencing laws in Alaska did not provide the flexibility needed to address these needs; a statutory change was critical. The pre-2012 statutory framework for felony sentencing still exists today. For example, a defendant in a serious (class "A" felony) drug or property crime offense faces a maximum of 20 years of jail time (Alaska Statute 12.55.125 (c), 2014 ed.). Even if the person has no prior felonies, the presumptive sentence must be between five and eight years. If the person has two or more prior felonies, the presumptive sentence becomes 15–20 years of jail time (Alaska Statute 12.55.125 (c) (3, 4), 2014 ed.). The sentencing law also provides that if a mitigating factor is shown by clear and convincing evidence, then the jail time can be reduced to as little as two and one-half years for a first offender, and seven and one-half years for an offender with two prior felonies (Alaska Statute 12.55.125(c) 2014 ed.). The pre-2012 law did not have a mitigating factor that applied to defendents with an FASD condition. The new FASD mitigating factor corrects this unfairness, at least for non-assaultive crimes to which the new mitigating factor applies.

Judge Michael I. Jeffery, an Alaska Superior Court Judge in Barrow for 32 years, experienced how the pre-2012 version of the sentencing laws did not allow for accommodation for a defendant diagnosed with an FASD condition. He presided over a felony case in 1990 in which the defendant had been evaluated and confirmed by a medical provider as having an FASD condition. The defendant had a prior felony and faced a presumptive term. His attorney requested consideration of a "non-statutory mitigating factor," arguing that it would be manifestly unjust to sentence the defendant without considering his FASD condition. Judge Jeffery agreed and directed referral to a "three-judge sentencing panel" (Alaska Statute 12.55.175 2014 ed.).

The three-judge sentencing panel did not agree and declined the opportunity to sentence the defendant without reference to presumptive sentencing and returned the case to Judge Jeffery for imposition of the presumptive term. Judge Jeffery recalls that the panel commented that the Legislature "had already spoken on the issue of mental conditions affecting sentencing and that FASD did not fit the existing law." The Alaska Court of Appeals affirmed the three-judge panel's decision (*State v. Leavitt* 1992).

Years later, as a member of the Alaska Criminal Justice Assessment Commission,[4] Judge Jeffery successfully advocated for one of the commission's 2000 legislative recommendations that "the legislature create a statutory mitigating factor for use at criminal sentencing, recognizing when the wrongful conduct was substantially affected by an organic brain disorder" (Alaska Criminal Justice Assessment Commission 2000). Passage of Senate Bill 151, the FASD mitigating factor bill introduced 12 years later, finally implemented this recommendation.

Both the Canadian Bar Association (CBA) and the American Bar Association (ABA), the national organizations representing attorneys and judges in both Canada and the USA, have recognized the need for such adjustments in the law. In 2010, the CBA passed a resolution at its annual meeting recommending the CBA support "access to justice for people with FASD and [to] urge all levels of government to allocate additional resources for alternatives to the current practices of criminalizing individuals with FASD." The resolution supported efforts to "prevent persistent overrepresentation of FASD-affected individuals in the criminal

---

[4]The Alaska Criminal Justice Assessment Commission (CJAC) was created in 1997 when the National Institute of Corrections selected the State of Alaska to participate in the federally funded Criminal Justice System Project. The Governor's Criminal Justice Cabinet sought participation in the project to study the problem of prison overcrowding. Its goal was to bring together legislators and representatives from across the criminal justice system spectrum and members of the public to participate in the review and to collectively develop recommendations to alleviate the problem.

justice system" and to "amend criminal sentencing laws to accommodate the disability of those with FASD" (Canadian Bar Association 2010).

This resolution inspired advocates at the ABA. During its August 2012 annual meeting, and with support from numerous national organizations, including its own judicial division, the ABA House of Delegates unanimously passed ABA Resolution 112B. The resolution stated that: "… the American Bar Association urges lawyers and judges, as well as bar associations and law school clinical programs, to … work with medical, mental health, and FASD disability experts to promote … applying FASD as a mitigating factor in the mitigation of juvenile justice and criminal sentencing … and [consider] alternatives to incarceration that reduce recidivism" (American Bar Association 2012).

The Hornby Zeller study indicated that the Alaska Department of Corrections (DOC) is the number one mental health care provider in the state (Hornby Zeller Associates 2014). Presentations and public testimony at legislative hearings since 2009 have suggested that DOC would benefit from providing alternative interventions for people with mental health and substance abuse disorders as a way to reduce recidivism. Former DOC Commissioner Joe Schmidt and Deputy Commissioner Carmen Gutierrez regularly reported to legislators that DOC had a "captive audience" and should be implementing evidence-based interventions that increase the likelihood of rehabilitation.

Overview of Alaska Legislation Establishing a New Mitigating Factor for FASD

In 2012, Alaska was the first U.S. state to explicitly recognize FASD as an available mitigating factor for sentencing in certain felony crimes when the burden of proof required by the law has been met. The bill's sponsor collaborated with stakeholders to craft a bill that would address the large number of people in the criminal justice system that are affected by FASD. The Alaska State Legislature unanimously passed Senate Bill 151, Fetal Alcohol Spectrum Disorder as a Mitigator, on April 15, 2012, and it became law on September 19, 2012.

A "mitigating factor" in the current Alaska sentencing law is a potential justification for lowering jail time below what would otherwise be mandated by statute. Like other mitigating factors, the FASD mitigator must be shown to exist by "clear and convincing evidence" (Alaska Statute 12.55.155(f)(1) 2014 ed.).

The FASD mitigating factor affords judges limited flexibility to reduce jail time for people with FASD in certain cases where there is clear and convincing evidence that the "defendant committed [an] offence while suffering from a condition diagnosed as a fetal alcohol spectrum disorder, the fetal alcohol spectrum disorder substantially impaired the defendant's judgment, behavior, capacity to recognize reality, or ability to cope with the ordinary demands of life, and the fetal alcohol spectrum disorder, though insufficient to constitute a complete defense, significantly affected the defendant's conduct."

The statute defines a fetal alcohol spectrum disorder as "a condition of impaired brain function in the range of permanent birth defects caused by maternal consumption of alcohol during pregnancy."

Alaska's law does not require a judge to use this or any other mitigating factor to actually reduce a sentence and does not automatically adjust a presumptive sentence, even when the defendant has met his or her burden of proof. When a minimum presumptive term is more than four years, the potential jail time reduction is up to one-half of the minimum presumptive term (Alaska Statute 12.55.155(a)(2) 2014 ed.). As discussed above, the potential jail time reduction for a serious drug or property crime conviction could be as much as 7.5 years.

Designing the Language of the Bill

The FASD mitigating factor legislation was developed by a dedicated legislator and a diverse working group of advocates. Alaska State Senator Kevin Meyer became interested FASD issues after adopting a daughter from Russia who was diagnosed with the disability. He regularly attended FASD rallies and events and introduced a resolution for many years in the Legislature proclaiming September 9th as FASD Awareness Day. In 2012, he introduced and saw passed, Senate Bill 127, which permanently established September 9th as FASD Awareness Day in Alaska.

Inspiration and critical support for this bill came from a coalition of individuals, organizations and state agencies organized by the Alaska FASD Partnership.[5] Formed in early 2010 and staffed by the Alaska Mental Health Board and Advisory Board on Alcoholism and Drug Abuse, the Partnership has an active Steering Committee that guides the organization, and a number of workgroups. In 2011, the FASD and the Legal System workgroup had determined the FASD mitigating factor issue to be one of its priorities for action. When Senator Meyer expressed interest in introducing legislation that would help Alaskans affected by FASD, the Partnership suggested the idea of establishing the mitigating factor in the felony sentencing law, and Senator Meyer agreed. The workgroup collaborated with the Senator's office to recruit representatives from a variety of disciplines to assist with drafting the law and advocating for it.[6] After excluding assaultive cases from the mitigating factor as it had done with several other statutory mitigating factors, the Legislature passed the bill unanimously and it became law.

Senator Meyer provided useful comments about effective methods to obtain such a result. He believed that the support of FASD advocates was key in helping

[5]Website is http://dhss.gov/adada/Pages/fasd.aspx.

[6]The workgroup for the FASD mitigating factor legislation included representatives from the Disability Law Center of Alaska, Advisory Board on Alcoholism and Drug Abuse, FASD Legal Issues Resource Center at the University of Washington, Washington State FAS Diagnostic and Prevention Network, an Alaska Superior Court, and others.

gain legislator approval. "As it went through the process, the phone calls, emails, the testimony, and those who could come down to Juneau, really provided us with a great wealth of information, and educated legislators on FASD. I think testifying on the bill itself and explaining some of the unknowns … really helped a lot," he said in the interview for this publication. While most legislators knew about FASD and were sympathetic to it, the education piece provided by advocates helped legislators understand the deeper complexities of the disability and make more informed decisions.

The senator noted his colleagues' concern about mitigating factors being "soft on crime." In an interview, Senator Meyer stated that "[o]ver the last couple years we've been changing our attitude a little on that, which I think is a great idea … The fact is, kids and adults with FASD don't necessarily process normally, and putting them in jail is not really helping them process right from wrong. In fact, it may be that you're making it worse … A lot of people with FASD just need the constant reminder. If we can do that instead of putting them in jail ... society is a lot better off and certainly the people with FASD are" (Meyer 2014).

Implementation of the FASD Mitigating Factor

Alaska's FASD mitigating factor provides an important tool for a cost-effective approach to sentencing in felony cases. The current sentencing laws incorporate a system of jail sentences applicable to even minor felonies and which become far more onerous, and expensive for the state, if a person has one or more prior felonies. As discussed above, a defendant with low adaptive functioning, poor impulse control, and little concern for consequences can easily generate a felony of multiple convictions. The defendant may be faced with a minimum presumptive term of 15 years in a serious drug or property sentencing. Application of a mitigating factor can reduce the sentence by as much as seven and one-half years, which can lead to earlier availability of cost-effective services, such as intensive case management and probation/parole supervision.

Why the FASD Mitigating Factor Has Been Rarely Proposed Since Its Passage

At the time of this writing, two years have passed since the FASD mitigating factor became law in Alaska. In July 2014, twelve Alaskan Superior Court judges representing each of the state's four judicial districts, and whose caseload includes sentencing in felony cases, were surveyed. The findings revealed that none of these judges had seen the mitigator argued. Doug Wooliver, Deputy Administrative Director for the Alaska Court System, said almost all the judges surveyed indicated they "suspected the reasons it is not used are that few attorneys are aware of it, and that few defendants have an actual diagnosis" (Wooliver 2014).

Judge Jeffery noted a felony case in his court in which the prosecuting and defence attorneys agreed to the mitigating factor as part of a plea bargain, but the case was a felony assault which is the type of crime that the Legislature had specifically excluded from application of the law.

One Juneau Superior Court Judge saw the FASD mitigating factor applied in one case, but expressed frustration about the lack of resources for referral outside the corrections system, and effectively "could not do much" with the mitigating factor (Pallenberg 2014).

Quinlan Steiner, Public Defender for the State of Alaska, noted concern among defense attorneys about the expense and difficulty of obtaining an FASD diagnosis. He stated that "if attorneys are unable to adequately identify the need to seek a diagnosis, the application of the mitigator will be limited to cases where a diagnosis has already been made. And due to [the] expense and complexity of this process, a preexisting diagnosis is not always available," he said.

Mr. Steiner also noted "potential concerns" with proposing the mitigator where support services are not available, or in cases where FASD is considered untreatable: "Some attorneys are concerned that their client will be treated more harshly at sentencing under the theory there is no rehabilitative program that can address their client's needs" (Steiner 2014). Rick Allen, Director of the Office of Public Advocacy, added that defense attorneys could be hesitant to introduce the mitigating factor fearing the judge may assume the client with an organic brain injury "is less likely to be rehabilitated and will continue to be a danger to the public" (Allen 2014).

John Skidmore, Director of the Criminal Division for the Alaska Department of Law, commented, "I understand their standpoint, they're saying if the judge doesn't know much about (FASD) they're not sure that the judge would be able to give it the weight that they think they should, and impose a sentence that they think would be appropriate." He noted that prosecutors don't assert mitigating factors and it hasn't come up in discussions with his colleagues (Skidmore 2014).

At the 2008 Whitehorse "Path to Justice" conference regarding FASD and the legal system, participants discussed that the limited time justice professionals spend with clients is an important factor inhibiting the use of more creative approaches for clients affected by FASD. The conference report reflected the extreme difficulties faced by persons affected by FASD when dealing with justice system professionals:

> It often takes more time than professionals have to clearly communicate with individuals with FASD to ensure they understand what is said to them. Also, some behavioral characteristics associated with FASD are sometimes seen by justice professionals as behavior that suggests disobedience, noncompliance or aggressiveness. If justice professionals are not aware how FASD may impact an individual, they may not want to get to know their client or work with them any more than the basic minimum to "get by" (Fraser 2008).

Clearly, there are complex reasons for not seeing the mitigator applied.

Strategies to Support Application of the Mitigating Factor

Rick Allen suggested that judges would likely be better persuaded to support the mitigator if criminal defense attorneys presented a plan that involved helping the defendant get housing, basic employment, treatment, and other supports. "If [the judge] knew … we were going to keep them occupied with healthy activities and keep them doing the things they need to be doing to stay out of trouble, I think that would be reassuring to everybody."

Allen recommended open dialog where "everybody is educating themselves" about FASD and appropriate interventions. He noted that while many judges understand the complexities of working with defendants with disabilities, there are others who have only practiced corporate law before coming to the bench and may not have the deeper understanding of the ways people with FASD (and other disabilities) can be accommodated for successful outcomes. "Lawyers need to feel comfortable that judges … care about this and have done the research," he said (Allen 2014).

Quinlan Steiner agreed, "Education is needed so judges and attorneys understand what services are available and what can be done to help people live productively" (Steiner 2014). The Public Defender Agency has arranged for a "social services specialist" staff person at its office in Bethel, Alaska to develop local alternatives to incarceration for adults and detention of juveniles.

California public defender William J. Edwards has defended and litigated on behalf of clients with FASD since 2000. He raises ethical issues about crafting sentences for persons affected by an FASD condition. He argues that people with FASD who so often have poor judgment and low adaptive functioning may also have a lower level of culpability. He suggests that instead of prison or jail, people with FASD, like people with other disabilities, need supportive services to remain stable, productive, and crime-free in the community.

Edwards recommends educating disability lawyers so they better understand the problem, and consider competency and litigation issues. "There are a lot of cases out there that talk about ineffective assistance of counsel for not identifying FASD as a mitigating factor," he said. "If it's in statute, I think we have a duty to raise it."

Edwards stresses the importance of identifying FASD as a developmental disability in statute so that people with FASD are offered appropriate, disability-informed consequences. He noted that courts and disability law organizations often don't consider FASD because "they're looking for people with IQs lower than 70." Most of Edwards' clients with FASD have an IQ higher than 70, he said, but still have low adaptive functioning, which leads to an inability to accomplish normal day to day life tasks, such paying rent, keeping a job, etc.

Reforming or Eliminating the Need for the FASD Mitigating Factor

Alaska's FASD legislated mitigating factor can serve as a model in many ways for other jurisdictions with similar presumptive sentencing laws. The credibility and political attractiveness of the mitigating factor are enhanced by a medically-correct legal definition of "fetal alcohol spectrum disorder" which includes a range of conditions. The law clarifies that only people with FASD conditions that are severe enough to warrant such protection are included, a requirement that the defendant show that he or she has an FASD condition of this severity, and that the condition "significantly affected" the criminal conduct in the incident under consideration.

The requirement of proof "by clear and convincing evidence" presents a difficult hurdle, but this burden of proof is embedded in Alaska's presumptive sentencing law and has little chance of being changed. Perhaps a lower burden of proof (by a preponderance of the evidence) would be more appropriate to at least allow the judge and the parties involved in the case to further evaluate whether application of the mitigating factor would meet the sentencing goals of the jurisdiction.

The provision banning consideration of the FASD mitigating factor in assault cases was a political necessity in order for the bill to pass the Legislature. But the dysfunctions caused by an FASD condition can lead to comparatively minor felony assaults, which nonetheless can trigger severe sanctions. The promise of an FASD mitigating factor would be enhanced by eliminating the blanket prohibition on consideration of the mitigating factor in all assaultive crimes to allow consideration of the applicability of the mitigating factor on a case-by-case basis.

But what if a jurisdiction did not need a mitigating factor at all? Mitigating factors are part of presumptive sentencing legislation that requires certain sentencing ranges for serious crimes. Then-Chief Justice Walter L. Carpeneti in his 2012 State of the Judiciary address compared the current situation in which prosecuting and defense attorneys reach plea bargains in almost all the criminal cases that include the jail sentence and may include the specific probation conditions. The judge simply approves the sentence or refuses to, and not approving a plea bargain is very unusual.

Chief Justice Carpeneti contrasted this system with "open sentencing," which allows active involvement of the judiciary in the sentencing process that he experienced while a practicing attorney before his appointment to the bench. He acknowledged that over the last thirty years, substantial research has become available about what response is most effective in actually reducing community problems and expense of recidivism. This research was not available when the presumptive sentencing law in Alaska was first implemented. Now we know that a "one size fits all" approach in sentencing is not the answer:

> We must give judges and others involved in sentencing the tools to distinguish those who pose the greatest risk of recidivism from those who don't, and we must empower them to respond accordingly. And I hope you will consider restoring the judge's proper role in sentencing decisions. When we have a talented and experienced judiciary that has been

> carefully selected to make the tough decisions in our justice system, and when we know that effective sentencing requires a close fit between the sanctions imposed and an individual offender's risks and needs, it makes little sense to set narrow ranges in advance and leave all the vital work of crafting appropriate sentences to others. Judges are a valuable resource, and they can and should play a strong role in implementing 'Smart Justice' principles (Carpeneti 2012).

The best result is a flexible sentencing system that reduces recidivism and provides more effective, humane treatment of all offenders so that an FASD mitigating factor statute is not needed. Instead, the persons involved in a sentencing hearing can simply present all relevant information (including the challenges from an FASD condition) to the judge for use in imposing the appropriate mix of jail time and community supervision for that particular offender.

Conclusion

People with an FASD condition or similar brain-based disorders have specialized needs apart from what people with Intellectual/Developmental Disabilities or Severe Mental Illness need. Angry outbursts, impulsivity, stealing, trespassing, committing assaults under the influence of alcohol or drugs, lack of appropriate hygiene, homelessness, criminal involvement, etc., are some of the behaviors legal professionals face when working with prenatally-exposed individuals. Yet experience has shown that with sufficient long-term (or even life-long) supports, these negative outcomes can be sharply reduced or even avoided altogether.

Communities and organizations that encourage trauma-informed FASD supports are more likely to see fewer crimes committed, and fewer public dollars spent on expensive prison beds and emergency services (Institute of Social and Economic Research (ISER) 2009).

As Alaskan lawmakers, agencies, and concerned individuals continue to support Smart Justice initiatives, as public advocates continue to hold awareness events and conduct organized trips to the Capitol to educate policymakers, as providers continue to learn the nuances of serving people with prenatal alcohol exposure and adapt their practices to be "trauma-informed" and to better serve clients individually according to the specifics of their disability, as parents and teachers understand the disability and make the appropriate accommodations to reduce the likelihood of developing secondary disabilities, and as courts and legal professionals become more informed about FASD and available services in the community, then Alaskans will be able to say they are making solid progress in reducing the barriers and increasing the successes for serving people with FASD.

Adequate and life-long supports for persons affected by FASD will forestall their involvement in the criminal justice system and avoid victimization with the inevitable heartaches and huge expense involved. This goal must be primary in addressing FASD.

However, when the criminal justice system must deal with a situation involving a presumptive sentence of a person affected by FASD, then a mitigating factor such as the statute in Alaska becomes vitally important. Experience has shown that the defense team must develop a plan for support services after any jail time in order to effectively advocate for an actual reduction of the jail time. When judges and attorneys feel confident about the availability of services and programs to adequately serve people with FASD, then we will likely see more frequent application of the FASD mitigating factor in Alaskan courts.

Furthermore, having more flexibility in sentencing by the elimination of presumptive terms (for all but the most serious cases) so that judges have more discretion is essential. This will eliminate the need for this type of statute in favor of presenting the individualized factors (including mental health conditions) regarding a defendant during the sentencing hearing to allow the judicial officer to craft a sentence that complies with legislative sentencing goals for the particular offender. Whether from a mitigating factor or from new legislation, allowing earlier jail release with adequate supports will implement Smart Justice practices and bring the many benefits to any jurisdiction using it.

References

Alaska Criminal Justice Assessment Commission. 2000. Final Report. www.hss.ajc.state.ak.us/reports/CJACFinal2004.pdf.

Alaska Department of Corrections. 2011. Alaska Prisoner Reentry Task Force, Five-Year Prisoner Reentry Strategic Plan, 2011–2016. http://www.correct.state.ak.us/TskForce/documents/Five-Year%20Prisoner%20Plan.pdf. Accessed Oct 2014.

Alaska Statute 12.55.125(c). 2014 ed.

Alaska Statute 12.55.155(a)(2). 2014 ed.

Alaska Statute 12.55.155(f)(1). 2014 ed.

Alaska Statute 12.55.175. 2014 ed.

Allen, R. 2014. Director of Alaska Office of Public Advocacy. (T. Tibbett, Interviewer).

American Bar Association. 2012. http://www.americanbar.org/content/dam/aba/administrative/mental_physical_disability/Resolution_112B.autcheckdam.pdf. Accessed Oct 2014.

Canadian Bar Association. 2010. hpps://www.cba.org/cba/resolutions/pdf/10-02-A.pdf. Accessed 10 April 2015.

Carpeneti, W.L. 2012. State of the Judiciary 2012. http://courts.alaska.gov/soj/state12.pdf. Accessed Oct 2014.

Edwards, W.J. 2014. Deputy Public Defender, Mental Health Branch, Los Angeles County Office of the Public Defender. (T. Tibbett, Interviewer).

Erickson, J. 2014. *Former Probation Officer and Case Manager for the Juneau Coordinated Resources Project*. Juneau, AK, USA.

Evensen, D. 2014. MA, Director, Fetal Alcohol Consultation and Training Services (FACTS). (T. Tibbett, Interviewer).

Fraser, C. 2008. Summary Report, Conference on the Path to Justice–Access to Justice for Individuals with Fetal Alcohol Spectrum Disorder. http://www.justice.gov.yk.ca/pdf/Path_to_Justice_Conference_Final_Report_Eng.pdf. Accessed 12 Oct 2014.

Gutierrez, C. 2015. 2015 Recidivism reduction plan. Alaska Department of Corrections. www.correct.state.ak.us/commish/docs/HB266%20-%20Recidivism%20Reduction%20Plan%20%20Report%20(2015).pdf. Accessed 13 April 2015.

Hornby Zeller Associates. 2014. Trust Beneficiaries in Alaska's Department of Corrections. http://www.mhtrust.org/layouts/files/documents/focus_DisabilityJustice/ADOC%20 TrustBeneficiaries%20May%202014%20FINAL%%20PRINT.pdf. Accessed Oct 2015.

Institute of Social and Economic Research (ISER). 2009. Research Summary: The Cost of Crime: Can the State Reduce Future Crime and Save Money By Expanding Education and Treatment Programs? http://correct.ak.us/TskForce/an2009/documents/ISERResearchSummary.pdf. Accessed 12 April 2015.

Jeffery, M.I. 2010. An Arctic Judge's Journey With FASD. *The Journal of Psychiatry and the Law* 38(Winter): 585–618.

Kelly, P. 2014. Fetal Alcohol Spectrum Disorders in Alaska report to Alaska State Senate Health and Social Services Finance Subcommittee. http://www.legis.state.ak.us/basis/get_documents.asp?session=28&docid=18982. Accessed Oct 2015.

Meyer, K. 2014. (T. Tibbett, Interviewer).

Office of National Drug Control Policy. 2011. Criminal Justice Brief Alternatives to Incarceration. Executive Office of the President. http://www.whitehouse.gov/sites/default/files/ondcp/Fact_Sheets/alternatives_to_incarceration_policy_brief_8-12-11.pdf. Accessed 14 April 2015.

Pallenberg, P. 2014. Judge. (M. Jeffery, Interviewer).

Skidmore, J. 2014. Director, Criminal Division, Alaska Department of Law. (T. Tibbett, Interviewer).

State v. Leavitt, MO&J No. 2453, 1992 WL 12153213 (Alaska App. 06 30, 1992).

Steiner, Q. 2014. Public Defender. (T. Tibbett, Interviewer).

Streissguth, A.B. 1996. *Understanding the Occurrence of Secondary Disabilities in Clients with Fetal Alcohol Syndrome (FAS) and Fetal Alcohol Effects (FAE)*. University of Washington: University of Washington. Seattle.

Wooliver, D. 2014. Deputy Administrative Director, Alaska Court System.

Additional Resources

Edwards, William J., and Stephen Greenspan. 2010. Adaptive behavior and fetal alcohol spectrum disorders. *Journal of Psychiatry and Law.* http://www.americanbar.org/content/dam/aba/migrated/2011_build/child_law/adaptive_behavior_and_fasd.authcheckdam.pdf.

American Bar Association Section of Litigation and the ABA Center for Continuing Legal Education. (2012). Fetal Alcohol Spectrum Disorders: What You Need To Know to Help Your Client. http://www.americanbar.org/content/dam/aba/publications/litigation_committees/childrights/112juncm1.authcheckdam.pdf.

Wartnik, Anthony P. 2007. Stopping the Revolving Door of the Justice Systems-Ten Principles for Sentencing Other Disposition of People with a FASD. http://www.uwcita.org/fetal-alcohol-spectrum-disorder.html.

FASD & Justice: The Ethical Case for Effective Training and Knowledge Mobilization Practices for Frontline Justice Professionals in Canada

Michelle Stewart

Introduction

This book represents a cross-section of chapters focused on legal and ethical issues in the area of Fetal Alcohol Spectrum Disorder (FASD). This chapter will contribute to the volume by focusing directly on the role of justice professionals and their experiences working with individuals that have FASD. Although FASD is first and foremost a health issue, it is a consideration in the justice system because individuals with FASD have contact as victims, witnesses and offenders. In what follows, I will present qualitative research findings from working with frontline members of the justice sector and community organizations. This chapter will draw on research to argue that there is an ethical need, and indeed an obligation, for appropriate training to better prepare police to work with clients that have FASD. By illustrating the ways in which many frontline workers misunderstand FASD, coupled with the stigma associated to FASD, this chapter argues that individuals with FASD have unique needs and the stakes in the justice system are simply too high to not have appropriate training in place to meet those needs.

Police agencies and researchers are turning attention to the encounters between frontline officers and individuals with mental health issues (Wilson-Bates 2008; Vancouver Police Department 2013) and cognitive challenges including brain injuries (Togher et al. 2004) and FASD (Stewart and Glowatski 2014). Similarly, researchers (Steadman et al. 2009) and policy-makers are becoming increasingly aware of prison populations that are filling up with individuals with cognitive challenges and brain injuries (Williams et al. 2010). Fast and Conry (2009) found that within a Canadian correctional setting 10 % of the population had FASD and 18 % was suspected to have FASD. Despite concerns raised about the rise in marginalized peoples having increased police and prison contact, there is relatively little research that focuses on how frontline police understand clients with different

M. Stewart (✉)
University of Regina, Regina, SK, Canada
e-mail: Michelle.Stewart@uregina.ca

M. Nelson and M. Trussler (eds.), *Fetal Alcohol Spectrum Disorders in Adults: Ethical and Legal Perspectives*, International Library of Ethics, Law, and the New Medicine 63, DOI 10.1007/978-3-319-20866-4_12

cognitive abilities. This research project draws on nearly 100 interviews with justice professionals and community workers with particular attention to the need for increased training in the area of FASD.

As a researcher focused on the ways marginalized individuals are understood in the criminal justice system, I see the legal and the ethical as deeply entangled. My goal in this chapter is to illustrate how these entanglements might be tackled, through collaborative work and effective practices to mobilize accurate information about FASD. FASD is a stigmatized condition that is often misunderstood—these misunderstandings can directly impact the lives of marginalized individuals through negative justice outcomes. This chapter presents an ethical case for increased training in the name of social justice for those living with FASD.

Background

FASD and the Justice System

As has been outlined in other chapters, FASD is an umbrella term used to describe a wide range of impacts that result from maternal drinking. Researchers argue that alcohol is a teratogen and, for expecting mothers, it can have negative impacts on the developing fetus (Guralnick 2000) including physical, behavioural and cognitive outcomes. Jones and Smith (1973) are most cited and recognized for their early work on the impacts of alcohol during pregnancy. Over the past 40 years, research on FASD has expanded beyond paediatrics (Jones and Smith 1973; Lemoine et al. 1968) to include research on more culturally-appropriate prevention practices (Salmon and Clarren 2011) and innovative practices in the area of women's health (Rutman 2013; Poole 2011). Other research has investigated the construction of FASD, associated risk categories, and their impacts on the lives of women (Drabble et al. 2011; Armstrong 2003).

Oldani's (2009) work in Manitoba Canada argues FASD can follow a "racialized script" in the areas of treatment and management. He argues that "white, mainstream Canadian children" are more likely diagnosed with ADHD whereas Aboriginal and First Nations children are more likely treated as having FASD. These differing practices have impacts in the area of medical treatment; I would extend his work to argue those different treatments can transfer to police settings as those with FASD are understood to be at a higher risk for negative contact with police. Here we also face a troubling situation in which a health condition, coupled with racialized practices, moves into the justice sector. Although researchers are investigating FASD and the justice system, much of this work takes place after arrest or negative contact (Fast and Conry 2004; Sinclair 2004; Cox et al. 2008; Fast and Conry 2009; Gagnier et al. 2011). Popova et al. (2011) conclude, "there is an urgent need to raise awareness not only about the prevalence of FASD in the criminal justice system and the disabilities associated with FASD, but also the appropriate responses necessary to reduce the pervasiveness of this disorder in this setting" (Popova et al. 2011, p. 339). There is a need to expand research in the justice sector and to work together to create critical capacity and better understanding of this disability from the frontline police through the courts, and into the prisons.

Stigma

FASD is a highly stigmatized condition in many sectors including justice. Similar to mental health or a cognitive disability, those with FASD experience discrimination in part because of misunderstandings of the condition and these misunderstandings can negatively impact the lives of individuals. Researchers argue that stigma can produce barriers when it comes to accessing care and services (Church et al. 2001; Smith and Marshall 2007). Canada FASD Research Network has argued for education and outreach campaigns that are aware of the underlying stigma (Canada FASD Research Network 2011[1]). Research on FASD and stigma has shown that individuals with FASD, and women who drink when pregnant, can experience stigma. This stigma can produce barriers (to accessing care) for expectant mothers who might need assistance leading one report to recommend: "Careful planning is required to avoid unnecessary repercussions for women who drink alcohol and for families affected by FASD" (Burgoyne 2005). The role of stigma cannot be overstated when discussing appropriate interventions, prevention and care practices for those with FASD. As a marginalized group there is a need to think carefully and creatively to provide the accommodation in service individuals are entitled to by law.

Research on Police Training

Prior to discussing the research findings I will offer a short overview of the literature on training which will be taken up again when analyzing interview material. Over the past 30 years, considerable attention has been paid to the *science* of training including texts focused on efficacy and evaluation (Salas and Cannon-Bowers 2001). Klinger (2012) discussed the role of police training as a mechanism to ensure transparency and accountability in police practice. In this chapter, the notion of accountability will be blended with the notion of ethical obligation. In addition to the need for accountability, there is also a need to establish training practices that are effective, and where efficacy can be tested through officer buy-in.

[1]http://fasdprevention.wordpress.com/2011/04/20/thinking-about-fear-based-vs-support-based-awareness-campaigns/.

Martinez (2012) discusses how "hands-on" training best resonates with officers. Despite the research on police training, Klinger (2012) argues "little research has been undertaken that carefully examines the effects of police training on officer's behavior in the field" (Klinger 2012, p. 120). Given the high costs of training (Cordner and Shain 2011) and the relative lack of research on police training, researchers should be mindful when officers indicate that their training does not resonate as it marks not only a loss of investment but also indicates that learned practices may not be implemented. This is an especially important consideration when discussing training practices that focus on training police to work with vulnerable subjects.

In most locations, police officers are subjected to a formal training process in a policing academy to understand normative police behavior (Conti 2011). As Karp and Stenmark (2011) discuss, police training blends new approaches to police work (innovation) with traditional policing expectations, which can be contradictory forces at times. Accordingly police must navigate the historical expectations of their organization and the new challenges they face in the community. Following this formal training period officers are expected to attend in-service trainings; this is coupled with the ongoing need for police to continue to professionally develop on their own (Kordaczuk-Was and Sosnowksi 2011). This obligation to ongoing training, coupled with the challenge to be innovative and voluntarily professionalize, will be taken up in the findings section. The obligation to self-train should be considered with recent research that found that the majority of officers reported an interest in professional development that included formal-training and self-development (Kordaczuk-Was and Sosnowksi 2011).

In their review of police perceptions of training, Schulenberg and Warren (2009) include both formal and informal mechanisms of training as central to policing practices. As most officers are expected to respond to situations as they happen, there is a need for informal networks, as well as effective feedback. Snook et al. (2012) reported that police officers are keen to receive critical feedback from supervisors as they become trained in particular skills. Research published by the Federal Bureau of Investigation (McDermott and Hulse 2012) indicates that junior police officers benefit from a blend of positive feedback when they are performing well and corrective feedback when they are underperforming.

It is well understood that police are often expected to engage in crisis management and to work with diverse populations with specialized needs. Bard and Berkowitz articulated in 1967, "[I]ndeed, it has been estimated that 90 per cent of a policeman's [*sic*] function is in activities unrelated to crime control or law enforcement" (Bard and Berkowitz 1967, p. 315). Bard and Berkowitz were promoting a police training program in family crisis intervention. This approach to train police in crisis management, they argued, was not intended to turn police officers into mental health workers. Rather, it was intended to help police be better equipped to do the work they were already doing at the front-line (Meyer et al. 2009). These findings were supported by similar work conducted by Pearce and Snortum (1983) in which officers trained in crisis intervention felt they had better handled cases following their training. More recently, research on crisis intervention training found that police who attended this specialized training reported being more equipped to manage someone with mental health issues after taking the specialized training (Ritter et al. 2010).

This chapter makes a similar case for specialized training on FASD for officers. This training could also be modified to have broader applications for police encounters with individuals with Acquired Brain Injuries and other cognitive challenges. A lack of training for this group of clients with particular needs might be seen as unethical if one considers the potential negative outcome they might suffer because of a cognitive disability. That being said, police must balance this particular training need relative to the many other training requirements. As this is a balancing act, my goal is to make an ethical case for more training which is informed by research with frontline justice professionals including police officers. The following section will present qualitative research methods and findings as they relate to stigma and need for further training.

Research Findings

Methods

Findings and recommendations in this chapter are based on nearly 100 interviews with justice professionals and community workers from a cross-section of metropolitan, medium, small and remote police stations in Saskatchewan, a Canadian province with a population of one million. Interview questions (see Appendix A) focused on how police (and other justice professionals) gain understandings of FASD. For the purposes of this chapter, I will focus primarily on the police interviews.

Interviews were conducted with municipal police and the Royal Canadian Mounted Police (RCMP). The RCMP are the national police force, and the cadet training centre is located in Regina, Saskatchewan. The RCMP are responsible for provincial policing in the province including most rural and small communities (of less than 15,000 people). In the larger cities, there are municipal police forces. To account for different policing experiences, both RCMP and municipal police were interviewed. For an overview of policing demographics in Canada see Table 1.

Permission to conduct research with police was secured through police leadership. This body facilitated access to police stations based on criteria provided by the researcher to include a cross-section of police offices including: rural, small, medium and urban settings. A total of eight locations were visited and 34 interviews were conducted with frontline police in these locations. Interviews were conducted on site in a private location. Although each officer was on shift, and a frontline worker, all 34 interviews were completed in one sitting. All officers were consented, and interviews were recorded. A semi-structured interview guide was used to collect comparable data. The average interview time was approximately 15 min.

Table 1 Populations and Policing in Canada and Saskatchewan

	Canada	Saskatchewan	Source
Total Population	34,880,500	1,080,000	Statistics Canada, (2012)
RCMP Officers	29,235	1,393	Royal Canadian Mounted Police, (2012); Government of Saskatchewan, (2012)
Municipal Police Officers	40,636	1,033	Burczycka, (2012)
Total Number of Officers	69,539	2,298	Statistics Canada, (2012); Burczycka, (2012)

Source: Glowatski, K (2013)

Regarding the demographics of the interviewees: all officers were classified as frontline staff and had 9.2 average years of service. Most officers were seasoned and our sample included officers with over 25 years of experience, and rookies who did the interview during their first day on the job. Most interviewees were men, which is reflective of the overall demographic in the profession. Burczycka (2012) reported that upwards of 80 % of Canadian police officers are men. Most of our interviewees were from Saskatchewan but some did have experience policing in other locations in Canada.

The following section will discuss key findings of the research as it relates to the legal and ethical issues surrounding FASD to include: (1) an overview of the baseline knowledge officers have of FASD along with their perceptions surrounding prevalence, (2) stigma and the negative connotation officers have of FASD, and (3) discussion about the need for dedicated FASD training and specific ideas officers have about training.

Officer Knowledge and Prevalence of FASD

Prior to conducting interviews we were cautioned that frontline police might have little knowledge of FASD. However, our interviews revealed that most officers had a baseline understanding of FASD. Most officers could describe elements of the condition including: FASD is the result of maternal drinking, impacts are lifelong and can including cognitive challenges. During the course of the interviews, we came to find out that many officers did not recall elaborate formal training on FASD, but rather acquired much of their understandings from "hands on" experiences and/or discussion with fellow officers. We were often told about informal exchanges after an incident with a client or casual conversations. One officer commented that "anything you come across with this, you kind of share, over a coffee or conversation." It is good that there is active discussion about FASD in the police office but the accuracy of this type of information exchange is unknown. Moreover, some officers see a value in their collective experience over formal training.

When asked how information about FASD is transferred to officers one interviewee responded:

> I mean we're all doing the same job so we all see the same thing and just through observation when we're on calls together, or just talking amongst each other. There's not a whole lot of scholastic conversation about FASD.

This officer highlights the suspicion many expressed when they discussed training. Namely that there was a presumption that a "scholastic conversation" would take place—here a scholastic conversation marks training that is academic and divorced from the realities of frontline work.

Most officers expressed faith in "hands on" experiences. For example, one interviewee commented that limited information was transferred to officers about FASD but that this was not really a concern because, "it's basically the people we deal with regularly or we know their issues." This comment is encouraging in one respect in that the officers are getting to know their clients and trying to best understand their "issues." However, in the absence of formal training, police are not in a position to best understand these clients' "issues" since FASD is experienced on a spectrum, and therefore "issues" are often changing. Therefore the "issues" for one person with FASD can be numerous and divergent.

Despite different levels of understanding about FASD, there was one thing that many officers seemed to agree on: FASD was a consideration in their daily frontline police work. Most also agreed that individuals with FASD required patience on the part of the officer because these clients do not process information in the same way which can be frustrating for police. It should be noted that many officers were compassionate about clients with FASD. When asked the advice they would give to fellow officers working with a client, the constant refrain was: patience. One officer outlines the experience of working with clients thusly:

> I think we need to be aware of it, and considering it in our investigations, we get… I guess, I don't know which word, maybe [be] empathetic towards it and try to understand it more so we don't lose our patience or don't forget about it and then not understand why someone is the way they are or fault them for it I guess – it's not their fault of their own I guess.

This officer's choice to use the language of empathy highlights the experience of working with clients that do not understand the frontline encounter. The officer's comment was echoed by other justice professionals we would later interview who expressed exasperation when discussing the ways their clients would misunderstand expectations when they were embroiled in the legal system with outcomes that could include: violations of probation, jail time, and/or loss of custody of their children.

The challenge of the frontline experience is that much more alarming if one considers the perceived prevalence of FASD relative to training. One officer commented, "I think a majority of our clients suffer from FASD whether they know it or they don't."

Stigma and Negative Connotations

Most officers could recall working with a client that they believed to have FASD and/or had disclosed that they had FASD. However, for many officers, FASD raised suspicion or concern as officers regularly reminded us that FASD could not be a "get out of jail free card." This level of suspicion about the condition may be a result of ongoing, sustained contact with clients that are challenging to work with. However, it is also reflective of the stereotypes and misunderstandings surrounding FASD. That being said, some officers were aware of the stigma surrounding the condition and the challenge the stigma can present to an individual:

> Officer: They don't understand—it has a terrible stereotype.
>
> Interviewer: Stigma?
>
> Officer: Yeah it's terrible, so.... I don't know how they're supposed to feel about saying they are diagnosed with it and I don't understand the disease enough to know if they understand what that diagnosis means.

This officer's comment shed light on the nuanced experience of frontline encounters in which some police are aware of the potential stigma that FASD can carry, and the stereotypes that can result from that stigma. Here, the officer is also raising the compounded challenge: a person with FASD can experience stigma because of the diagnosis, which then raises concern about how much this person understands their diagnosis *and* the potential associated stigma when they are diagnosed.

During one interview, an officer was particularly candid. After stating that there was little to no training on the topic, they continued:

> You get terms thrown around and that kind of stuff. Like I said a lot of the time we deal with the same people. You tend to get the understanding or belief that they are suffering from FASD and it becomes very negative. Oh that "FASD bastard" or whatever. In that kind of sense, it's never a positive reflection or a beneficial discussion about it.

These negative understandings of FASD are framed here as a result of ongoing contact. And from that negative contact, and perhaps frequent contact, the discussions about FASD that follow are "never a positive reflection." In the absence of formal training and in light of ongoing and sustained contact with troubled clients, police are left to navigate the terrain of a brain injury with limited resources. An officer with eight years experience, when asked about training stated matter-of-factly, "there really hasn't been anything. We haven't received any new training, we haven't received any refresher training."

Training

When asked what resources would best assist them in their frontline work, most officers answered education and/or training. Officers were quick to caution against so-called scholastic discussions of FASD and indicated an interest in a "non-clinical explanation … of what to expect" when they encountered someone with FASD. Officers expressed a need for training that would be applicable in frontline encounters and were quite detailed in the type of training they believed would be most effective. In what follows, I will offer an overview of the suggestions made by officers. I offer these suggestions in part because they represent what officers believe will be helpful at the frontline.

In speaking with officers about resources they thought would be most useful at the front-line, the majority answered that education was crucial. Unfortunately, some felt that the obligation fell on the individual. One officer commented:

> educate yourself. Unfortunately we don't receive nearly enough of it...and its not just FASD it is other mental issues that we deal with that we don't have virtually any training in. But specifically with FASD we need to take it upon ourselves to educate ourselves because unfortunately we don't receive it from anywhere else.

It was clear that self-education and informal networks were the primary mechanism of information. It should be noted that a few officers did mention online training requirements. Officers rejected this form of training as it is often done in conjunction with their expected regular police work. This could result in a two-window approach whereby officers would complete police work in one window screen and "participate" in a training session in the other window. This type of Internet based training was deemed ineffective. One officer commented "95 % of us cheat."

There were alternatives to mandatory online training and self-education by "googling it." Some officers commented that a brochure or handouts would be helpful, as it could be something they could put in their duffle bags. However, one officer was quick to point out the limited value of brochures:

> Everything we're given gets thrown in the bag. We got enough stuff on our belt when we walk down the street. I'll be honest with you, I'm not going to pull out the brochure when I am talking with someone. But if we get that training in the classroom, with the live scenarios, where we are actually in a controlled situation...with professional people that do the acting and so we deal with mental health scenarios and armed robbery scenarios all that kind of stuff and that's how cops learn....if you give us something where its hands-on and you force us to think about it that'll come back in our training. Our motto is you are only as good as your training. How you act in your training is how you will act on the street.

This officer believes that a practice-based, "hands on" approach will have the most impact. One might further extrapolate that if a police agency takes the time to create a scenario-based model the participating trainees will see that the issue is understood to have some importance in the area.

Some officers expressed an interest in training that would be conducted off-site (for example at the police college). However, the time and resources available to agencies makes it quite challenging to send officers "away" for training aside from annual certifications. For local training, some officers expressed concern about the role of experts coming into the police station (without a background in policing) alongside the challenge of permeating all sections of the police force through short briefings held in the police station. Moreover, officers cautioned about bringing in outside "experts" to discuss FASD as front-line officers can quickly discredit them because these experts do not understand frontline police work. Some shared stories of advocates coming to the police station and having unrealistic expectations of what police can do at the frontline and/or naive understandings of what frontline work entails. Officers were quick to remind the research team that they are neither doctors nor social workers and, as such, they need information and strategies appropriate for police not highly medicalized descriptions. A key example of this would be to pay less attention to the language and medical elements of the cognitive disability and place the emphasis on the potential behaviours it can produce. With behaviors in mind, officers can imagine scenarios, or draw on previous experiences, and rethink the encounter if they understand a particular behavior is related to a disability rather than evidence of being openly defiant or aggressive.

Recommended Strategies

It is important to identify when front-line police officers make pointed arguments about how to effectively deliver material and ways to ensure that material permeates the front-line. One officer we interviewed enthusiastically endorsed partnered training in which a representative from the community or community-based organization would present material with another police officer. Here the collaboration between that officer and the community member might assist with getting frontline officers to engage with the session. Others suggested the possibility of having someone with FASD speak to the police as part of a simple Question and Answer session. This raises many concerns for advocates as a so-called high-functioning individual with FASD could give police a misconception about FASD and how it is experienced on a spectrum. Although police had many great ideas there is a need to balance them out with realistic approaches that will help further understanding about FASD without accidentally doing a disservice to those living with FASD who do encounter the justice system. In the section that follows, I will draw on officer's ideas for training practices that could allow for effective knowledge mobilization that balances out the potential negative impacts of the proposed idea.

Short Briefings

In the interviews conducted, officers not only expressed frustration with clients with FASD but also articulated the need to offer these clients different forms of service. While most officers were aware that clients with FASD can have a cognitive challenge, they also could see the need for more patience with these clients. However, in the absence of formal training many expressed frustration and argued for the need for more information about FASD and strategies to better work with these clients. The targeted suggestions offered by interviewed officers could be translated into training practices that could directly impact frontline practice.

Officers had inexpensive suggestions about how to transfer knowledge about FASD to all involved in the police agency, and how to best secure buy-in.

From the discussions with police it seems that a short shift briefing could be one of the first steps in taking up the issue of FASD with front-line officers. Additionally, to address the challenges of officers feeling like they are expected to become social workers or accept excuses for behavior, it might be best if officers are involved in delivering the briefing. Each officer comes to policing with a wide-range of skills and experiences, and many police officers are entering this line of work as a second or even third career path. These previous career paths can include complementary field experiences including health, social work and education. It also means officers can have university or professional degrees. Some officers have served in multiple communities and can also draw on that rich background of diverse community needs. With this in mind, it seems that each police office could survey the skill-set of the officers in a unit and identify key individuals that have training or experience working with individuals with FASD. These officers could then be put in contact with a local FASD expert from a community organization or even local parents who would be willing to come and speak to police. The police officer and the community member could then work together to create a short briefing. The officer could help formulate material in a way that would resonate with officers and the community member could complement and expand the officers' understandings of FASD. This partnership would allow police in the station to identify that officer as a resource but also would have the opportunity to ask a local expert questions during the briefing.

Short pre-shift briefings are already an established practice in many police organizations. The recommendation here is to partner in-house expertise with an agency to deliver the briefing. It might take a bit of time and work to have the police liaise with the organization and adapt the material so that it feels more appropriate for the frontline; it will also take a bit of time to cycle through all the shifts.

Scenario-Based Training

As discussed earlier, scenario-based training is a proven practice and one that police officers find rewarding. Scenario-based trainings allow officers to act out a response to a particular situation and includes corrective feedback from trainers and colleagues. Scenarios can involve officers acting out the roles with a script that drives them to resolve a situation (from an active shooter to a mental health client encounter). In some scenarios, officers learn verbal strategies and in other scenarios they learn weapons-based responses. Of note, officers that receive these trainings undergo a shared group experience and often must troubleshoot a situation as an individual and/or a group. Scenario-based trainings on FASD could employ professional actors to portray a client in crisis and officers could learn different strategies to assist that individual. Similar to briefings, these could be collaborative efforts that draw on the expertise in communities.

Officers expressed frustration with lengthy trainings and endless power point slides. In a short 15-min briefing, much material could be covered including the key elements of FASD and how these translate into specific behaviors officers might encounter. Officers nearly unanimously argued that patience was a key factor when dealing with FASD clients, as such it is important to discuss strategies to help maintain patience and strategies for effective communication with FASD clients in a group setting where specific tactics could be explored. In short collaborative briefings or scenario trainings officers could be given small nuggets of information about FASD and then have time to discuss that particular characteristic of FASD relative to frontline experiences which would allow for accurate information about FASD to be parlayed to officers while also giving them time to think about FASD within the real-world setting they experience.

Returning to the literature on crisis training, officers did express reservations about being turned into social workers, but most also understood that much of their work involves dealing with individuals in crisis. Therefore, it is likely not surprising that officers are requesting training to deal with these clients in crisis; training that might leave them better equipped to manage the situations they find themselves in on a regular basis. These crisis situations are often with the same clients who lack appropriate community supports and therefore are having increased contact with police. A short briefing or innovative scenario-based practice could speak to the needs expressed by officers for more training.

Based on research findings, and in light of increased attention for the need to address the needs of marginalized individuals in the justice system, this chapter recommends that police agencies undertake collaborative efforts to create appropriate training programs about FASD. Officers are requesting training, and there is a need for training. The closing section will make the ethical case for police agencies to undertake mandatory training.

The Ethical Case for Training

Research presented here indicates that officers have empathy for clients with FASD but also harbor frustrations that can fuel stigma and stereotypes. In the absence of formal training, officers are not best equipped to work with clients that have FASD. Officers expressed the need for more training on FASD and also had ideas about to effectively execute that training. Since officers argued that FASD was a regular part of their frontline experience, they see a need for more training to work effectively with clients. In what follows, I will amplify that need by highlighting other considerations when thinking about FASD and the justice system.

Many individuals with FASD are likely to not fully understand the legal process, and therefore can be susceptible to victimization at various times in the justice system. Fast and Conry (2004), drawing from the work of Perske, argue that individuals with FASD share similar challenges as those experienced by individuals with a mental disability including "a desire to please people in authority, an inability to understand abstract concepts, confabulating [mixing facts and fiction], taking blame quickly to 'get out of the situation and go home', and being victimized by the plea bargaining of an accomplice" (Fast and Conry 2004, p. 163). Fast and Conry call on all aspects of the justice system to be more accountable to this group of vulnerable subjects. I would add that police are the first point of contact into the justice system and training must start there. Without training to recognize these characteristics, officers may inadvertently attribute guilt to an individual because they cannot recognize the characteristics of the disorder—this misunderstanding might limit their critical capacity to also recognize victimization.

Contemporary training practices for police include language of "client-centered" approaches, problem-solving strategies for officers and inclusion of content that discusses root causes of criminal activity. Police are expected to make connections in their community, in part, to understand their community's needs. This attention to making police officers more embedded in their community is an effort to deliver a more effective version of the justice system. This chapter argues that people with different abilities, such as FASD, require special consideration so that the justice system is both effective but also *just*. For that effective and just system to be executed, there is an obligation on the part of the state and its agents to undertake appropriate training to best meet the needs of all citizens—anything less is unethical.

Conclusion

Individuals with FASD can have unique needs. When some of these individuals encounter the justice system, their needs and abilities must be recognized and met. Individuals with FASD experience the condition on a spectrum and as such there is not a one-size-fits-all solution when discussing FASD and frontline policing. However, one thing is for certain: for those individuals that experience a cognitive disability as part of their FASD there must be an *awareness* of the challenges they are facing when they encounter the justice system. An individual with a brain injury will understand the justice system differently and that difference must be accounted for—anything less is unacceptable because the stakes are too high in the justice system.

Research indicates that when police receive training on issues that stigmatize a population, that there is a change in police perception and this can impact the lives of those who are marginalized (Beletsky et al. 2005; Rhodes et al. 2006). The goal of this chapter is to highlight the need for more effective training and knowledge mobilization practices within police agencies and the justice system more broadly. The feedback of officers was used to outline the need and desire they have to acquire more understanding of FASD. Innovative collaborations between researchers and police agencies are being undertaken to create new training programs on FASD. Although this work is spotty, it serves as a starting point for innovative approaches to FASD training that can grow to include other marginalized populations.

Appendix A: Interview Guide

1. What is your understanding of Fetal Alcohol Spectrum Disorder?
2. How prevalent is FASD in your area?
3. How did you acquire this knowledge or understanding?
4. Is FASD considered to be an important consideration in policing in your area?
5. What if any local agencies are called upon to assist with FASD client?
6. How is FASD information or knowledge transferred to local officers?
7. Is FASD training conducted in your police office? If yes, please explain the training.
8. Is there a particular local network of FASD expertise in the area?
9. Are there any resources offered by province or federal government to assist police with local FASD networks or practices?
10. What, in your opinion, is the role of police in FASD client care and management?
11. Is there someone in your community that you believe you can turn to for information if needed?
12. If there were one suggestion you could make to have an impact on those who work with FASD clients what would it be?
13. In addition: What resources do you think would be helpful to assist officers who do come into contact with individuals with FASD?
14. Can you say how long you have been on the force?
15. During your time on the police force can you think of a time when you had to deal with a client with FASD? If yes, please explain how you knew and if this effected how you engaged with them?

References

Armstrong, Elizabeth M. 2003. *Conceiving risk, bearing responsibility*. Baltimore: The Johns Hopkins University Press.

Bard, M., and B. Berkowitz. 1967. Training police as specialists in family crisis intervention: A community psychology action program. *Community Mental Health Journal* 3:315–317

Beletsky, Leo, Grance E. Macalino, and Scott Burris. 2005. Attitudes of police officers towards syringe access, occupational needle- sticks, and drug use: A qualitative study of one city police department in the United States. *International Journal of Drug Policy* 16: 267–274.

Burczycka, M. 2012. Police resources in Canada, 2010. http://www.statcan.gc.ca/pub/85-225-x/2010000/part-partie1-eng.htm. Accessed 4 July 2013.

Burgoyne, Wendy. 2005. What we have learned: Key Canadian FASD awareness campaigns. Best Start: Ontario's Maternal Newborn and Early Child Development Resource Centre and Public Health Agency of Canada.

Canada FASD Research Network. 2011. Should awareness campaigns be fear-based or support-based? http://fasdprevention.wordpress.com/2011/04/20/thinking-about-fear-based-vs-support-based-awareness-campaigns/. Accessed 15 Aug 2014.

Church, S., Henderson, M., Barnard, M., and G. Hart. 2001. Violence by clients towards female prostitutes in different work settings: Questionnaire survey. *British Medical Journal* 322:524–525. doi:10.1136/bmj.322.7285.524

Conti, N. 2011. Weak links and warrior hearts: a framework for judging self and others in police training. *Police Practice and Research* 12:410–423. doi:10.1080/15614263.2010.536726

Cordner, G., and C. Shain. 2011. Editorial: The changing landscape of police education and training. *Police Practice and Research* 12:281–285. doi:10.1080/15614263.2011.587673

Cox, L.V., D. Clairmont, and S. Cox. 2008. Knowledge and attitudes of criminal justice professionals in relation to Fetal Alcohol Spectrum Disorder. *Journal of Population Therapeutics and Clinical Pharmacology* 15(2): e306–e313.

Drabble, L.A., N. Poole, N. Magri, N.M. Tumwesigye, Q. Li, and M. Plant. 2011. Conceiving risk, divergent responses: Perspectives on the construction of risk of FASD in six countries. *Substance Use and Misuse* 46(8): 943–958.

Fast, Diane K., and J. Conry. 2004. The challenge of fetal alcohol syndrome in the criminal legal system. *Addiction Biology* 9: 161–166.

Fast, Diane K., and J. Conry. 2009. Fetal Alcohol Spectrum Disorders and the criminal justice system. *Developmental Disabilities Research Reviews* 15(3): 250–257.

Gagnier, Karina Royer, Timothy E. Moore, and Justice Melvyn Green. 2011. A need for closer examination of FASD by the criminal justice system: Has the call been answered? *Journal of Population Therapeutics and Clinical Pharmacology* 18(3): e426–e439.

Guralnick, Michael J. (ed.). 2000. *Interdisciplinary clinical assessment of young children with developmental disabilities*. Baltimore: Brookes.

Jones, K.L., and D.W. Smith. 1973. Recognition of the fetal alcohol syndrome in early infancy. *The Lancet* 302:999–1001

Karp, S., and H. Stenmark. 2011. Research article: Learning to be a police officer. Tradition and change in the training and professional lives of police officers. *Police Practice and Research* 12:4–15. doi:10.1080/15614263.2010.497653

Klinger, David A. 2012. Police training as an instrument of accountability. *Saint Louis University Public Law Review* 32(1): 111–121.

Kordaczuk-Wąs, M., and S. Sosnowski. 2011. Police in-service training and selfeducation in Poland. *Police Practice and Research* 12:317–324. doi:10.1080/15614263.2011.563967

Lemoine, P., H. Harousseau, J.P. Borteyru, and J.C. Menuet. 1968. Les enfants de parents alcooliques–anomalies observées: à propos de 127 cas. *Ouest Médical* 21: 476–482.

Martinez, L. E. 2012. The police response to critical incidents in academic institutions. *Journal of Police Crisis Negotiations* 12: 69 –77. doi:10.1080/15332586.2012.646900

Meyer, J.F., Paul, R.C., and D.R. Grant. 2009. Peacemaking criminology: A special issue – part II: Peacekeepers turned peacemakers: Police as mediators. *Contemporary Justice Review* 12:331–344. doi:10.1080/10282580903105897

McDermott, P.J., and D. Hulse. 2012. Corrective feedback in police work. *FBI Law Enforcement Bulletin: Federal Bureau of Investigation* 13–17. Retrieved from http://www.fbi.gov/statsservices/publications/law-enforcement-bulletin/june-2012/focus-on-training

Oldani, Michael. 2009. Uncanny scripts: Understanding pharmaceutical emplotment in the aboriginal context. *Transcultural Psychiatry* 46(1): 131–156.

Pearce, J.B., and J.R. Snortum. 1983. Police effectiveness in handling disturbance calls: An evaluation of crisis intervention training. *Criminal Justice and Behavior* 10:71–92. doi:10.1177/0093854883010001005

Poole, Nancy. 2011. Bringing a women's health perspective to FASD prevention. In *Fetal Alcohol Spectrum Disorder: Management and policy perspectives of FASD*, ed. Riley, E.P., S. Clarren, J. Weinberg, and E. Jonsson. Weinhem: Wiley-Blackwell.

Popova, Svetlana, Shannon Lange, Dennis Bekmuradov, Alanah Mihic, and Jurgen Rehm. 2011. Fetal Alcohol Spectrum Disorder prevalence estimates in correctional systems: A systematic literature review. *Canadian Journal of Public Health* 2011: 336–340.

Rhodes, Tim, L. Platt, A. Sarang, A. Vlasov, L. Mikhailova, and G. Monaghan. 2006. Street policing, injecting drug use and harm reduction in a Russian city: A qualitative study of police perspectives. *Journal of Urban Health: Bulletin of the New York Academy of Medicine* 83(5): 911–925.

Ritter, C., Teller, J.L.S., Munets, M.R., and N. Bonfine. 2010. Crisis intervention team (CIT) training: Selection effects and long-term changes in perceptions of mental illness and community preparedness. *Journal of Police Crisis Negotiations* 10:133–152. doi:10.1080/15332581003756992

Rutman, Deborah. 2013. Voices of women living with FASD: Perspectives on promising approaches in substance use treatment, programs and care. *First Peoples Child & Family Review* 8(1): 107–121.

Salas, E., and J.A. Cannon-Bowers. 2001. The science of training: A decade of progress. *Annual Review of Psychology* 52:471–499. doi:10.1146/annurev.psych.52.1.471

Salmon, A., and S.K. Clarren. 2011. Developing effective, culturally appropriate avenues to FASD diagnosis and prevention in northern Canada. *International Journal of Circumpolar Health* 70(4): 428–433.

Schulenberg, J.L., and D.M. Warren. 2009. Content and adequacy of specialized police training to handle youth-related incidents: Perceptions of trainers, supervisors, and frontline officers. *International Criminal Justice Review* 19:456–477. doi:10.1177/1057567709340403

Sinclair, Bob. 2004. Commentary on 'The challenge of fetal alcohol syndrome in the criminal legal system' by D. K. Fast & J. Conry. *Addiction Biology* 9(2): 167–168.

Snook, B., House, J.C., MacDonald, S., and J. Eastwood. 2012. Police witness interview training, supervision, and feedback: A survey of Canadian police officers. *Canadian Journal of Criminology and Criminal Justice* 54:363–372. doi:10.3138/cjccj.2011.E.13

Smith, F.M., and L.A. Marshall. 2007. Barriers to effective drug addiction treatment for women involved in street-level prostitution: a qualitative investigation. *Criminal Behaviour and Mental Health* 17: 163–170.

Steadman, H., F. Osher, P. Robbins, B. Case, and S. Samuels. 2009. Prevalence of serious mental illness among jail inmates. *Psychiatry Services* 60(6): 761–765.

Stewart, Michelle, and Krystal Glowatski. 2014. Front-line police perceptions of Fetal Alcohol Spectrum Disorder in a Canadian province. *Police Journal: Theory, Practices and Principles* 87(1): 1–17.

Togher, L., S. McDonald, and C. Code. 2004. Training communication partners of people with traumatic brain injury: a randomized controlled trial. *Aphasiology* 18(4): 313–335.

Vancouver Police Department. 2013. *Vancouver's mental health crisis: An update report.* Vancouver: Vancouver Police Department.

Williams, W., G. Cordan, A. Mewse, J. Tonks, and C. Burgess. 2010. Self-reported traumatic brain injury in male young offenders: A risk factor for re- offending, or mental health and violence? *Neuropsychological Rehabilitation* 20(6): 801–812.

Wilson-Bates, Fiona. 2008. *Lost in transition: How a lack of capacity in the mental health system is failing Vancouver's mentally-ill and draining police resources*. Vancouver: Vancouver Police Department.

Part IV
Social and Policy Issues

Social Care and Fetal Alcohol Spectrum Disorders in the UK

Alison McCormick

How effective is the UK Social Care system in meeting the needs of families with children who have been prenatally exposed to alcohol? In this chapter I will briefly review the legal framework and how local authorities are reported to have met their responsibilities in delivering their Social Care role for these families.

Substance misuse is a key feature in social work with children and families across the UK. FASA, a family support charity in Belfast, have found that 70 % of their "Looked After Children" are living away from home as a direct result of parental substance abuse. Of the different forms of substance misuse, alcohol misuse is arguably the least understood area, especially in terms of its long term effects on families.

Hidden Harm (Advisory Council on the Misuse of Drugs (ACMD) 2003), published by the Home Office, was the first major UK attempt to look at the impact of drug and alcohol misuse on children. It has been estimated that while between 250,000 and 350,000 children in England and Wales lived in families where one or both parents had a serious drug problem (ACMD 2003), around 900,000 children are thought to be affected by alcohol misuse (Brisby et al. in Forrester et al. 2006). The *Munro Review of Child Protection* (Munro 2010, p. 12) emphasizes the importance of Social Care professionals maintaining a focus on such children:

> The biennial reviews of SCRs [serious case reviews] report recurrent problems in practice, e.g. children being invisible to professionals because the focus is on the parents, inadequate assessment of the dangers of parental problems of substance misuse, domestic violence, and mental illness, and fixed judgments not being challenged and revised.

There appears to be a discrepancy in the level of Social Care response for children affected by family drug misuse compared with those living in families who misuse alcohol; those cases involving alcohol typically come to the attention of social services later and follow a different pathway through Social Care (Adamson

A. McCormick (✉)
GSCC, Huntingdonshire, UK
e-mail: fasawareukeast@btinternet.com

M. Nelson and M. Trussler (eds.), *Fetal Alcohol Spectrum Disorders in Adults: Ethical and Legal Perspectives*, International Library of Ethics, Law, and the New Medicine 63, DOI 10.1007/978-3-319-20866-4_13

and Templeton 2012). The facts that (1) misused drugs are often illegal and alcohol is legal, and (2) the recommendations for screening and assessment in *Models of Care for Alcohol Misusers* (Department of Health and National Treatment Agency for Substance Misuse 2006) do not extend to their children (Cleaver et al. 2011), perhaps account for this. However, Forrester and Harwin (2004, p. 126) found that:

> Children are far more likely to have already experienced harm at the point of allocation if their parents misused alcohol and, worryingly, the harm was often more serious in nature. By contrast, most children affected by parental drug misuse were identified at an early stage and strong and co-ordinated protective action was taken before harm had set in.

Research in the USA found that cases involving alcohol misuse were given more chances of rehabilitation, although they were less likely to succeed than any other type of case (Murphy et al. 1991). A retrospective review of the social work cases in the study exposed a worrying trend of leaving vulnerable children at home in an over optimistic belief that the mother would overcome her difficulties. Placing the child out of home was more typical in situations of drug misuse.

Children With Fetal Alcohol Syndrome Disorders (FASD)

The United States Department of Health and Human Services report to Congress (Substance Abuse and Mental Health Services Administration (SAMHSA) 2002, p. 46) states that: 'Of all the substances of abuse, including heroin, cocaine and marijuana, alcohol produces by far the most serious neurobehavioral effects in the fetus'. Based on a widely accepted prevalence for Fetal Alcohol Spectrum Disorders (FASD) of one in 100 live births (Sampson et al. 1997), it is estimated that over 7000 babies with FASD are born in the UK every year to families where the mother drinks during pregnancy. However, this figure is believed to be conservative (May 2009).

Many of these children need permanent substitute families. Alarmingly, findings from the Foetally Affected Children's Team (FACT) study of children placed by Parents for Children (a specialist FASD adoption, fostering and family support agency now merged with TACT: www.tactcare.org.uk) (see Williams, Chap. 8, this publication) highlighted huge areas of unmet need and a lack of understanding of these children's functional deficits (Brocklesby et al. 2009). The project found that many children placed by Social Care had never had their needs comprehensively assessed in the context of their probable, hidden, prenatal alcohol damage. Local authorities appeared reluctant to fund assessments for these children, perhaps fearing, with some justification, that the identification of long-term support needs for the children could prove to be costly and complex. Families had not been given essential background information and so were struggling to access resources from education, health and social services systems. Prospective carers had been told they had to 'live with uncertainty' without any explanation of what this 'uncertainty' might mean or what services were in place to help. As the complexity of the child's problems emerged, they were often left struggling to cope alone.

As Cousins and Wells (2005, p. 381) state: 'It is unfair to expect families to take on the challenges and responsibilities of caring for a child affected by FAS [fetal alcohol syndrome] without a clear idea of what this may involve.'

The lack of knowledge of FASD among Social Care professionals was the most consistent feedback from parents involved in Dumont's (2011) research into strategies developed by parents to support their adopted children with FASD. Parents felt unsupported by Social Care's lack of understanding of the complexities of FASD, reporting that they had had to waste energy and time continuously fighting for their children's diagnosis, for professionals' awareness and support, and for adequate resources to meet their children's needs. Most parents stated that they had to explain to professionals what FASD was, and that they were not always believed. Some were told by a range of professionals that there was nothing wrong with their children, and that they were fantasizing. Others explained that their children's so-called 'bad behaviour' had been perceived as due to poor parenting rather than as organic brain damage. While involvement with professionals who had an understanding of FASD led to better support for their children, most participants expressed anger and disappointment at the lack of provision and support for their children, as well as at the lack of awareness of how FASD impacted on their children and their families.

Similar findings emerged from Turney et al.'s (2011) study, 'Social work assessment of children in need: what do we know?', and McCormick (2009) reported that, among 42 parents responding to her *Adoption Today* article, '91 per cent of adoptive parents felt professionals lacked awareness of their child's FASD disability and services required.' This is a situation that needs urgently addressing.

The Duties of the UK Social Care System and Children With FASD

Research among families of children with FASD has uncovered apparent failure by Social Care to follow legislation, regulations and policy, including The Children Act 1989, The Adoption Act 2004 and The Adoption Support Services Regulations, and has found discrepancies between placing and receiving local authorities (Brocklesby et al. 2009).

Local authorities, via Social Care, have a duty to safeguard and promote the welfare of children under the Children Act 1989 and to recognize the needs of carers under the Carers (Recognition and Services) Act 1995. According to The Children Act 1989 (Section 17 (10)), a child is taken to be 'in need' if they require services to achieve or maintain a reasonable standard of health or development and would suffer significant impairment without them. Once a child is identified as 'in need', they have an initial meeting with a social worker, following which the child and family are assessed and offered services (e.g. from Social Services, Health, Education or voluntary organisations) to help them resolve any problems or difficulties they are experiencing.

Section 10 of The Children Act 2004 states that services should:

- Enable families and carers to effectively support their children
- Intervene before a crisis necessitates statutory intervention
- Urgently resolve weak accountability and poor relationships between agencies
- Value, train and support all those working with children and young people so they can carry out their work competently and confidently.

For those children who need to be placed with adoptive families, Regulation 3 of *The Adoption Support Regulations 2005* outlines a comprehensive range of support services to be offered to meet the needs of adoption. Regulation 4 states that a local authority must carry out an assessment of their needs for adoption support services. Despite this, 90 % of 42 families who responded to McCormick's (2009) *Adoption Today* article reported that their children with FASD had not been properly assessed or had been placed for adoption without a diagnosis or an appropriate support package.

The tragic death of Victoria Climbié (age 8 years) at the hands of her relatives in London in the year 2000, as well as the subsequent Laming inquiry (Laming 2003), highlighted the failure of the services involved to communicate with each other and to protect her. Her horrific abuse within a private fostering family arrangement went unnoticed and unchecked; her death was however, subsequently responsible for enormous changes to safeguard children. Developed in response to the inquiry findings, the Every Child Matters strategy (Department for Education and Skills 2004) was intended to fuel a radical reform of public services in England, and to address the inadequacies revealed in how different services dealt with children at risk or in need. In England, the strategy was quickly enshrined in the Children Act 2004, and was described as 'one of the most significant changes in local children's services in living memory' (Lownsbrough and O'Leary 2005, p. 11). It prompted the development of integrated, accessible, personalized, child-centred services. The Government's aim was for every child from birth to age 19 years, whatever their background or their circumstances, to have the support they needed to make progress against five key outcomes: to be healthy, stay safe, enjoy and achieve, make a positive contribution and achieve economic well-being.

Although *Every Child Matters* offered a sweeping vision about children and young people's entitlements, full accountability for the delivery of the services that enable children, young people and their parents/carers to achieve these entitlements is delegated to local public services (Munro 2010). Difficulties arise due to the different laws and policies across counties in the UK, and prevent uniform practice and entitlement to services.

The Need for Professional Training

Unbelievably, there is currently no requirement for social workers studying for a social work degree to learn about substance misuse (Forrester and Harwin 2004), although the awareness of this need is growing. Galvani and Hughes (2010), in an exploration of social work students' knowledge and attitudes about substance use, found that 96 % of students thought that training on substance use was very or extremely relevant to their practice, but that 69 % had not received any. They concluded that alcohol and drug education needs to be included in social work training. (For training resources developed, see: Galvani and Forrester 2008, 2010; Forrester et al. 2006; McCarthy and Galvani 2010.) The Office of the Children's Commissioner of England (Adamson and Templeton 2012) also noted the lack of, and necessity for, pre- and post-qualification training.

However, while there is plentiful information in the Social Care literature on illegal drugs and how they may compromise parenting, there is comparatively little about alcohol misuse in pregnancy, the prevention, assessment and awareness of fetal alcohol spectrum disorders (FASD), and the complex impact that FASD may have on children and their carers. There is also little information regarding good social work practice specific to FASD, whether this be policy interpretation, management strategies or ways to enhance practice, assessment or support of families.

The FACT project (Brocklesby et al. 2009) noted a lack of understanding of FASD and all its complexities, implications and stages among professionals, and the families involved disclosed that social workers, midwives, health visitors and teachers did not have enough insight or training regarding the long-term consequences of alcohol misuse. Forrester and Harwin (2004) in their study of children exposed to parental substance misuse found that social workers were often isolated and ill prepared, and that 71 % of these families had no professional trained in substance misuse working with them. This resulted in the use of inadequate conceptual frameworks and a lack of understanding of how to help families.

The FACT study found that many social workers working with families of children with FASD:

- Undertook only superficial case histories
- Accepted information given by birth parents about maternal prenatal alcohol consumption at face value even when contradicted by other evidence on file (e.g. alcohol induced antisocial behaviour and witness statements)
- Wrongly attributed children's complex problems to other difficulties (e.g. attachment disorders, post-traumatic stress disorders or poor parental coping)
- Had not identified developmental discrepancies (e.g. between a child's expressive and receptive language) and their likely impact.

Assessment and Diagnosis

The significance of obtaining accurate information to maximize the child's outcomes in life for both diagnosis and prognosis is poorly understood (Brocklesby et al. 2009). Families of children with FASD are often passed from professional to professional, accumulating diagnoses, none of which fully addresses the young person's complexity.

Good assessment has a critical role to play in early intervention strategies, contributing to the effective and timely targeting of interventions. As Turney et al. (2011, p. 4) write:

> The importance of timely assessment is reflected in the current concern with very early or 'earlier' intervention… The rationale for this approach – that it is better to intervene before difficulties become established and potentially more severe – is hard to challenge.

A good Social Care assessment can improve home stability for children by preventing delay and helping to ensure the provision of appropriate and adequate support (Wade et al. 2010). However, there are many ways that a child with FASD can 'slip through the net' of commonly used assessment tools, and their support needs are often not picked up. For the Common Assessment Framework to have effective service provision outcomes for children with FASD, those assessing will require an in-depth understanding of the FASD-associated 'hidden' developmental disabilities (Armistead and McCormick 2011); also many FASD issues and complexities cannot be identified through generic 'tick box' formats (Cleaver et al. 2004, 2007; Mitchell and Sloper 2008; Munro 2010). FASD professionals, parents and researchers agree that most local authorities are not in a position to accurately assess and identify the needs of a child with FASD (Armistead and McCormick 2011; Dumont 2011; McCormick 2009; Turney et al. 2011). Turney et al. (2011, p. 14) caution:

> Barriers to quality in assessment…can operate at a number of levels…including whether or not the practitioner is competent, has the appropriate knowledge and confidence to carry out the required task and has the scope to do so within their individual caseload.

There are also concerns about the usefulness of existing assessment formats for recording the views of young people with FASD (Mitchell and Sloper 2008). While these young people may appear to the uninitiated to be articulate, intelligent and able to converse appropriately, this often masks developmental and cognitive delays and a very uneven neurological profile (Brocklesby et al. 2009). Their language and comprehension difficulties impact on their social functioning and capacity for judgment (Brocklesby et al. 2009). An assessor unsensitized to FASD may not be able to recognize or gather evidence pointing to the young person's lack of understanding, inability to differentiate between fact and fantasy, or the fragmented logic they use to interpret the world. With an unusually strong desire to please, many young people with FASD hide their poor memory recall by confabulating—inventing very inaccurate stories according to what they think the assessor wants to hear. They have no understanding of the consequences of what they may be saying. For example, they may unwittingly trigger Child Protection alarm bells. Inexperienced professionals may find it difficult to comprehend FASD issues, and may even hold parents responsible for the child's difficulties (Armistead and McCormick 2011; Brocklesby et al. 2009; Dumont 2011; McCormick 2009; Turney et al. 2011).

The FASaware UK (Armistead and McCormick 2011) survey of 38 parents (a 71 % response rate from 52 approached) revealed their perceptions and experiences of the Social Care system:

- Lack of appropriate understanding of FASD and the associated, complex, invisible impact on families
- Unwillingness to seek information from parents, FASD charities or medical personnel
- Refusals to assess children
- Stressful and time-consuming assessment processes
- Serious inaccuracies in records and assessments leading to unhelpful recommendations
- Unwillingness to correct inaccuracies.

Inaccurate recording and assessments have serious implications. All the studies reported in this chapter found significant errors in children's files and no means of amending them. The *Every Child Matters* drive towards joined-up services has led to the indiscriminate sharing of confidential and sensitive information via joined up databases. If this information is inaccurate, it can damage innocent people seeking help for their disabled child.

In the course of the FACT study (Brocklesby et al. 2009), researchers designed an assessment which took account of the issues and needs associated with FASD. As part of this assessment, a specialist social worker with knowledge of the family and child undertook a comprehensive assessment backed up by validated psychometric screening tests appropriate to the child's age and development. Wherever possible, the child was assessed in a familiar and local environment either at home or at school. The services of an occupational therapist, a psychiatrist, a paediatrician, a speech and language therapist, an education consultant and an advocate were also engaged where appropriate. A summary report was then sent to the main carers, and also to the local authority or the court. The multidisciplinary assessment was linked to an annual review and FACT's ongoing support system involving specialists suitable for the child's needs. The FACT multidisciplinary assessment provided ample evidence of the need for specialist services for children and young people with FASD, particularly special school placements, respite care and more information and support for carers.

Families' Experiences of Service Provision

Increasingly, children fetally affected by alcohol are permanently placed out of their family or find themselves living with extended family members, including aging grandparents. This can be an issue. Children and young adults with FASD have a spectrum of complex hidden needs and disabilities, and can be very challenging and exhausting to care for, sometimes resulting in parent/carer exhaustion and burnout. Brocklesby et al. (2009) found very high levels of carer stress and distress due to huge barriers in identifying and supporting fetally affected children, and to the extreme levels of supervision and care-giving skills they demand—emotionally, physically and mentally—24 hours a day, 7 days a week, 52 weeks a year. In some situations, the demands of children with FASD may make them vulnerable to further abuse and secondary disabilities.

The British Medical Association report on FASD (2007, p. 32) unequivocally identifies the service implications:

> Treatment for FASD requires the implementation of tailored management programmes and specialized support in the provision of healthcare, education and social services.

However, from Armistead and McCormick's (2011) survey of 38 UK parents/carers with parental responsibility (i.e. adoptive, birth and extended families only; no feedback was requested from carers of accommodated children for 57 young people with a diagnosed FASD, it was clear that many parents found themselves battling for services. These parents reported: fraught and hostile relationships with agencies; failure by Social Care to provide support and respond to parents' requests for help; and unwillingness by Social Care to seek information from parents, FASD charities or medical personnel. The survey highlighted the following issues:

- Evidence of enormous disparity between Government agencies' understanding of FASD as a 'disability' in relation to Government definitions. Only three of 57 children met Social Care's disabled children's services threshold criteria, yet 51 out of 57 children had been awarded high rate Disability Living Allowance (DLA). Twenty received some form of support via Social Care, mainly via Direct Payments
- None of the children had received respite care provided directly by Social Care post-adoption, even though some had previously received it when fostered
- Only one family (two children out of 57), who did not live on the UK mainland, reported receiving Social Care support services suitable for their needs.

When Social Care support was provided, parents/carers reported that it was often inadequate; for example, insufficient Direct Payments for families to organize their own appropriate help. One family was offered 90 minutes respite funding a month—equating to approximately £15—to be shared between their three children with FASD (Armistead and McCormick 2011). The family declined the payment saying that this would not cover even 1 hour's care for one child and, rather than supporting the family, would hinder them due to the time taken to manage the Direct Payments. In the FACT study (Brocklesby et al. 2009), two families each with three children with FASD reported that the only resolution offered by Social Care to the repeated challenging behaviour of one of their children, was to place them back into care without trying out any practical support or innovative thinking. While this may have alleviated the families' immediate stress, in the long term it would have compounded the relationship and attachment issues for their already complex adopted children.

Adopters were often expected to navigate their own way through disability and medical services to get their own and their child's needs met (McCormick 2009). Families reported that their child with FASD fell between two categories: respite care was judged by Social Services to be inappropriate and was usually inaccessible via adoption support budgets; yet, because the children lacked a 'visible disability', they were rarely classified as disabled therefore adoptive families were unable to access the short-break care funded through *Aiming High* (Department for Children, Schools and Families 2007).

Both Armistead and McCormick (2011) and Brocklesby et al. (2009) reported cases where requests by parents/carers for respite care had been misinterpreted by local authorities and child and adolescent mental health services (CAMHS) resulting in parents/carers being blamed and isolated. In some cases, professionals had questioned parents'/carers' commitment or ability to parent their children, and a few had culminated in Child Protection investigations. Four families from four different areas, each caring for three children with FASD, described how they had been accused by social workers (who were unqualified in medicine or mental health) of fabricating their situation and of having unresolved mental health issues themselves (Armistead and McCormick 2011). The allegations made by Social Care professionals against one mother following 'unreasonable requests for help' resulted in extended court proceedings, a major psychiatric assessment, financial hardship and nearly cost her home and marriage. This needless, traumatic process ended with her being judged as fit to adopt the three children with FASD who had been placed with her family.

An informal survey of responses from 42 adoptive families following the author's *Adoption Today* article (McCormick 2009; see also Table 1) revealed that:

- 80 % of families responding had not been able to access respite care despite their requests, even when this had been in place previously for their child's foster carers.
- More than half the adoptive parents/carers had either had children removed or feared their removal.
- More than a third had had child protection allegations made against them, allegedly due to Social Care's misunderstanding of the nature of FASD (i.e. children's concrete thinking, tendency to fantasize or to confabulate memories to please listeners, and their lack of understanding of consequences).
- Half of the carers confided they had had a relationship breakdown or partial breakdown within their own family which they attributed to the stresses of the placement and lack of support and understanding from agencies.

Finally, the studies found a lack of understanding within Social Care about the development of FASD across the lifespan. This has led to failures by Social Care professionals to put in place, prior to children's adoption, appropriate long term intervention and support plans which would prevent the later serious secondary disabilities associated with FASD (Brocklesby et al. 2009). While very young children may present with near typical development at the point of adoption, as these children grow older their difficulties become increasingly acute and challenging. McCormick (2009) found that most adoptive carers responding to her article had not been offered any post-adoption support because their child had been classified as having 'no problems' early on. There has to be acknowledgement by Social Care that families of children with FASD may need to access services

Table 1 Perceptions and experiences of the 38 parents of 57 children with FASD of the social care support they received (Armistead and McCormick 2011)

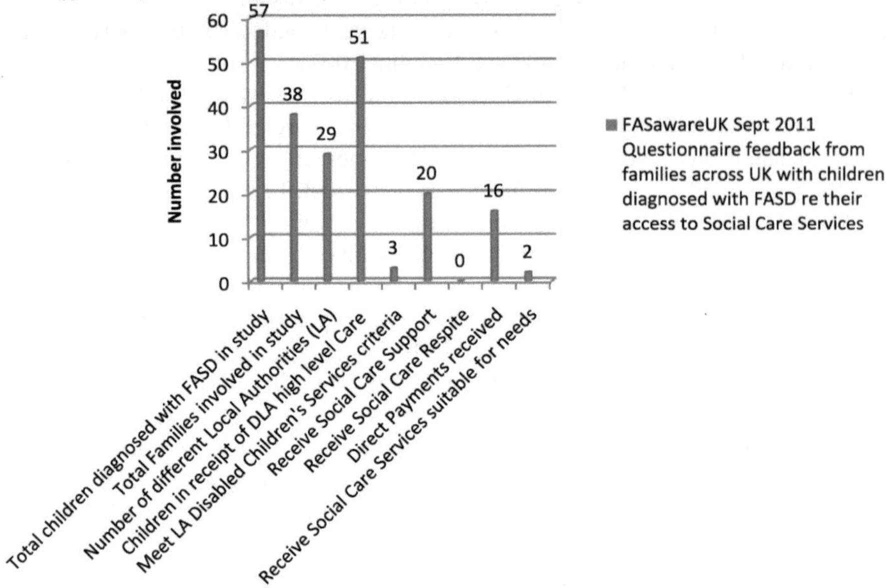

post-adoption. Currently, once an adoption order is granted, funding or access to an appropriate assessment is near impossible (Brocklesby et al. 2009). Families of children with FASD should not be abandoned with the sole, permanent responsibility for managing their children's acute problems for which most have been woefully ill-prepared (Table 1).

Summary and Conclusion

Most people with FASD feel like a 'square peg in a round hole', no service or facility fits their requirements, and they are grossly misunderstood. People with FASD, due to their inherent invisible brain dysfunction and disability, place demands on each of their environments—family, school and the wider community. If their needs are not understood and met, their behaviours can become out of control, bizarre, desperate, unpredictable, impulsive and dysfunctional. Many find themselves in corrective institutions and/or with poor mental health (Streissguth et al. 1997).

The compelling evidence outlined in this chapter suggests that the UK's Social Care model is not robust enough to prevent an avalanche of innocent children being born every year with FASD or to provide adequately educated professionals and appropriate support services for fetally affected individuals and their families. The families of all studied children reported Social Care's resistance to recognizing FASD as a disability according to the Government's definition and the Children in Need criteria, and to providing appropriate services and support as laid down by *Every Child Matters*. This is contrary to both national guidance and accepted good practice. Many families have had to go to significant lengths and through great crisis to care for their complex disabled child with minimal support.

It is vital that Social Care practitioners, management and strategic decision makers address the systemic denial of FASD and take action on alcohol misuse seriously, especially in pregnancy where there is a chance to alleviate the risks of having a child, or further children, with FASD and the impact of possible secondary disabilities for the person, their agency and Society. Social Care needs to take a positive lead on FASD—obtaining fully evidenced facts and recording accurate data; sending out strong messages regarding alcohol misuse; maximizing training on FASD; and developing effective and supportive policies (Munro 2010). They need to improve outcomes for families by: raising awareness of how to prevent and detect FASD; obtaining accurate multidisciplinary assessments; accessing funding for effective personalized support packages; and understanding the enormous lifelong impact this disability has on the person, their family, carers and society.

Alcohol consumption is a major problem for Social Care. Many cases tend to be 'top end' involving Care Proceedings or Child Protection. The costs are high, complex and involve a large amount of interagency liaison (often by professionals with insufficient FASD training and awareness) with very vulnerable and often disabled children (Forrester and Harwin 2004). In the Yukon Territory, Canada, the Government found that children with FASD from the care system are approximately four times more costly to care for, both financially and in care time, and are significantly more likely to require additional education, health and support than other adoptive children who do not have FASD (Alton and Evenson 2006).

However, if the Government does nothing to support families of children with FASD, the costs are likely to spiral in adulthood. In an American study, Streissguth et al. (1997) found that children with FASD were likely to be vulnerable to life events and to develop secondary disabilities in adulthood: out of a cohort of 415 individuals with FASD, 90 % had needed long-term support in the transition to adulthood; 60 % were or had been confined in mental health and criminal justice situations; and 50 % exhibited some form of sexually inappropriate behaviour. Children with FASD are therefore a group whom it is worth understanding and making an investment in at an early stage to reduce costs to Society.

This is a large agenda, especially in a time of great cutbacks, but alcohol misuse in the UK is out of control. It is one of the most important challenges and areas of future expenditure facing Social Care's policy makers, budget holders and practitioners this century. Tackling the UK's drinking culture will need an engaged, sustained and appropriately complex response to change attitudes, knowledge and behaviour (Stead et al. 2009). Young women who binge drink play 'Russian Roulette' with their pregnancies, while children with preventable and lifelong disabilities are born in enormous numbers every year. Multi-targeted actions at policy and service level are needed to support people in making changes.

In the UK there is still no legal requirement to notify and centrally record if a child is diagnosed with FASD (Adamson and Templeton 2012). This lack of central recording is failing our children by preventing them getting the support they rightly require and feeding systemic denial in services. Until this and our alcohol culture are addressed, children will continue to be born with a devastating and preventable disability. Ensuring a good childhood is hard enough (Layard and Dunn 2009), but for the increasing numbers of children born fetally affected in 'Binge Britain', Social Care need to take action to optimize the long-term wellbeing of an ever increasing number of innocent people and to provide services for a hidden section of society crying out for practical help with one of the most frustrating and complex conditions to manage.

References

Adamson, J., and L. Templeton. 2012. *Silent voices: Supporting children and young people affected by parental alcohol abuse*. London: The Office of the Children's Commissioner.

Advisory council on the misuse of drugs (ACMD). 2003. *Hidden harm: Responding to the needs of children of problem drug users*. London: Home Office.

Alton, H., and D. Evenson. 2006. *Making a difference: Working with students who have fetal alcohol syndrome*. Whitehorse, Yukon: Yukon Education, Government of Yukon.

Armistead, G., and A. McCormick. 2011. FASawareUK study of social care support (unpublished).

British Medical Association. 2007. *Fetal alcohol spectrum disorders: Aguide for healthcare professionals*. London: British Medical Association.

Brocklesby, E., J. Le Vaillant, A. McCormick, D. Mandelli, and M. Mather. 2009. Substance misuse in pregnancy: An unrecognised and misdiagnosed problem for a child. *Seen and Heard* 19(1): 22–32.

Cleaver, H., I. Unell, and J. Aldgate. 2011. *Children's needs—parenting capacity. Child abuse: Parental mental illness, learning disability, substance misuse and domestic violence,* 2nd ed. London: The Stationery Office.

Cleaver, H., S. Walker, and P. Meadows. 2004. *Assessing children's needs and circumstances: The impact of the assessment framework*. London: Jessica Kingsley.

Cleaver, H., D. Nicholson, S. Tarr, and D. Cleaver. 2007. *Child protection, domestic violence and parental substance misuse: Family experiences and effective practice*. London: Jessica Kingsley.

Cousins, W., and K. Wells. 2005. "One more for my baby": Foetal alcohol syndrome and its implications for social workers. *Child Care in Practice* 11(3): 375–383. http://ulster.academia.edu/WendyCousins/Papers/271217/One_More_for_My_Baby_Foetal_Alcohol_Syndrome_and_Its_Implications_for_Social_Workers. Accessed 6 Oct 12.

Department for children, schools and families. 2007. *Aiming high for disabled children: Better support for families*. London: HM Treasury.

Department for education and skills. 2004. *Every child matters: Change for children*. Annesley: DfES Publications.

Department of health and national treatment agency for substance misuse. 2006. *Models of care for alcohol misusers (MoCAM)*. London: Department of Health.

Dumont, L. 2011. *Exploring strategies developed by parents to support their adopted children with foetal alcohol spectrum disorder*. London: Goldsmiths University.

FASA. 2008–2012. Family matters. http://www.fasaonline.org/young-people/services/family-matters/. Accessed 1 Oct 12.

Forrester, D., and J. Harwin. 2004. Social work and parental substance misuse. In *Children exposed to parental substance misuse: Implications for family placement*, ed. R. Phillips. London: British Association for Adoption and Fostering.

Forrester, D., J. McCambridge, S. Rollnick, J. Strang, and C. Waissbein. 2006. *Child risk and parental resistance: Can motivational interviewing improve the practice of child and family social workers in working with parental alcohol misuse?*. London: Alcohol Research UK.

Galvani, S., and D. Forrester. 2008. *What works in training social workers about drug and alcohol use?: Asurvey of student learning and readiness to practice.* Coventry: University of Warwick/University of Bedfordshire.

Galvani, S., and D. Forrester. 2010. *Social work and substance use: Teaching the basics.* Southampton: Higher Education Academy. Subject Centre for Social Policy and Social Work (SWAP).

Galvani, S., and N. Hughes. 2010. Working with alcohol and drug use: Exploring the knowledge and attitudes of social work students. *British Journal of Social Work* 40(3): 946–962.

Laming, Lord. 2003. *The Victoria Climbié Inquiry: Report of an inquiry by Lord Laming.* Norwich: HMSO.

Layard, R., and J. Dunn. 2009. *A good childhood.* London: The Children's Society.

Lownsbrough, H., and D. O'Leary. 2005. *The leadership imperative: Reforming children's services from the ground up.* London: Demos.

May, P. 2009. Prevalence and incidence internationally. In *Fetal alcohol spectrum disorder (FASD): Across the lifespan (proceedings from an IHE consensus development conference 2009)*, ed. E. Jonsson, L. Dennett, and G. Littlejohn. Institute of Health Economics: Alberta, Canada.

McCarthy, T., and S. Galvani. 2010. *Alcohol and other drugs: Essential information for social workers.* Luton: University of Bedford.

McCormick, A. (2009) 'Caring for children with foetal alcohol syndrome' *Adoption Today.*

Mitchell, W., and P. Sloper. 2008. The integrated children's system and disabled children. *Child and Family social work* 13(3): 274–285.

Munro, E. 2010. *The Munro Review of child protection: Part 1—a systems analysis.* Norwich: The Stationery Office.

Murphy, J.M., M. Jellinek, D. Quinn, G. Smith, F.G. Poitrast, and M. Goshko. 1991. Substance abuse and serious child mistreatment: Prevalence, risk, and outcome in a court sample. *Child Abuse and Neglect* 15(3): 197–211.

Sampson, P.D., A.P. Streissguth, F.L. Bookstein, R.E. Little, S.K. Clarren, P. Dehaene, J.W. Hanson, and J.M. Jr Graham. 1997. Incidence of fetal alcohol syndrome and prevalence of alcohol-related neurodevelopmental disorder. 56(5): 317–326.

Stead, M., R. Gordon, I. Holme, C. Moodie, G. Hastings, and K. Angus. 2009. *Changing attitudes, knowledge and behaviour: Areview of successful initiatives.* London: Joseph Rowntree Foundation.

Stratton, K., C. Howe, and F. Battaglia. 1996. *Fetal alcohol syndrome: Diagnosis, epidemiology, prevention, and treatment.* Washington, DC: Institute of Medicine, National Academy Press.

Streissguth, A., H. Barr, J. Kogan, and F. Bookstein. 1997. Primary and secondary disabilities in fetal alcohol syndrome. In *The challenge of fetal alcohol syndrome: overcoming secondary disabilities*, ed. A. Streissguth, and J. Katner. Washington, DC: University of Washington Press.

Substance Abuse and Mental Health Services Administration (SAMHSA). 2002. *Report to congress on the prevention and treatment of co-occurring substance abuse disorders and mental disorders.* Washington, DC: United States Department of Health and Human Services.

Turney, D., D. Platt, J. Selwyn, and E. Farmer. 2011. *Social work assessment of children in need—What do we know?: Messages from research.* Bristol: School for Policy Studies, University of Bristol.

Wade, J., N. Biehal, N. Farrelly, and I. Sinclair. 2010. *Outcomes for children looked after for reasons of abuse or neglect: The consequences of staying in care or returning home, report to department for education.* York: University of York.

FASD in a Canadian Aboriginal Community Context—An Exploration of Some Ethical Issues Involving the Access to FASD Service Delivery

Lori Vitale Cox

Prenatal exposure to alcohol, PEA, has been shown in both animal and human research studies to have a deleterious effect on many aspects of development particularly neurodevelopment.[1] Brain injury from PEA can affect learning, memory, attention, language and social behaviour.[2] Recent research also indicates that pre-natal exposure to alcohol can result in changes to hormonal functioning and disruption of HPA system that affects cortical production and stress response regulation.[3]

Fetal Alcohol Spectrum Disorder (FASD) is the name given to describe the spectrum of effects due to pre-natal alcohol exposure. Primary diagnostic conditions along the FASD spectrum include Fetal Alcohol Syndrome (FAS) Partial Fetal Alcohol Syndrome (pFAS) and Alcohol Related Neurodevelopmental Disorder (ARND).[4] Secondary conditions or disabilities related to pre-natal exposure to alcohol were first recognized in the research literature by Streissguth.[5] These are exacerbated by the absence of diagnosis and subsequent accommodations for the FASD disability at home and in the community. Socio-economic factors and environmental adversity play a significant role in the clinical expression of both the primary and secondary conditions.[6,7,8,9] FASD is recognized as a permanent and significant disability that can affect functioning throughout the lifespan.

[1]Chudley et al. (2005).

[2]Riley Edward et al. (2003).

[3]Hellemans et al. (2010).

[4]Chudley et al. (2005).

[5]Streissguth and Kanter (1997).

[6]Abel (1995).

[7]Abel and Hannigan (1994).

[8]May and Gossage (2011a, b).

[9]Bingol et al. (1987)

L.V. Cox (✉)
Elsipogtog First Nation, New Brunswick, Canada
e-mail: lsdcox@nbnet.nb.ca

M. Nelson and M. Trussler (eds.), *Fetal Alcohol Spectrum Disorders in Adults: Ethical and Legal Perspectives*, International Library of Ethics, Law, and the New Medicine 63, DOI 10.1007/978-3-319-20866-4_14

Older youth and adults, especially those who lack FASD diagnostic health services and disability accommodation, often have problems with depression, suicide, panic disorders, addictions, incarceration, and institutionalization.[10] Children's problems include learning and behaviour dysfunction that (in the absence of recognition and support) often lead to school disruption: suspension, expulsion, early school leaving and addictions and trouble with the law.[11]

Despite these findings many babies born in Canada are still being exposed to alcohol in utero. Research indicates that approximately 15 % of women in all of the Canadian provinces reported that they consumed alcohol during their pregnancy.[12] This rate may be higher since 62.4 % of women report that they drank in the 3 months prior to pregnancy recognition.[13] Overall rates of alcohol consumption among aboriginal women seem to be lower than in the general population but binge drinking, associated with greater fetal risk, is reported to be more prevalent among aboriginal women who do drink.[14] Published research on prenatal alcohol consumption in aboriginal communities in Canada is limited but a few community studies have indicated that prenatal alcohol use may be problematic among some populations.[15,16]

Epidemiological data regarding the incidence of FASD in the Canadian population is lacking. We have not yet established a mechanism for surveillance or national capacity for screening and diagnosis. The incidence in Canada was previously approximated at 1 % of the population based on US data of 9.4 per 1000.[17] New US data from a large population study indicates considerably higher rates. Researchers reported an updated incidence estimate of 3.6 %.[18] Actual incidence rates of FASD among Canadian aboriginal populations are unknown but using the new general population estimates many thousands of aboriginal people are affected and lack services (Table 1).

---

[10]Ibid.

[11]Ibid.

[12]Health Canada. Canadian Perinatal Health Report, 2003. Ottawa: Minister of Public Works and Government Services Canada, 2003. Table F1.3 Rates of maternal alcohol consumption by province, p. 159.

[13]Public Health Agency of Canada. What Mother's Say: The Canadian Maternity Experiences Survey. 322 Ottawa 2009.

[14]Statistics Canada. 2001 Aboriginal Peoples Survey See also Statistics Canada. Canadian Community Health Survey—2000/01. Retrieved from:
http://www.statcan.ca/english/freepub/82-221-XIE/00502/tables/html/2156.htm In Nunavut it is reported that 51 % of women drink alcohol compared to 74 % of women in the general Canadian population. 30 % of those 12 years and over who do drink in Nunavut, binge drink 5 or more drinks per occasion compared to only 20.1 % binging in this way in general Canadian population.

[15]Muckle et al. (2003).

[16]Potyrala and Brewster (2009).

[17]May and Gossage (2001).

[18]May et al. (2014a).

Table 1 Projected rates of FASD

Population	Population	Prevalence rate (%)	Population FASD
Canadian[a]	35,540,400	3.6	1,279,454
Canadian births[b]	383,822	3.6	13,818
Aboriginal[c]	1,400,685	3.6	50,424
Aboriginal births[d]	33,616	3.6	1,210

[a]Retrieved from: Statistics Canada: http://www.statcan.gc.ca/daily-quotidien/140926/dq140926b-eng.htm
[b]Ibid
[c]Aboriginal Population 2011: Data retrieved from Statistics Canada: http://www12.statcan.gc.ca/nhs-enm/2011/dp-pd/aprof/details/page.cfm?Lang=E&Geo1=PR&Code1=01&Data=Count&SearchText=Canada&SearchType=Begins&SearchPR=01&A1=All&B1=All&Custom=&TABID=1
[d]Aboriginal Birth rate Information: Retrieved from Statistics Canada : http://www.hc-sc.gc.ca/fniah-spnia/pubs/aborig-autoch/stats-profil-atlant/index-eng.php#a621 In 2011 the aboriginal population was 1,400,685. According to Stats Canada the crude birth rate is 24 births/1000 population. We can project that there are roughly 33,616 babies born a year

In much of Canada access to FASD health service delivery is limited but in aboriginal communities, access to FASD health service delivery is especially problematic. There are no multi-disciplinary diagnostic FASD teams in all of Nunavut and limited access in Yukon and NWT although team diagnosis is recommended by Health Canada and the Canadian Guidelines for Diagnosis.[19] A report in 2007 on Inuit Public Health published by the National Collaborating Centre for Aboriginal Health at the University of Northern British Columbia noted[20]:

> FASD has been identified as a pressing health issue facing Inuit communities. Interviews with representatives from each of the northern regions identified FASD as one of the major issues they were concerned about in their region.

There also may be some health inequalities within indigenous populations. Stakeholders in Inuit communities reported that Inuit communities receive a disproportionately lower portion of FASD Federal funding than other aboriginal groups in Canada for FASD. They noted the general lack of Inuit-specific planning, programming, and resources for FASD specific activities.[21] Across all

[19]Salmon and Clarren (2011). There are diagnostic teams in Whitehorse and Yellowknife but most of Canada's north continues to lack FASD service delivery.

[20]National Collaborating Centre for Aboriginal Health, "State of the Knowledge: Inuit Public Health" University of Northern British Columbia, 2007; Prince George, BC.

[21]Ajunnginiq Centre, Fetal Alcohol Spectrum Disorder: An Environmental Scan of Services and Gaps in Inuit Communities. Ottawa: Ajunnginiq Centre, National Aboriginal Health Organization. 2006.

regions the following priorities were identified in regard to FASD health service delivery[22,23]

- Lack of diagnosis
- Lack of follow-up and coordination following diagnosis
- Lack of intervention and family support services after diagnosis to assist diagnosed individuals and their families
- Lack of alcohol counselling and harm reduction services for women
- Lack of training of health professionals and lack of health staff in the communities
- Lack of staff training for community members to assist them in implementing FASD-related programs and services
- Lack of services and staff for FASD-affected offenders in Corrections including adult diagnosis
- Lack of culturally specific FASD strategy and approach to diagnosis, prevention and intervention

Without diagnosis, appropriate interventions in educational and social systems are limited or lacking. This would suggest an increase in associated FASD secondary disabilities. Research on the incidence of secondary FASD disabilities in aboriginal communities is unavailable—since this data depends upon primary recognition and diagnosis and this continues to be a problem in all regions.

According to Statistics Canada (2011) there are 94,400 people living in the Atlantic provinces of Canada (Newfoundland-Labrador, Nova Scotia, New Brunswick, and Prince Edward Island) who are aboriginal.[24] Using the recent prevalence general population estimate of 3.6 %,[25] we can project that 3398 aboriginal people in Atlantic Canada need FASD health services at this time. In 2011, there were only 10 diagnostic slots in Atlantic Canada.[26] At the present time, with the opening of the New Brunswick FASD Centre of Excellence, capacity increased to approximately 30; but most of these diagnostic slots do not serve the aboriginal population. Ninety-nine percent of the aboriginal population affected by FASD in Atlantic Canada do not have access to FASD health services in terms of diagnosis and disability accommodations treatment.

The lack of access to FASD health service delivery in aboriginal communities may be negatively affecting the health and well-being of indigenous people living in these communities. It also is contrary to Canada's ethical and legal obligations as expressed by international covenants as well as by the Canadian Charter of Rights and Freedoms.

---

[22]National Collaborating Centre for Aboriginal Health, "State of the Knowledge: Inuit Public Health" University of Northern British Columbia, 2007; Prince George, BC.

[23]Results of Asset Mapping and Medicine Wheel Community Development Workshop—Elsipogtog First Nation, NB (2004).

[24]Statistics Canada: Retrieved from: www.statcan.gc.ca/daily-quotidien/140926/dq140926b-eng.htm.

[25]May et al. (2014b).

[26]Clarren et al. (2011).

In 1948, Canada was a signatory to the UN Declaration of Human Rights affirming in Article 25 the rights of everyone to adequate medical care as well as support in the event of a disability.[27] In 1966, Canada was a signatory to the UN International Covenant on Economic, Social and Cultural Rights, ICESCR, which also recognized the right of everyone to medical care. In Article 12 of the covenant there was special mention of services to ensure: prevention and treatment, the healthy development of the child, and the creation of conditions that would assure medical services.[28] The ICESCR was ratified by the Canada in May of 1976 committing the government in law as well as in principle to the human rights principles of the covenant.[29] In December 1991 Canada also ratified the Convention on the Rights of the Child (CRC).[30] Article 24 of the CRC specifically mentions the responsibility of states to 'ensure that no child is deprived of his or her right of access to…health care services' that further 'states shall strive to ensure that no child is deprived of his or her right of access to such health care services (and) shall pursue full implementation of this right. The CRC also recognized its obligations to mothers to 'not only take appropriate measures….to ensure the provision of necessary medical assistance and health care to all children…..(but also) to ensure appropriate pre-natal and post-natal health care for mothers.'[31]

Research has determined that FASD is a brain based disability but outcome can be improved by early diagnosis, a stable and supportive home and appropriate treatment and accommodations. In Article 23 of the CRC Canada also affirms its obligations to provide all children with a disability the special accommodations that are needed without charge if parents lack resources and that services 'should be designed to ensure that the disabled child has effective access to and receives education, training, health care services, rehabilitation services, preparation for employment and recreation opportunities.'[32]

---

[27]UN Declaration of Human Rights (1948) Retrieved from: http://www.un.org/en/documents/udhr/. Everyone has the right to a standard of living adequate for the health and well-being of himself and of his family, including food, clothing, housing, and medical care and necessary social services, and the right to security in the event of unemployment, sickness, disability, widowhood, old age or other lack of livelihood in circumstances beyond his control. (Article 25).

[28]UN International Covenant on Economic, Social and Cultural Rights (1966): Retrieved from: www.who.int/hhr/Economic_social_cultural.pdf
Recognizing "the right of everyone to the enjoyment of the highest attainable standard of physical and mental health" (including) provision for the…….a. healthy development of the child; b. prevention, treatment and control of epidemic, endemic, occupational and other disease (and) c. creation of conditions which would assure to all medical service and medical attention in the event of sickness (Article 12).

[29]Canada Heritage-Reports on UN human rights treaties and other documents: ICESCR. Retrieved from: http://www.canadianheritage.gc.ca/eng/1357136859881/1357137099645.

[30]Canada Heritage-Reports on UN human rights treaties and other documents: Convention on the Rights of the Child (CRC). Retrieved from: http://www.canadianheritage.gc.ca/eng/1357136859881/1357137099645.

[32]Ibid.

[31]Ibid.

Like many developmental disabilities FASD is permanent and requires life-long support and accommodation. Section 15 of the Charter, signed into law 1982, clearly indicates Canada's legal responsibility to provide access and accommodation for people with physical or mental disabilities of all ages without discrimination.[33] The nature of this commitment is more specifically articulated in the UN Convention of Rights of Persons with Disabilities—signed by Canada in 2007 and fully ratified in 2010. This Convention affirms the rights of all people to access, accommodation and support in event of disability. Specific mention in the convention is made of many areas of life including: Justice (Article 13), Education (Article 24) Health (Article 25)[34]: The convention clearly states provision should be made of services that work to prevent secondary disabilities:

> Provide those health services needed by persons with disabilities specifically because of their disabilities, including early identification and intervention as appropriate, and services designed to minimize and prevent further disabilities, including among children and older persons;

The Convention also notes 'the difficult conditions faced by persons with disabilities who are subject to multiple or aggravated forms of discrimination on the basis of race, colour…… language……indigenous or social origin……age or other status'.

The disadvantage faced by an Aboriginal person with an FASD disability is exacerbated by the already significant gap they have to traverse in health, education, housing, and all other determinants of health. According to the 2014 UN report of the Human Rights Council[35]:

> The health of First Nations, Inuit and Métis people in Canada is a matter of significant concern….significant gaps still remain in health outcomes of aboriginal as compared to non-aboriginal Canadians, including in terms of life expectancy, infant mortality, suicide, injuries, and communicable and chronic diseases such as diabetes. The health situation is exacerbated by overcrowded housing, high population growth rates, high poverty rates and the geographic remoteness of many communities, especially Inuit communities in the north.

[33]Canadian Charter of Rights and Freedom of the Constitution Act. Retrieved from: http://laws.justice.gc.ca/eng/Const/Const_index.html. Section 15. (1) Every individual is equal before and under the law and has the right to the equal protection and equal benefit of the law without discrimination and, in particular, without discrimination based on race, national or ethnic origin, colour, religion, sex, age or mental or physical disability.

[34]Convention on the Rights of Persons with Disabilities. Eighth session, New York, (August 2006) Retrieved From: http://www.un.org/disabilities/convention/conventionfull.shtml.

[35]United Nations General Assembly-Human Rights Council 27th Session (July 2014) Report of the Special Rapporteur on the right of indigenous people: The situation of indigenous people in Canada; Retrieved from: http://www.ohchr.org/EN/Issues/IPeoples/SRIndigenousPeoples/Pages/SRIPeoplesIndex.aspx.

Socio-economic status, SES, is one of the biggest risk or protective factors to the clinical manifestation of the FASD condition.[36,37] In 1987 a group of researchers compared the offspring of high SES and low SES women with controls. The study population all had chronic drinking problems and similar prenatal alcohol consumption when pregnant yet the incidence of FAS in offspring was 4.5 % for upper-middle class mothers vs. 70.9 % for low SES mothers.[38] There were significant differences in rates of intellectual deficiency but more importantly in areas that could not be attributed to post-natal environment, such as a significant reduction in head circumference and growth, as well as increased congenital malformations.

In 2010, Canada became one of the signatories of the United Nations Declaration on the Rights of Indigenous Peoples.[39] In doing so Canada affirmed its indigenous peoples' 'equal right to the enjoyment of the highest attainable standard of physical and mental health'. Canada also agreed that they would 'take the necessary steps with a view to achieving progressively the full realization of this right' (Article 24). Special note was taken in the document of the need to protect the rights of people with disabilities. 'Particular attention shall be paid to the rights and special needs of indigenous elders, women, youth, children and persons with disabilities in the implementation of this Declaration (Article 22)'.[40]

Despite its ethical and legal responsibilities lack of access to FASD diagnostic services means lack of identification and subsequent lack of eligibility for provision of disability accommodation in school and community. The 2014 UN report of the Human Rights Council notes that despite conditions funding levels in aboriginal communities for education, health and social services continue to lag far behind.[41]

---

[36]Abel and Hannigan (1994). See also Abel (2004).

[37]May and Gossage (2011a, b).

[38]Bingol et al. (1987).
'There was highly significant difference in the incidence of FAS offspring between upper middle and lower class alcoholic mothers, 4.5 % versus 70.9 % respectively. Mean weight, length, and head circumference at birth in children of upper middle class alcoholic women was -ISD, those of lower class alcoholic women fell into -2SD. All other parameters, congenital malformation rate, failure to thrive, mental retardation were also significantly greater in children of lower class alcoholic women (p less than or equal to 0.01)'.

[39]United Nations General Assembly-Human Rights Council 27th Session (July 2014) Report of the Special Rapporteur on the right of indigenous people: The situation of indigenous people in Canada; Retrieved from: http://www.ohchr.org/EN/Issues/IPeoples/SRIndigenousPeoples/Pages/SRIPeoplesIndex.aspx.

[40]Canada became a signatory to the Declaration in 2010; but in 2007 when the Declaration passed by a vote of 143 countries in favour to 4 against Canada was one of only four countries in the world that originally voted against it. The other countries are: Australia, New Zealand and the United States—countries that have large historically disadvantaged indigenous populations.

[41]United Nations General Assembly-Human Rights Council 27th Session (July 2014) Report of the Special Rapporteur on the right of indigenous people: The situation of indigenous people in Canada; Retrieved from: http://www.ohchr.org/EN/Issues/IPeoples/SRIndigenousPeoples/Pages/SRIPeoplesIndex.aspx.

The challenging behaviors of youth and young adults who may be affected by prenatal exposure are often perceived to be intentional because they have not been identified through the medical system as having a disability. FASD behaviors can become more challenging as individuals grow into adulthood without support and accommodation. Secondary disability research indicates that without early diagnosis and support 60 % of individuals with FASD begin to get into trouble with the law.[42] Research indicates that many individuals with FASD end up in the revolving door of the criminal justice system where they are punished for behaviors that have less to do with criminality than with their disability.[43]

The criminal justice system in Canada has no process in place to screen or assess FASD. Forensic psychiatrists are not often trained nor always available to diagnose FASD, so unless the diagnosis is brought into the court individuals with FASD do not get appropriate consideration or accommodation within the criminal justice system for their disability.[44]

To remedy the situation regarding the lack of FASD service delivery Health Canada implemented FASD initiatives beginning in the early 2000's.[45] First Nation Inuit Health or FNIH's FASD Program and PHAC's (Public Health Agency of Canada) FASD National Strategic Projects Fund are the basic components of Health Canada's FASD Initiative introduced in the 1999 Budget.

PHAC's FASD National Strategic Fund has been responsible for advancing a number of important FASD projects and Canada has become well-known throughout the world for its ground-breaking work in advancing FASD knowledge, research and practice.[46] They funded a wide range of projects including development of: Clinical Practice Guidelines for Diagnosis of FASD,[47] Alcohol Use and Pregnancy Clinical Practice Guidelines for physicians[48]; a National Screening Toolkit for FASD,[49] Prevention statements, Adult diagnosis statements, development of a national prevalence plan, economic analysis and research funding through Canada's FASD Research Network and NeuroDevnet.

---

[42]Streissguth and Kanter (1997).

[43]Conry and Fast (2000).

[44]This may change with the recent addition in the DSM-5 of the condition, ND-PEA, Neurodevelopmental Disorder-Prenatal Exposure to Alcohol.

[45]PHAC's FASD Initiative is described on their website: http://www.phac-aspc.gc.ca/hp-ps/dca-dea/prog-ini/fasd-etcaf/funding-fonds/index-eng.php FNIH FASD program is described on the Health Canada website: http://www.hc-sc.gc.ca/fniah-spnia/finance/index-eng.php.

[46]Ibid. The objective of the FASD initiative is to develop: 'practical tools for community-based programs and health and allied professional (with the goal) to reduce the number of alcohol-affected births and to improve the outcomes for those affected by FASD….(by providing) knowledge and evidence, knowledge translation and exchange, leadership, co-ordination, and collaboration, systems capacity development, and evaluation.'

[47]Chudley et al. (2005).

[48]Carson et al. (2010).

[49]CAPHC (2012).

Despite the success of PHAC's National FASD initiative, actual service delivery at the community level in Canada has not improved significantly. In 2011, Clarren and Lutke report that the clinical capacity in Canada to diagnosis was 2288 while the number of people projected to be affected by FASD was 1,279,454.[50,51] Canada's specific legislation dealing with healthcare, the Canada Health Act (1985)[52] shares the responsibility of running the Canadian medical system between levels of government. The provinces/territories have been given the primary responsibility for health service delivery while the federal government through Health Canada provides direction, research and support.[53] So despite PHAC's direction, there was no real change in provincial health policy and practice.

Health services for aboriginal people are provided through more complex arrangements between the provinces or territories, the federal government and aboriginal communities. Often there is confusion as to the responsibility of each of these in relation to the funding and the delivery of particular health services. The federal department that has the specific mandate for aboriginal health is FNIH (First Nation and Inuit Health). While an Aboriginal right to health care has never been specifically acknowledged by the federal government or decided in the courts, the courts have clearly recognized the fiduciary responsibility of the government towards aboriginal people. FNIH's mandate is to[54]:

- Ensure the availability of, or access to, health services for First Nations and Inuit communities;
- Assist First Nations and Inuit communities address health barriers, disease threats, and attain health levels comparable to other Canadians living in similar locations; and
- Build strong partnerships with First Nations and Inuit to improve the health system.

Most of the health services for aboriginal people, however, are delivered by the provinces or territories where the people reside whether off or on-reserve. FNIH's

[50] Clarren et al. (2011).

[51] Universal access to medical care is not a specifically mentioned as a Charter right. Canadian Charter of Rights and Freedom of the Constitution Act Retrieved from: http://laws.justice.gc.ca/eng/Const/Const_index.html Sect. 7 of the Charter has been used in successful legal challenges to guarantee access to health services.

[52] The Canada Health Act (1985) Retrieved from: http://www.laws-lois.justice.gc.ca/eng/acts/c-6/page-1.html.

[53] Health Canada: Retrieved from: http://www.hc-sc.gc.ca/hcs-sss/index-eng.php
`Health Canada's mandate is to help Canadians maintain and improve their health….. Health Canada's responsibilities for health care include setting and administering national principles for the health care system through the *Canada Health Act* …..Working in partnership with provinces and territories, Health Canada also supports the health care system through initiatives in areas such as health human resources planning, adoption of new technologies and primary health care delivery.

[54] FNIH Retrieved from: http://www.hc-sc.gc.ca/ahc-asc/branch-dirgen/fnihb-dgspni/index-eng.php.

'un-insured benefits program' is supposed to fill in the gaps by covering health services that are not provincially insured i.e. dental services, eyeglasses or medicine. This program could provide the framework to fund FASD diagnostic services but at this point in time there is only project support (not core funding) for FNIH's FASD program. The actual focus of the program has been on prevention through mentoring services and support for women, although the stated objectives of their program are broader:

> (1) Reduce the number of babies born with FASD; and (2) support children (age 0–6) who are diagnosed with FASD and their families to improve their quality of life.[55]

When FNIH's FASD program was first launched in the early 2000s there were three eligible priority areas: Asset Mapping, Diagnosis (Formation of Multidisciplinary team), and Mentoring (PCAP, Parent Child Assistance Program).

Amidst this context, one community based FASD multi-disciplinary diagnostic team, the Eastern Door, was established in Atlantic Canada through the FNIH's call for proposals in the early 2000s. The Eastern Door Team, is an aboriginal community based diagnostic team, that's been operating since 2005-6; it was used as a model for the New Brunswick FASD Centre of Excellence that opened in 2012.[56] The Eastern Door is actually the result of successful tri-partite collaboration between the federal government, the provincial government and the community. It would not have been possible without the funding provided by FNIH's FASD Program for training and team coordination, but the in-kind provincial services of physician and OT and the special needs accommodation services offered post-diagnosis by the FN on-reserve school were also essential.[57] The Eastern Door is unique in its development of culturally based tools for screening, diagnosis and intervention and the participation of traditional healers in the diagnostic process. After the Eastern Door was operating for a few years a satisfaction survey was sent to the families who had accessed services; the satisfaction rate was over 90 %.[58] One of the birth mothers who was able to access FASD services in the community speaks of her experience with diagnosis and then after as accommodations were put in place and the child began to learn[59]:

[55]Health Canada Archived content: Retrieved from: http://www.hc-sc.gc.ca/fniah-spnia/finance/agree-accord/prog/index-eng.php.

[56]NB FASD Advisory Committee Final Report 2010. The NB Centre of Excellence was first operated by Moncton Family Service and is now being managed by VON Canada. When they are fully operational they are able to do about 20 diagnosis a year and the Eastern Door team can do about 10 diagnosis.

[57]The community school provides the psychologist, speech pathologist, and social worker for the team. The school also provides accommodation services in the school after diagnosis. The school program is funded by AANDC, Aboriginal and Northern Development Canada (Indian Affairs) through its special needs funding. This funding was not available until the children were diagnosed with FASD.

[58]EHWC Evaluation 2008.

[59]Interview with Bev Sanipass-2010—Eastern Door Video.

> When she was a baby I didn't understand why she would you know when she would get angry she would bang her head on the floor.....like constantly.....she didn't know how to say how she was feeling...so just for...for not to get hurt I would put my hand on her head, where she was pounding her head, and so she wouldn't get hurt......
> At first I dreaded going and didn't know....didn't want to know....but I put all those feelings aside and I said this is for her...... She asked me a lot of questions and I just answered them in my honest way as best as I could...what I know...but she helped me figure out a lot of stuff in my head...about it...just talking to her it was a relief....for me...that it had a name....the doctor told me that this is what it is and then what they were going to do and every word he was saying after that...it was lifting my spirits and then things started going more easier after that and both of us...and she started to learn and when she first started reading a book I cried...I never thought she would read a book...... and I was so happy...when she graduated high school—all that, because of that, the diagnosis, it made my life easier—

Despite its success, the Eastern Door funding from FNIH has been reduced to a point that it now does not even cover the salary of the team coordinator. The original plan was to begin to offer adult diagnosis and to expand in order to allow nearby FN communities to access diagnostic services as they have requested.[60] Instead, one of the health center administrators recently had to restrict Eastern Door services because of inadequate funding from FNIH. FNIH removed diagnosis as one of the funded components in their FASD program since they consider medical diagnostic services to be the responsibility of the province.

Because of the limited funding in the provinces for the development of multidisciplinary diagnostic teams an individual's access to FASD diagnosis in aboriginal and non-aboriginal communities has depended upon the practice of individual physicians. These health professionals are funded through provincially insured health plans and they face a number of challenges in providing diagnosis. Twenty years ago the Public Health Agency of Canada commissioned a survey of pediatricians and family doctors to explore their level of knowledge and attitudes to FASD in light of the lack of access to FASD diagnostic services. One of the most important issues for them was lack of training. Over half reported that they lacked the training to diagnose FASD. A few had never even heard of the condition.[61] Some of the physicians reported that they had considered FAS as a diagnosis but also admitted they never made reference to the condition in their case notes because they were uncomfortable with the issue.[62] Physician lack of FASD training and knowledge continues to be a problem. Physicians also report they are uncomfortable asking about alcohol use in pregnancy and they do not have the resources needed to deal with the issue. [63]

---

[60]NB FASD Advisory Committee report. 2009. A consultation meeting was held with FN health center administrators and community health professionals. There was consensus that they wanted the Eastern Door Center made into an aboriginal center of excellence to serve their communities.

[61]Nanson et al. (1995) (2 % of physicians had never even heard of FAS. Of the 98 % remaining who had some knowledge of the condition less than 50 % felt able to diagnose it).

[62]Ibid.

[63]Carson et al. (2010).

In 2007, the *Lancet* published an article written by British pediatricians entitled 'Fetal Alcohol Syndrome: law and ethics' that suggested that diagnosing FASD poses a conflict for physicians.[64] They write that they are caught between 'the right of the individual to know the likely cause of their disability against emphasising the mother's potential sense of guilt, thereby probably damaging the relationship of mother and child'.[65] Despite this conflict the authors conclude that ethically it is only by 'reliably making the diagnosis can more effective interventions be offered to affected individuals'. But there are practical barriers to FASD diagnosis that affect physicians offering FASD diagnostic services.

FASD diagnosis is time-consuming. It requires interviews with the birth mother, information from other specialists and coordination of care. Physicians report that there are no provincial FASD billing codes that would enable them to charge for the services that are required.[66]

The provincial government however, could choose to supply FASD billing codes for individual physicians as well as resources for the coordination and development of FASD diagnostic teams. Health officials justify the continuing lack of FASD service delivery in terms of a risk-benefit or cost-effective analysis (i.e. the cost of FASD diagnosis is high; there is no cure for FASD, and FASD interventions are expensive to provide.[67]) This economic approach to health service delivery would appear to have no ethical basis in light of the principles that health professionals and policy makers agree should guide health care policy and practice. Beauchamp and Childress first articulated these in the 1970s and they are now widely accepted by most health professionals and their regulatory bodies[68]:

1. Beneficence—Policy and practice that has the best interest of the other in mind
2. Maleficence—Policy and practice that above all 'does no harm'
3. Justice— Policy and practice that is fair and that promotes equality among individuals
4. Autonomy— Policy and practice that allows for the right of each individual to have freedom of choice

In our society cost-effectiveness/cost-benefit analysis in healthcare has become identified with the ethical principle of beneficence. This is based on an interpretation of John Stuart Mill's philosophy of utilitarianism. Mill proposed that decisions, by individuals or the state, should be made on the basis of what will do the

---

[64]Mukherjee et al. (2007).

[65]Ibid.

[66]Gander FASD conference. 2000 Physician Session facilitated by Dr.Ted Rosalas. Goose Bay physicians' session at the hospital (2001).

[67]Notes from Big Cove Wellness Committee Inc. In attempt to get diagnostic services in FN community as well as in the province members of the Wellness Committee met with representatives from the department of health and hospital medical services in a series of meetings from 2004–2009.

[68]Beauchamp and Childress (2008).

Table 2. Risks and benefits associated with FASD diagnosis/service delivery

Risks	Benefits
Lack of appropriate interventions after diagnosis	• Interventions provided anyway—diagnosis means more appropriate ones
	• Diagnosis leads to shift in focus of interventions
	• Diagnosis leads to better resource allocation
	• Diagnosis needed in research to help improve outcome
Stigma and labeling	• Diagnosis leads to understanding of disability
	• Shift from individual being a problem to having a problem
	• More effective interventions
	• More autonomy as individuals with FASD are supported and empowered
Diagnosis and interventions are expensive in a time of scarce resources	• Research indicates diagnosis and intervention works to prevent or ameliorate secondary disabilities
	• Secondary disabilities lead to great economic costs
	• Appropriate interventions mean less recidivism in CJS
	• Providing mothers with FASD services and support for their disability means improving the lives of their children
FASD diagnosis is not cost-effective. Health resources could be used on health conditions that would serve more people and have more impact	• Diagnosis and support means improved relations to self, to family, world—Mi'gmaw ethical concept 'Nogemag'
	• Increased well-being for individual, family and community—Greek ethical concept 'Eudaimonia'

greatest good for the greatest number of people.[69] In Utilitarian terms, 'Beneficence' would then be equated with an action that brought the greatest benefit to the greatest number of people. In making policy decisions the theory is that there is a need to balance the economic benefits of a certain course of action with the risks it entails to achieve results.

Given this framework, the decision to provide access to FASD diagnosis and intervention would be based on analyzing and weighing the probable risks and benefits that would be incurred. Health care decisions are often guided by tools such as QALY (Quality Adjusted Life Year), that are used in cost effective analysis to calculate the value for money of a medical service—in terms of the quality and the quantity of life lived after the service/intervention. Below is a chart outlining the risks and benefits associated with FASD diagnosis (Table 2).[70]

[69]Mill (1993).

[70]Cost effective analysis and Cost benefit analysis are distinct. Cost benefit analysis has to do with monetary savings and cost effective approach with what appears to be a more qualitative approach. This distinction is not important for the purposes of this analysis.

The economic implications of FASD have been studied extensively in recent years, and the research indicates that using a cost-benefit analysis it would actually be more effective to deliver FASD health services despite the initial expense.[71] Despite this research, the political will that drives policy is lacking at the provincial level. As Michael Foucault pointed out, ethical principles are socially constructed.[72] Who decides what benefits are of value and what risks are worth taking? Who decides what "quality" means to an individual or group who may have entirely cultural different values. Who decides the groups that should benefit by the decisions that are made? For instance, we would never consider refusing health services to children with inoperable brain cancer, although by cost effectiveness analysis the QALY score would be low. Extensive health services for Autism Spectrum Disorder have been provided in most regions after much lobbying by upper middle-class parents, yet services for FASD (a similar developmental condition that is perceived to affect only the aboriginal population or the marginalized) are non-existent. Public health policy and professional practice are affected by prevailing politics and perceptions and interests.

Policy makers would have a different set of priorities if the issue of FASD service delivery was instead considered using the framework of Aristotelian ethics where 'Eudaimonia', happiness, harmony and well-being, is the 'end' or purpose of human life and of virtuous and ethical action.[73] This concept is similar in a sense to the M'igmaw aboriginal concept of Nogemag—well-being in relationship to self-others-creator. From a Mi'gmaw perspective health and right action should be considered in context of the balancing and harmony of the whole.[74]

One of the most important issues that need to be addressed when considering the provision of FASD services in an aboriginal community context has to do with a certain resistance within the aboriginal community to the way FASD is framed in the present medical model. Often this is posed as an issue about lack of culturally sensitive norms or assessment tools, but the issue is deeper than this. Noel Milliea, one of the traditional healers who are part of the Eastern Door Diagnostic team, suggested a different way of looking at FASD by framing it in relation to a circle (rather than a grid) and considering it as a system disorder as well as an individual one. In traditional terms the disorder we now call FASD might be considered as 'a disordering'. There is a breakdown in relationship between mother and child that has to do with the breakdown of social relationships to self, family, community and larger society.

When the condition of FASD is framed in this way there is no stigma involved to the affected individuals, the mothers, families and communities. This is not the

---

[71]Stade et al. (2007).

[72]Foucault (1984). See On the Genealogy of Ethics: An Overview of Work in Progress, pp. 340–372. Also Politics and Ethics, pp. 373–380.

[73]Aristotle (1941). The word *Eudaimonia* (εύδαιμον-ία) literal English translation in Liddell and Scott's Greek lexicon is 'prosperity, good fortune, opulence, happiness' but the word as a concept used by Aristotle holds much more meaning than the simple English translation.

[74]Discussions with elder Noel Milliea The English translation of the word *Nogemag* is 'all my relations' but the concept holds much more meaning than the simple English translation.

case when FASD is considered through the lens of the present medical model. In the medical paradigm FASD as it is constructed tends to be perceived as the 'fault' of the mother. One particular article discussing FASD and ethics published in a physicians' journal reported[75]: 'by exposing the fetus to a teratogen the mother is causally, and arguably morally, responsible for the outcome.'

This way of thinking stigmatizes the mother and so leads to resistance in terms of accessing FASD diagnostic services for her child or disclosing pre-natal alcohol use (even if she drank before she knew she was pregnant or was told by her physician that moderate drinking would not harm her child). In Canada, the aboriginal population as a whole now carries the 'stigma' of FASD because in Canada FASD is widely believed to be primarily an aboriginal problem.[76] The ethical implications of this kind of mis-information are serious as it has hindered progress and empowerment by generating resistance and discomfort within the aboriginal community itself. The Ontario Federation of Indian Friendship Centres reports that the stigma associated with FASD and aboriginal people is working to silence aboriginal voices about FASD.[77]

> This erroneous characterization has not only negatively affected the mainstream understanding of Aboriginal health and community, but has also inhibited the progress of FASD research and strategy development…….

The reality is that adequate FASD service delivery is a serious health problem in both aboriginal and non-aboriginal communities in Canada; however, aboriginal communities (and individuals living in them) are particularly disadvantaged because of the 'disproportionate hardship that they experience in relation to the social determinants of health' and the lack of access to the services they need:

> Colonization and assimilation practices have resulted in high rates of mental illness, addiction, homelessness, poverty, unemployment, incarceration, poor physical health, and community and familial dysfunction….Many individuals do not have access to the holistic services that would help them to cope with the social isolation and experiences of violence and abuse that can arise from these issues. Together, these consequences of a shared history of trauma embody the underlying, or ultimate, causes of FASD in Aboriginal communities, and must be addressed with culturally appropriate methods that foster cultural reintegration, personal and collective healing, and positive change. FASD is a product of socioeconomic status rather than race or culture and as long as a national or provincial strategy does not exist…….Aboriginal people affected by FASD will go without the culturally appropriate, wholistic, and self-determined supports and resources that they need to ensure positive health outcomes and an optimal quality of life.[78]

---

[75]Mukherjee et al. (2007).

[76]When we were lobbying in NB to get provincial FASD diagnostic services I was asked to give an FASD 101 presentation to a group of provincial ministers including Health, Education, Social Welfare and Aboriginal Affairs. During discussion period that followed one of them commented that they thought FASD was only a problem of aboriginal people—he wasn't aware that it was a health issue for others.

[78]Ontario Federation of Indian Friendship Centres Toronto, Ontario (2013) Fetal Alcohol Spectrum Disorder: a Position Paper. Retrieved from: http://www.ofifc.org.

[77]Ontario Federation of Indian Friendship Centres Toronto, Ontario (2013) Fetal Alcohol Spectrum Disorder: a Position Paper. Retrieved from: http://www.ofifc.org.

Therefore, Canada's ethical obligation to provide resources for system based FASD diagnostic and intervention services in aboriginal communities is clear, as is its legal obligation to accommodate for the FASD disability. However, there are further considerations. Naomi Adelson pointed out that system inequities in Canada in regard to aboriginal health are the root cause of health disparities.[79] In this light FASD service delivery to aboriginal people-whether diagnosis, prevention or intervention-implies a system focus and also identifies a system responsibility. This is true for both indigenous and nonindigenous populations in Canada. Prevention and healing of FASD has much to do with Canada's ethical and legal responsibility to actively work together at all levels of government to create healthy systems. This would, in turn, help provide access to culturally informed FASD diagnosis services and to the social, economic, physical and spiritual accommodations that are needed by affected individuals and families to deal with the disability.

References

Abel, E. 1995. An update on the incidence of FAS: FAS is not an equal opportunity birth defect. *Neurotoxicology and Teratology* 17: 437–443.

Abel, E.L. 2004. Paternal contribution to fetal alcohol spectrum disorder. *Addiction Biology* 9(2): 127–133.

Abel, E., and J. Hannigan. 1994. Maternal risk factors in fetal alcohol syndrome: Provocative and permissive influences. *Neurotoxicology and Teratology* 17: 437–443.

Adelson, N. 2003. The embodiment of inequity: Health disparities in Aboriginal Canada. *Canadian Journal of Epidemiology and Community health* 57: 254–258.

Aristotle. 1941. Aristotle. Nichomachean ethics. In *The basic works of Aristotle,* ed. Aristotle, R. e. McKeon, 935–1126. New York: Random House.

Beauchamp, T.L., and J.F. Childress. 2008. *Principles of biomedical ethics*, 6th ed. Oxford: Oxford University press

Bingol, N., C. Schuster, M. Fuchs, S. Iosub, C. Turner, R. Stone, and D. Gromisch. 1987. The influence of socioeconomic factors on the occurrence of fetal alcohol syndrome. *Advances in Alcohol and Substance Abuse* 6: 105–118.

CAPHC. 2012. *National screening tool kit*. FASD Toronto: CAPHC.

Carson, G., L. Cox, J. Crane, P. Croteau, L. Graves, S. Kluka, and R. Wood. 2010. Alcohol use and pregnancy consensus clinical guidelines. *JOGC* S1–S32.

Chudley, A., J. Conry, J. Cook, C. Loock, T. Rosalas, and N. LeBlanc. 2005. Fetal alcohol spectrum disorder: Canadian guidelines for diagnosis. *CMAJ* 172(S suppl).

Clarren, S., J. Lutke, and M. Sherbuck. (2011). The Canadian guidelines and the interdisciplinary clinical capacity of Canada to diagnose FASD. *Journal of Population Therapeutics and Clinical Pharmacology* 18(3): e494–e499.

Conry, J., and D.K. Fast. 2000. *Fetal alcohol syndrome and the criminal justice system.* Vancouver: Fetal Alcohol Syndrome Resource Society.

Foucault, M. 1984. *Foucault reader* (ed. P. Rabinow). New York: Pantheon Books.

Hellemans, K.G., J.H. Sliwowska, P. Verma, and J. Weinberg. 2009. Prenatal alcohol exposure: Fetal programming and later life vulnerability to stress, depression and anxiety disorders. *Neuroscience & Biobehavioral Reviews* (2010 May) 34(6): 791–807. doi: 10.1016/j.neubiorev.

[79]Adelson (2003).

May, P.A., and J.P. Gossage. 2001. Estimating the prevalence of fetal alcohol syndrome. *Alcohol Research and Health,* 25(3):159–167.

May, P.A., et al. 2014. Prevalence and characteristics of fetal alcohol spectrum disorders. *Pediatrics*. American Academy of Pediatrics. Published online October 27, 2014.

May, P.A., A. Baete, J. Russo, A.J. Elliott, J. Blankenship, W.O. Kalberg, D. Buckley, M. Brooks, J. Hasken, O. Abdul-Rahman, M.P. Adam, L.K. Robinson, M. Manning, and H.E. Hoyme. 2014. Prevalence and characteristics of fetal alcohol spectrum disorders. *Pediatrics* 134(5): 855-66. doi: 10.1542/peds.2013-3319.

May, P.A., and J.P. Gossage. 2011a. Maternal risk factors for fetal alcohol spectrum disorders: Not as simple as it might seem. *Alcohol Research & Health* 34: 15–26.

May, P.A., and J.P. Gossage. 2011b. Maternal risk factors for fetal alcohol spectrum disorders: Not as simple as it might seem. *Alcohol Research and Health* 34: 15–26.

Mill, J.S. 1993. *Utilitarianism, on liberty, considerations on representative government.* London: Everyman.

Muckle, G., et al. 2003. Smoking, alcohol, and drug use during pregnancy among Nunavik Inuit Women. In *Poster Presentation, 12th International Congress on Circumpolar Health.* Nuuk, Greenland.

Mukherjee, R., N. Eastman, J. Turk, and S. Hollins. 2007. Fetal alcohol syndrome: Law and ethics. *The Lancet* 369: 1149–1150.

Nanson, J., R. Bolaria, R.E. Snyder, and B.A. Morse. 1995. Physician awareness of fetal alcohol syndrome: A survey of paeditricians and general practitioners. *CMAJ* 152: 1071–1076.

Potyrala, M., and J. Brewster. 2009. The Anaaana Project: Maternal Health Survey in the Olkiqtani Region of Nunavut. In *Government of Nunavut Workshop Paper Presentation 14th International Congress on Circumpolar Health*, Yellowknife, NWT.

Riley Edward P., N. Mattson Sarah, and T.K. Li, et al. 2003. Neurobehavioural consequences of prenatal alcohol exposure: An international perspective. *Alcoholism, Clinical and Experimental Research* 27(2): 362–373.

Salmon, A., and S. Clarren. 2011. Developing effective, culturally appropriate avenues to FASD diagnosis and prevention in northern Canada. *International Journal of Circumpolar Health* 70: 4.

Stade, B., W.J. Ungar, B. Stevens, J. Beyen, and G. Koren. 2007. Cost of fetal alcohol spectrum disorder in Canada. *Canadian Family Physician* 53(8): 1303–1304.

Streissguth A.P., and K. Janter, ed. 1997. The challenge of fetal alcohol syndrome: Overcoming secondary disabilities. Seattle: University of Washington Press.

FASD and the Concept of "Intellectual Disability Equivalence"

Stephen Greenspan, Natalie Novick Brown and William Edwards

Introduction

The term "Intellectual Disability (ID) equivalence" refers to accommodations that are made by legal and other governmental entities when they provide services, supports or protective arrangements to people who—because of brain impairment—function as if they have ID but fail to qualify for the ID label which is needed for access to many programs because their IQ scores are a few points too high. Fetal Alcohol Spectrum Disorder (FASD) is a logical candidate for such an accommodation as (a) it (like ID) involves brain impairment, (b) people with FASD have adaptive deficits and support needs that are identical to those with ID, and (c) while many people with FASD do qualify as having ID, the majority do not, because full-scale IQ scores are typically too high.

ID-equivalence accommodations are an attempt to free the human services field from the straight-jacket that has been imposed by over-reliance on full-scale IQ ceiling scores, which functioned as a gate-keeper and barrier for developmental services eligibility. In this chapter, using specific examples, we review some of the forms that these ID-equivalence accommodations have taken, point out problems with those forms, and suggest improvements for implementing this concept. First, we provide a brief historical overview of the role of intelligence in intellectual disability and the growing belief that an IQ score provides an inadequate basis for determining whether someone may or may not have that disorder.

S. Greenspan (✉)
University of Connecticut, Storrs, USA
e-mail: stephen.greenspan@gmail.com

N.N. Brown
University of Washington, Seattle, USA
e-mail: drnataliebrown@gmail.com

W. Edwards
LA Mental Health Court, Los Angeles, USA
e-mail: WEdwards@pubdef.lacounty.gov

M. Nelson and M. Trussler (eds.), *Fetal Alcohol Spectrum Disorders in Adults: Ethical and Legal Perspectives*, International Library of Ethics, Law, and the New Medicine 63, DOI 10.1007/978-3-319-20866-4_15

Limitations of IQ and IQ Cut-Offs

Intellectual Disability, under various names and most of them now offensive, has been around for centuries, as seen in references to it in Egyptian, Greek, Roman and other ancient documents (Scheerenberger 1983). Until the advent of intelligence testing in the early twentieth century, individuals with ID were identified by how they functioned in the everyday world or what has come to be termed "adaptive behavior". Specifically, people with ID were differentiated from the general population by the perception that they lacked ability to survive on their own. With the invention of the IQ statistic, diagnostic emphasis shifted from everyday functioning as seen by others in one's environment to "intelligence" as measured on a one or two hour test of mainly academic problems, administered by a psychologist who typically did not know the child or adult being tested.

The original Binet-Simon scale, on which most subsequent tests were modeled, took progressively more advanced items from the educational curricula. Thus, as was argued by Anastasi (1983), there was essentially no difference between a measure of aptitude (IQ) and a measure of academic achievement, as seen in the extremely high correlations among the two types of instruments and extremely good predictive validity between IQ and academic performance. When there was divergence between aptitude and achievement scores, as in former but now largely abandoned operational definitions of "learning disabilities," these often reflected differences in test reliability as well as the fact that achievement profiles were typically differentiated in most people, including those with brain impairments. The question thus becomes: "how appropriate is it to place central reliance on a measure essentially of academic potential as a gateway to receiving developmental services, subsidies and entitlements?" The answer, it seems to us and many others, is "not much."

The IQ metric became overwhelmingly popular in North America in the second quarter of the twentieth century, as it proved useful to eugenicists such as Lewis Terman (author of the Stanford-Binet, a translation and extension of the original French Binet-Simon) in promoting their racist political agenda (Blumenfeld 2011). That agenda included shutting down immigration to the United States from Southern and Eastern Europe, placing individuals with below-average IQ in gender-segregated institutions before subjecting them to involuntary sterilization, and otherwise preventing those termed "morons" (many of whom today would likely not be considered to have ID) and "imbeciles" (people who today would be considered to have mild ID) from transmitting their "diseased germ plasma." The implementation of that agenda depended on being able to persuade legislators and the public through quasi-scientific or outright fraudulent studies such as Henry Goddard's (1912) book on the "Kallikak" family, that low-IQ people posed a threat to society and to the continued superiority of the "white race." Much evidence that Goddard faked his Kallikak data can be found in Smith (1985).

The basic idea being promoted at the time was that IQ represented a precise snapshot of the brain that was concrete, immutable, inherited, and incorporative of all that we view as "intelligence." Now of course we understand that there are discrete cognitive capacities such as "social intelligence" that are not well captured by an IQ score. We also recognize that some cognitive measures such as "executive functioning" are better indices of intelligence than IQ, and that cognitive profiles across batteries of IQ and other tests are more useful than a single index score (McGrew and Flanagan 1998). It is also accepted that there are environmental as well as other biological (e.g., brain injury) contributors to poor intellectual functioning (Ceci and Williams 1997), and that instruments which purport to measure IQ are not equally reliable or valid and are affected by well-known psychometric variables, such as experience with prior tests, obsolescence of test norms, problems with test construction and statistics, and normal variation in performance that is random or due to examiner incompetence or bias (Greenspan and Olley 2015).

The idea of using IQ ceilings to demarcate a dividing line between impairment and "normality" came to the fore a little over half a century ago when the American Association on Mental Deficiency (later, the American Association on Mental Retardation and today, the American Association on Intellectual and Developmental Disability) published a diagnostic manual. In this manual, they established a three-prong definition of mental deficiency (later, "mental retardation" and today "intellectual disability") as a condition in which the first prong, intellectual impairment, was reflected in an IQ score that fell at least one standard deviation (SD) below the population mean (a standard score equal to or less than 85, or the 17th percentile of the population distribution). The conventional belief, not particularly based on research, was that the incidence of ID was 3 % of the population. Therefore, a standard that took in the bottom 17 % of the population was obviously too high as it created many "false positives" of individuals labeled ID who should not have been. This problem was exacerbated by a failure for over a decade to use the second diagnostic prong of adaptive behavior, the intended purpose of which was to bring the actual incidence down to 3 %. To correct for this problem, the criterion for prong two was changed three decades ago from one standard deviation (IQ = 85, 17th percentile) below the mean to two standard deviations (IQ = 70, 2nd percentile) below the mean (Greenspan and Switzky 2006). However, the minus 2 SD standard was just as arbitrary as the minus 1 SD standard had been two decades earlier. No apparent rationale was provided for using the minus 2 SD cut-score other than the superficial elegance of using the statistical convention of standard deviation units. Just as the earlier use of the minus 1 SD criterion was too easy a hurdle to clear, this new reliance on a minus 2 SD criterion was too difficult a hurdle, with the result being an excessive number of "false negatives" of people who deserved ID eligibility but were wrongly denied it. Most of the definitional reforms essayed over the past several decades were motivated by an attempt to rectify the problem of false negatives caused by setting the qualifying IQ score too low.

To date, four solutions have been proposed to overcome the problem of false negatives. The first attempted fix in the 1980s (Grossman 1983) involved encouraging diagnostic evaluators and agencies to take into account the confidence interval of the IQ test (5 points at the 95th percent of confidence) when making ID diagnostic determinations. When this admonition was largely ignored, the second attempted fix, enacted in the 1990s (Luckasson et al. (1992), was to take the suggested five-point confidence interval and make it the new ceiling standard, later modified (Lucksson et al. 2002) as the IQ range "70–75." More recently, this second fix has involved using 75 as the cutoff point (Schalock et al. 2002), which is also what DSM-5 adopted. Part of this fix was to insert the word "approximately" before the number 70 or, more typically, the term "approximately minus two standard deviations below the mean," with "approximately" meaning that one should take into account the unreliability or standard error of the IQ statistic. When changing the ceiling was typically ignored, a third fix was to encourage clinicians to change the number itself by correcting for norm obsolescence, a phenomenon known as the "Flynn effect" (Gresham and Reschly 2011). The Flynn correction process involves subtracting 0.3 IQ points per year of elapsed time between the date of norming and date of test administration to correct for the fact that norms are toughened by 3 years per decade of norm obsolescence to adjust for changes in population performance in the interim. This practice has become commonplace when determining eligibility exemptions in the US under the *Atkins v Virginia* standard but has not typically been used for school, residential or other less catastrophic purposes. The fourth fix is the one with the most promise for eventually solving the false negative problem and the most relevant for the ID-equivalence issue. It involves approaching the first diagnostic prong of "intellectual functioning" as a broad construct that is tapped by various cognitive measures of which IQ is but one data point. This is the position underlying the ID section in DSM-5 (American Psychiatric Association 2013), which actually states that neuropsychological tests, particularly of "executive functioning," are typically more useful diagnostically than an IQ score. This provision reflects a step away from the "disability" (arbitrary, numbers-based) view of ID to one that views the condition as a brain-based "disorder" determined clinically rather than psychometrically. In such a conceptualization, the key is to look at the individual as a whole person and not just through the narrow lens of a single numeric score. This position will be explored more fully in the concluding section of this paper, but first we will look at how various government entities have put in place ID-Equivalence provisions intended to get around our slavish and inappropriate reliance on arbitrarily-instituted full-scale IQ ceilings.

Forms of ID-Equivalency Accommodation

Various attempts have been made over the years to address the inequitable obstacles to ID service eligibility posed by the continuing rigid reliance on full-scale IQ cutoffs. People with FASD are among those most victimized by the current practice and therefore stand the most to gain from efforts to develop more flexible frameworks. In the following pages, we describe some of these ID-Equivalency approaches, and point out both their strengths and limitations.

Building Policy Around an Individual-Driven Lawsuit

Neal Fahlman is a young man in British Columbia, Canada, of First-Nations ethnicity who was adopted at age five weeks. As a child he had three diagnoses: FASD, ID and what today would be termed ASD (Autism Spectrum Disorder), in addition to ADHD (a common co-morbid problem for people with FASD). As a child and pre-adolescent, Neal qualified for developmental disability (DD) services as his full-scale IQ was below the minus two standard deviation threshold used in BC by Community Living British Columbia (CLBC, a Crown Corporation) to determine eligibility for residential DD services. He also met the other two criteria for a diagnosis of ID: onset in the developmental (pre-age-18) period and significant deficits in adaptive functioning. Starting at age 15, Neal was funded by CLBC for living in foster homes and eventually in his own small home due to behavior outbursts, with one-on-one daily supports but with continued involvement by his adoptive parents. The program was costly ($77,000 per year) but less expensive than institutional alternatives (such as prison), given Neal's very poor judgment and impulse control problems (e.g., lashing out at others).

When Neal reached age 19, his eligibility for CLBC services needed to be re-established. However, his newly-obtained IQ score of 79 was now a few points above the approximately minus two standard deviations level, which was cited as the basis for denial of services even though his adaptive behavior scores were still in the moderate (minus three standard deviations) range. In many other states or provinces, once someone is in the system and begins receiving services, eligibility remains intact, regardless of any change in IQ. Neal's family sued CLBC and a panel of judges ruled that the use of a specific IQ cut-off to determine continued eligibility, which had never been discussed in the agency's authorizing legislative history, was arbitrary and unfair. The agency was ordered to continue serving him and to devise a more equitable, and less IQ-driven, formula for making eligibility determinations for other applicants.

After considering a framework with broader applicability, the BC Ministry of Community Living Services (the government agency over CLBC) came up with a formula that appears to have been very narrowly tailored to people exactly like Neal Fahlman. Essentially, they established two pathways to residential services: (a) the traditional one, for people who qualified for a diagnosis of ID, which required an IQ below 70 along with adaptive deficits and onset in childhood; or (b) an alternative pathway, only available to people who—like Neal—had ASD or FASD. In addition to having onset in the developmental period, successful applicants needed to have deficits in adaptive and intellectual functioning, but IQ scores could be above the traditional 70–75 ceiling. However, to ensure that applicants were sufficiently deserving of services, prong two (adaptive functioning) required a score that fell at least three standard deviations below the population mean on a standardized rating measure. This happened to be the precise profile characterizing Neal Fahlman.

There are three problems with the ID-Equivalence solution in British Columbia that was devised in response to the Neal Fahlman lawsuit. The first problem was that the BC "fix" continued to be based on artificial and arbitrary numeric criteria, i.e., the numbers were adjusted to make the judges in the Fahlman case happy. The underlying issue raised by the court—namely the need to look at the whole individual—was not really addressed at all.

A second problem with the BC solution was that adaptive deficiency was defined as minus three standard deviation units, which was far too stringent a requirement as it included only a fraction of the first percentile of the population. This was also grossly inconsistent with the usual prong two criterion, which at minus two standard deviation units (at or below the second percentile) was more in line with the population for whom ID-equivalence relief was being sought. Our guess, which has been confirmed by conversations with experts and applicants in BC, is that adaptive behavior informants such as parents and even evaluators such as psychologists are then motivated or perhaps forced under the BC solution to exaggerate the severity of an applicant's deficits. There is obviously something wrong with such a state of affairs, as it is inequitable to reward those savvy or dishonest enough to exaggerate, while punishing those too unsophisticated or honest to do so.

The third problem with the BC solution is that the ID-equivalency pathway was made available only to people with one of the two disorders—ASD and FASD–which applied specifically to Mr. Fahlman. What about people with the dozens if not hundreds of other brain-based developmental syndromes who have the same problem of significant adaptive needs but whose IQs straddled the minus two standard deviation demarcation line? For example, do individuals with Prader-Willi syndrome or Dandy-Walker malformation, where there is an ID-equivalent need for services but where IQs sometimes fall above 70, have to file their own lawsuits when they are denied services? A more equitable solution would have been to use general language, such as "persons with brain-based neurodevelopmental disorders" rather than "persons with autism or FASD." Consideration of equity applies here as well, as it is unfair to give relief to very politically influential advocacy groups (such as parents of autistic children) and deny it to equally deserving clients (such as individuals with, say, Dandy-Walker syndrome, where IQs also fall on both sides of the 70–75 barrier) whose disorders are less known and thus lack comparable visibility or influence.

Building ID-Equivalency Around Prader-Willi Syndrome

Just as FASD is granted ID-Equivalency in British Columbia (as long as the minus three standard deviation criterion on prong two is met or faked), other specific syndromes have been specified in various jurisdictions for ID-Equivalency. A difference is that for these disorders, eligibility is sometimes met automatically if the syndrome diagnosis is established without any, let alone a more severe, adaptive behavior deficit finding. Perhaps the most prevalent of these exemption examples includes Prader-Willi syndrome (PWS), a rare (1:10,000 to 1:25,000 live births) developmental disorder caused by a deletion or partial expression of several genes on chromosome 15. Among the symptoms of PWS are cognitive deficits typically, but not always, involving IQ below minus two standard deviations, significant adaptive deficits, and a compulsive need to eat constantly often resulting in morbid obesity.

A statute in Connecticut gives automatic ID-Equivalency status exclusively to applicants for ID services who have PWS, regardless of IQ. According to a 2006 website description of the Connecticut Department of Mental Retardation (since renamed the Department of Developmental Services): "An application for eligibility determination may be made by… any person who is a resident of Connecticut at the time application is made, or by someone on the person's behalf, and who is, appears to be, or believes him/herself to be a person with mental retardation, as defined in Connecticut General Statutes 1–1 g or Prader-Willi Syndrome…" Later, the eligibility criteria are spelled out thusly: "…The process by which the Department examines information relative to an applicant for Department to determine if the applicant meets the statutory criteria for mental retardation or Prader-Willi Syndrome," and PWS is described as "a neurobehavioral genetic disorder that has been diagnosed by a physician utilizing medically appropriate criteria." Such a diagnosis, which until the 1990s was based mainly on clinical criteria (e.g., hypotonia, hypogonadism, inability to achieve food satiation), is now derived from extremely reliable DNA-based genetic testing methods.

The relevant aspect of the Connecticut exemption is that applicants with PWS do not have to meet any other criteria for eligibility such as level of adaptive behavior impairment. Apparently it is just assumed that an individual with PWS, a spectrum disorder with varying levels of severity, has significant enough service needs to merit DD services whether or not his/her IQ score falls below 70–75. Several other states also list PWS in their eligibility criteria but embed it within a slightly broader list of neurodevelopmental disorders and also specify that an individual still has to meet prong two, the adaptive functioning deficits. An example is Wisconsin, which specifies that developmental services are for people who have "…a disability attributable to brain injury, cerebral palsy, epilepsy, autism, Prader-Willi syndrome, intellectual disability, or another neurological condition closely related to an intellectual disability or requiring treatment similar to that required for individuals with an intellectual disability, which has continued or can be expected to continue indefinitely and constitutes a substantial handicap to the afflicted individual…." Similarly, Massachusetts affords ID-Equivalence status to people with "… cerebral palsy, epilepsy, autism, Prader-Willi or other condition other than mental illness or an emotional disturbance; closely related to mental retardation because the condition results in impairment of general intellectual functional or adaptive behavior similar to those with mental retardation; manifests before 22 years of age; [is] likely to continue indefinitely; [and] results in substantial limitations in three or more of major life activities: self-care, understanding and use of language, learning, mobility, self-direction, [and] capacity for independent living."

PWS is not the only chromosome abnormality syndrome given special ID-Equivalency status, but it is the most widely-noted. In Manitoba, special mention is given to Patau syndrome (a trisomy on chromosome 13) and Edward syndrome (a trisomy on chromosome 18). An advantage of the Manitoba legislation is that it is worded as "chromosome disorders *like* [emphasis added] Patau syndrome and Edward syndrome," so that the conditions are illustrative of a broader class of disorders. In contrast, Connecticut mentions only PWS. Patau syndrome has about the same incidence as PWS, while among the chromosomal disorders, Edward syndrome is more prevalent (1:6,000 live births), second only to Down syndrome (1:1,000 live births). Although one of the most common biological causes of ID, Down syndrome (DS) is not mentioned in ID-Equivalence statutes—except in Arkansas–perhaps because there is rarely any straddling of the 70–75 divide and therefore virtually all people with DS have no problem qualifying for ID services. Yet, we have found specific mention of Patau or Edward syndromes as ID-Equivalency triggers only in Manitoba, while PWS is mentioned in many places. The likely explanation for this discrepancy is that PWS parent support and advocacy groups are found virtually everywhere, while the "International Trisomy 18/13 Alliance" is much less visible. However, it does have an active presence in Western Canada, which may explain why it is mentioned in the Manitoba legislation.

According to a PWS researcher with first-hand knowledge (Dykens 1996), the legislation that established automatic ID-Equivalence for applicants with PWS in Connecticut occurred as a result of successful legislative lobbying of the state legislature by PWS parents and advocates. Not long after this legislation was enacted, a request was made for similar special treatment (i.e., ID-Equivalence) by parents of children with autism which is also a spectrum disorder straddling both sides of the 70–75 IQ divide but with virtually all diagnosed individuals exhibiting severe adaptive deficits and service needs. However, this request was turned down for three related reasons: (a) autism is diagnosed clinically without absolute certainty of a biological (DNA) test as in PWS; (b) the diagnosis of autism is made too freely (which reportedly is why DSM-5 eliminated Asperger disorder) and consequently includes many individuals unlikely to ever need disability services; and (c) even if an autism diagnosis was reliable and valid, the numbers of affected individuals would be enormous (perhaps as high as 1 %) in comparison to PWS; thus granting automatic ID-Equivalency to all autistic individuals would likely have severe fiscal consequences for any state or provincial human services budget. Unfortunately, ID definitions and diagnoses are in part driven by political and economic considerations, much as we might wish that were not the case.

By implication, therefore, the political initiative of PWS advocates on behalf of automatic ID-Equivalency status was more likely to be successful because diagnosis of PWS was highly reliable, virtually all affected individuals needed and deserved DD services, and—last but not least—the prevalence rate for PWS was low enough that such a scheme was financially viable. Still, one could make the same inequity argument against singling out PWS in Connecticut as we made against singling out FASD (and Neal Fahlman's other diagnosis, autism) in British Columbia.

Moreover, the case against PWS could be made even more strongly, as the lawsuit-driven solution in BC still required significant evidence of adaptive deficits. In fact, this was so excessive as to almost ensure fraud, while such (or any) evidence of adaptive deficits was not a requirement in Connecticut for PWS eligibility. Looking at this history and convoluted bases for "fixes" to the eligibility criteria, it seems questionable to us that a single genetic disorder is singled out for special ID-Equivalency treatment when there are so many other syndromes (Patau and Edward, to name but two) that are equally deserving. It is clear then that the broader problem goes beyond the naming of a specific ID-qualifying disorder and involves the somewhat outdated emphasis on underlying specific medical categorization itself.

Developmental Disabilities Solution

A major way in which government entities increase the population eligible for ID services beyond those with IQ scores at or below 70–75 is to use the broader term "Developmental Disabilities" (Administration for Community Living 2013; Disability Law Center, undated; National Council on Disability 2012; Zaharia and Moseley 2008). This term owes its origins to pioneering Kennedy-era legislation in the United States which authorized various government-funded disability-related human rights, research, training and facility-building initiatives (Gettings 2011). The term is now widely used in many state and provincial eligibility statutes. The original legislation—the "Mental Retardation Facilities and Community Mental Health Centers Construction Act of 1963"—defined the population addressed by these programs as people with "mental retardation." Seven years later, the law—renamed the "Developmental Disabilities Services and Facilities Construction Amendments of 1970"– was revised, with the term "Developmental Disabilities" (DD) substituted for "mental retardation." However, DD was still defined categorically as in this expanded list: "mental retardation, epilepsy, cerebral palsy, and other neurological conditions originating before the age of 18." In 1975, the legislation was revised again, and DD was again defined categorically to include mental retardation plus these conditions "closely related to mental retardation": cerebral palsy, epilepsy, autism and dyslexia, again with a pre-18 age of onset, which were expected to continue indefinitely and that constitute a substantial handicap." The term "other neurological conditions" was dropped for some reason. In our opinion that was a mistake as such a loophole along with qualifying language such as "producing service needs similar to those needed by people with ID", has since been added in many places to keep IQ-Equivalency from being unfairly limited to people in only a few, and somewhat arbitrarily chosen, diagnostic categories.

A major revision in 1978, termed the "Developmentally Disabled Assistance and Bill of Rights Act," raised the age-of-onset from 18 to 22 and switched from a categorical to functional definition of DD as a "severe, chronic disability…attributable to a physical or mental impairment…likely to continue indefinitely" that resulted in "substantial functional limitations in three or more areas of major life activity." A final revision in 2000 [note: the unusual reversal of typical order of Roman numerals is reported here as it actually was in the statute] defined DD as "a severe, chronic disability of an individual that (i) is attributable to a mental or physical impairment or combination of mental and physical impairments; (ii) is manifested before the individual attains age 22; (iii) is likely to continue indefinitely; (iv) results in substantial functional limitations in 3 or more of the following seven areas of major life activity: (I) Self-care, (II) Receptive and expressive language, (III) Learning, (IV) Mobility, (V) Self-direction, (VI) Capacity for independent living, (VII) Economic self-sufficiency; and (v) reflects the individual's need for a combination and sequence of special, interdisciplinary, or generic services, individualized supports, or other forms of assistance that are of lifelong or extended duration and are individually planned and coordinated." The 2000 law also clarified the application of the DD definition for children from birth through age nine, by stating that a child could still be considered to have DD without meeting all of the above criteria if "the individual, without services and supports, has a high probability of meeting these criteria later in life."

The intent of this functional and arbitrary formulation appears to have been an attempt to capture the adaptive limitation profiles of individuals who functioned as if they had ID in spite of having IQs that fell above the 70–75 IQ ceiling. However, at least two of the skills, language and mobility, were not specific to ID. Likewise, one also could argue that self-direction was not specific to ID-equivalency. It is not clear where this list came from as there is no science to support it. Certainly, a limitation in this list is that none of the items address deficits in social functioning, which many people (and virtually all family members) consider to be at the top of any list of reasons why people with ID need protections and supports.

Another curious aspect of this formulation is that it was never intended to be a diagnostic framework; rather, it was intended to clarify the scope and focus of various federally-funded programs (Developmental Disabilities Councils, Offices of Protection and Advocacy, University Affiliated training centers) and related facilities. Yet, it morphed over time into something approaching a diagnostic framework in spite of the questionable and slap-dash nature of the list of seven life functions. Today, there are many jurisdictions in both Canada and the United States that use the Developmental Disabilities construct, and several that use the 3-out-of-7 life skills areas specified in the 2000 legislation described above. One place that mentions FASD as an IQ-equivalent condition example is Minnesota, whose statute 252.27 (2012) notes several "related conditions," defined as: "a condition that is found to be closely related to a developmental disability, including but not limited to, cerebral palsy, epilepsy, autism, fetal alcohol spectrum disorder, and Prader-Willi syndrome." Like most statutes, the one in Minnesota excludes mental illness, but states that autism (at one time thought to be a form of childhood schizophrenia but now listed as a neurodevelopmental disorder) is not considered a mental illness and thus is not excluded.

An obvious advantage of using terms like "such as" or "not limited to" is that ID-Equivalence is then not limited to the few disorders listed but can be much more expansive. Minnesota is one of a very small list of jurisdictions where FAS is specifically included in an expanded disorder list. However, Minnesota's eligibility document then goes on to state that even if one has a qualifying medical underlying disorder, the condition must still cause "substantial functional limitations," as established by deficits in three out of the seven adaptive life activities.

The "Similar Services" Solution

As noted in the previous section, a number of jurisdictions have a mixed categorical/functional approach to ID-Exemption, but these generally lack an explicit statement that ID is the core construct upon which they are basing the expanded category list as well as the functional life skills impairment list. Such a statement would be a useful addition, as it would indicate the real purpose of an expanded list, which is to enable agencies to do the right thing, namely grant DD services and protections to people who clearly deserve them but are denied them solely because their full-scale IQ is above the arbitrary 70–75 full-scale IQ ceiling. Such a statement also would enable eligibility determiners to avoid being limited to the disorders in an expanded list, as adding four or five such categories (e.g., autism, FASD, Prader-Willi, etc.) still fails to include dozens of other disorders involving full-scale IQ scores that straddle the 70–75 barrier known to cause or be strongly associated with ID and serious adaptive deficits.

A good example of an ID service eligibility framework that went from an expanded categorical definition with a functional overlay to adding a similar-services component is California's vast regional center system. The regional centers are state-funded agencies that act as the gateway through which applicants for ID services in the state must pass (Disability Rights California 2012). The authorizing legislation, the Lanterman Developmental Disabilities Act (AB 846), which is widely known as the Lanterman Act, was initially proposed in 1973 and passed in 1977. The Act significantly expanded upon the Lanterman Mental Retardation Services Act (AB 225), initially proposed in 1969. Although originally created to serve people with mental retardation, the regional centers were later mandated to serve persons with four conditions: mental retardation (today termed ID), cerebral palsy, epilepsy and autism. This list obviously was directly modeled after Kennedy-era federal Developmental Disability legislation. To be eligible for services under the Lanterman Act, a person had to have a "substantial disability." Thus, people with epilepsy, who are at risk for cognitive impairment but often have superior intelligence, would not be eligible automatically for ID services.

In 1976, the Lanterman Act was amended to establish the right to treatment and habilitation services for persons with developmental disabilities. In 2003, the definition of "substantial disability" was amended to require the existence of significant functional limitations in "three or more…areas of major life activity…."

Previously, having a "substantial disability" only required the existence of a significant functional limitation in one area of major life activity. In these "major life activities" were self-care, receptive and expressive language, learning, mobility, self-direction, capacity for independent living and economic self-sufficiency. It is noteworthy that this is the same list (and 3-out-of-7 formula) previously developed in federal Kennedy-era legislation. As mentioned, that legislation was originally intended mainly to authorize programs (such as state DD councils), and the life skills list and formula were not necessarily intended to be a guide to diagnosis or service eligibility. However, practical application of the legislation has morphed into that in spite of the fact that the list and the numeric formula do not appear to have any convincing empirical or theoretical rationale.

Further delimiting the scope of the regional centers, the Lanterman Act specifically excluded "conditions that are considered solely a learning disorder, solely a psychiatric disorder or solely a physical disorder" (Cal. Code Regs., Tit. 17, Sec. 54000[c][3]). This obviously could be a source of confusion, as people with ID often have psychiatric and motor problems. A major innovation in the Lanterman Act, which keeps it from being unworkable with its very outmoded and overly limited list of four disorders, is what has been referred to colloquially as California's "fifth category" (Disability Rights California 2012). The fifth category is made up of "conditions similar to mental retardation, or conditions that require treatment similar to the treatment required for individuals with mental retardation." The "similar conditions" and "similar treatment" formulations make it possible in theory for people with a wide range of brain-based conditions that produce adaptive needs similar to those found in people with ID to receive the services to which they are entitled. The legislation does not specifically refer to a too-high IQ as the reason for this Fifth Category loophole, but obviously that is its justification, as is the case with any ID-equivalency provision.

In practice, attaining service eligibility for a too-high IQ applicant under the Fifth Category has proven to be anything but easy, as the application process often involves initial disapproval followed by one or more appeals and even a lawsuit before, in some cases, a successful resolution. Undoubtedly, the reason for this difficulty reflects insufficient funding to serve a large and growing pool of applicants, but it also may reflect the insidious survival of an IQ-ceiling-driven mindset, in spite of legislation which seems to explicitly allow IQ ceilings to be ignored when appropriate. Two court cases have served to clarify the intent of the Fifth Category provision and force the regional centers to more fully live up to their obligation to serve people who fall within the purview of the Fifth Category: *Mason* v. *Office of Administration Hearings* (89 Cal.App.4th 1128, 2001), referred to as "Mason," and *Samantha C. v. DDS* (185 Cal.App.4th 1462, 2010), referred to as "Samantha C."

The issue in Mason was whether having a "condition similar to mental retardation" meant one had to have an IQ in the same 70–75 IQ range as found in regular old-fashioned MR/ID. Of course, such an interpretation was a perversion of the ID-Equivalency purpose of the Fifth Category, but it was made possible because the Act failed to specifically mention IQ-ceiling waiver as a major reason for the provision. This position is reflected in the following exchange involving Dr. Bob Chang, a psychologist employed by the plaintiff in the lawsuit, and the Inland (San Bernardino area) Regional Center (IRC). When Dr. Chang was asked in a hearing if a person could fall in the Fifth Category with an IQ score in the low-average range, he answered "No. Low-average intelligence is not a condition similar to mental retardation. It is statistically significantly different. Low-average general intelligence is very different than somebody who is mentally retarded." So much for the idea of ID-Equivalency as a way out of the straightjacket imposed by rigid adherence to IQ ceilings!

The 2010 Mason case involved an appeal of a turn-down by IRC of ID services for a child who had a seizure disorder at birth and actually was institutionalized for a while in the now-shuttered Lanterman Developmental Center (earlier known as Pacific State Hospital). The IRC refused Mason for ID services because of an IQ score that was a little too high in spite of substantial adaptive deficits. In addition to arguing that the Fifth Category did not apply to applicants like Mason with above-75 IQs, the IRC also argued that the Fifth Category was unconstitutionally vague. The court rejected the IRC's position, stating that the wording of the Fifth Category was sufficiently clear, and that it made no sense to use as a criterion for the Fifth Category the same exact criterion that was used to define ID. However, while a victory for Mason in providing a broader interpretation of the Fifth Category, the decision of the appellate court was a loss for him in that it did not insist that he qualified for ID services, even under an expanded approach to eligibility. The reason given by the court was that Mason had a diagnosis of Learning Disability (LD) and consequently could be educated in a regular classroom. Plus, there was no evidence that a child with LD, defined in part by having an IQ score above the ID range, needed services similar to those provided to children with ID. Thus, in spite of the intention that the Fifth Category would provide an alternative to IQ-based eligibility determination, the Mason decision illustrated just how pervasive and persistent such views continued to be and how little weight was given to adaptive behavior scores in eligibility determination decisions.

The 2011 Samantha C case represented a leap forward in interpretation of the Fifth Category, particularly in clarifying the meaning of the "similar treatment" clause. The better outcome for the plaintiff in this case likely reflected advances in ID services philosophy and understanding of cognitive disabilities in the intervening 9 years, but undoubtedly it also reflected differences in the impairment severity profiles of the two plaintiffs. The basic facts of the case were that Samantha was deprived of oxygen for 30 min during the birth process and experienced significant cognitive and adaptive problems as a result. The plaintiff in this case, Harbor (Long Beach area) Regional Center (HRC), argued that Samantha was not entitled to state ID services because she had been labeled LD and ADHD. The court sided with an expert who said that Samantha's problems were better attributed to significant anoxia (oxygen deprivation) at birth and that her need for services similar to those required by people who were labeled ID because of low-enough IQ made her clearly eligible under the Fifth Category. HRC was ordered by the court to provide Samantha with appropriate ID services.

As the focus of this book is on legal issues involving people with FASD, it is important to include a brief account of a recent case–with some similarities to Mason and particularly Samantha C—involving a service applicant with FASD. The case is not cited in any court ruling because it was successfully resolved through the application appeal process before court intervention had to be sought. Nonetheless, the case is somewhat known in the literature as a profile of the petitioner—whom we refer to as "Lisa"—was contained in an article on adaptive functioning and FASD, which two of the authors published in a special FASD issue of the *Journal of Psychiatry and Law* (Edwards and Greenspan 2011). The account that follows is basically an update to the earlier one.

Lisa, the daughter of a woman who abused alcohol and lived in Los Angeles, was diagnosed with FASD as a young adult. She had two siblings who were diagnosed with FASD during childhood. Lisa suffered very severe abuse from her father, who went to prison as a result. She and her brother, who became a regional center client, were discovered by police living unclothed in dog cages when she was age four. Lisa was placed in a series of foster and institutional settings, where she received a great many psychiatric diagnoses. Lisa had a temper, and as a young adult, she got into a dispute with a caregiver that led to her setting a small fire outside a group home from which she had been expelled. While in jail for 2 years following the arson incident, she attempted suicide (one of several attempts) and almost succeeded. When her sentence was completed, she was committed to a secure psychiatric facility, where the state sought to have her permanently incarcerated.

Due to the efforts of her new attorney, Lisa eventually received a diagnosis of FASD from a team of knowledgeable experts. Using this diagnosis, and based on the experts' opinions that with proper care Lisa would no longer be a threat to herself or others, the attorney was able to persuade a judge to release her from confinement if a suitable residential and case management plan could be developed. An application was made to a regional center for ID services under the Fifth Category, but the application was turned down on the basis that Lisa's full-scale IQ (in the 90s) was too high and that her problems were mainly psychiatric and therefore not covered under the Lanterman Act. A disability rights group took Lisa on as a client. They used the fact that she was living temporarily in a different and better funded jurisdiction to make a new application to a different regional center, one which they believed was more likely to be sympathetic. This time, the application was successful on the basis that Lisa had a neurodevelopmental disorder (FASD) that produced cognitive deficits. These included major problems in executive functioning that impaired her judgment and made it necessary for her to live and work in supervised settings and receive case management services similar to clients with ID, many of whom also had psychiatric difficulties. As of today, Lisa is happy and thriving in an environment that is much more supportive, skilled and loving than would have been the case had she continued to be served through California's mental health and correctional systems. Lisa's case is relevant and potentially important on two related grounds: (a) she qualified for ID services with an IQ score that was essentially normal, and (b) the outcome establishes, at least in California, a *prima facie* case that people with FASD should be served by the ID service system if they have sufficient service needs that can be shown to be congenitally brain-based.

Adaptive Behavior Substitution for IQ

Adaptive Behavior, also known as Adaptive Functioning, plays an important role in determining ID eligibility even if IQ meets the minus two standard deviation criterion. When using this more conventional route, having low adaptive behavior by itself is not sufficient for ID eligibility if the IQ criterion is not met. In the Developmental Disabilities solution with singled-out diagnostic categories such as autism or cerebral palsy, service eligibility still requires significant deficits in adaptive behavior, even if it the criterion is not as stringent as that used in British Columbia. In the Similar Solutions approach described in the preceding section, adaptive behavior plays an important role, as seen in the case of Lisa. That young woman, who had a clear-cut diagnosis of Partial Fetal Alcohol Syndrome (PFAS), was granted ID-Equivalence status through California's so-called "fifth category" only because she has very significant support needs which were reflected in substantially impaired levels of adaptive behavior. An even more explicit emphasis on adaptive behavior as a basis for ID equivalence can be found in Colorado's recent reformulated eligibility standard (Block 2013).

Colorado revised its eligibility criteria in April 2014 for the usual reason: too many children and adults with brain-based developmental disorders (such as autism) and very significant support needs were being denied service eligibility because of IQ scores over 70. The solution devised is that a developmental disability "means a disability that is manifested before the person reaches 22 years of age, which constitutes a substantial disability to the affected individual, and is attributable to mental retardation or related conditions which include cerebral palsy, epilepsy, autism or other neurological conditions when such conditions result in impairment of general intellectual functioning or adaptive behavior similar to that of a person with mental retardation. (C.R.S. 27-10.5-102 11 (a), as amended)." The innovative thing about this provision is that for ID-Equivalency qualifying purposes, adaptive behavior now has a status equal to that of IQ, and too high IQ score is no longer an impediment to receiving ID services. Mentioning "neurological conditions" is also an important feature, in that it reinforces the critical (but rarely stated) notion that ID reflects a failure of brain development.

The Colorado emphasis on similar "impairment" differs from the California emphasis on similar support needs in that the former is more grounded in psychometrics, while the latter is more grounded in qualitative and subjective perceptions of need. Specifically, the Colorado statute specifies that a deficit in General Intellectual Functioning is to be determined by an IQ score below 70 (for the first route of MR/ID), while a deficit in Adaptive Behavior is to be determined by deficits on a standardized instrument of adaptive functioning, such as the Adaptive Behavior Assessment System, second edition (ABAS-2). According to Block (2013), who analyzed the revised Colorado standards while still in the proposal stage, the adaptive behavior standard was to be met by significant deficits in two of the eleven adaptive skills in the old (2000) DSM-4TR definition and the very old (1992) AAIDD definition of ID. This antiquated view of adaptive behavior/adaptive functioning is obviously problematic, in part because the taxonomy on which it is based lacks content validity (Zigler et al. 1984). The content validity of adaptive behavior is a topic outside the scope of this chapter, but suffice it to say that the absence of an adequate model or measure of adaptive behavior has obvious relevance for devising alternative (ID-Equivalence) eligibility pathways. This is because impaired functioning in the various community contexts (i.e. school, social relationships, independent living, etc.) is used as a prime reason for giving less weight to IQ, and adaptive behavior is, in theory, a way of getting at community functioning. A brief discussion of the DD behavioral phenotype—indicating why many people with FASD and related brain-based disorders deserve services or other accommodations—is therefore in order and will be provided later.

FASD and ID-Equivalence in Civil and Criminal Contexts

As noted, FASD is one of the brain-based syndromes mentioned for ID-Equivalency consideration in DD service systems in some jurisdictions. An example is Alaska (Title 47, chapter 20, Alaska Stat. § 47.20.290, 2012), where the term "Developmentally Delayed or Disabled" is used to describe children eligible for early intervention services. Developmentally delayed is defined as "… functioning at least 15 % below a chronological or corrected age or 1.5 standard deviations below age appropriate norms in one or more of the following areas: cognitive development, gross motor development, sensory development, speech or language development, or psychosocial development, including self-help skills and behavior, as measured and verified by appropriate diagnostic instruments and procedures or through systematic observation of functional abilities in a daily routine by two professionals and a parent, developmental history, and appropriate assessment procedures."

The term Developmentally Disabled is defined as "…having an identifiable physical, mental, sensory, or psychosocial condition that has a probability of resulting in developmental delay even though a developmental delay may not be exhibited at the time the condition is identified…" This list of such conditions is then supplied (note: we have changed the format slightly to fit it into a paragraph style): "…(a) chromosomal abnormalities associated with delays in development, such as Down's syndrome, Turner's syndrome, Cornelia de Lange syndrome, or fragile X syndrome; (b) other syndromes and conditions associated with delays in development, such as fetal alcohol syndrome, cocaine and other drug-related syndromes, metabolic disorders, cleft lip, or cleft palate; (c) neurological disorders associated with delays in development, such as cerebral palsy, microcephaly, hydrocephaly, spina bifida, or periventricular leukomalacia; (d) sensory impairment, such as hearing loss or deafness, visual loss or blindness, or a combination of hearing and visual loss, that interferes with the child's ability to respond effectively to environmental stimulus; (e) congenital infections, such as rubella, cytomegalovirus, toxoplasmosis, or acquired immune deficiency syndrome; (f) chronic illness or conditions that may limit learning or development, such as cystic fibrosis, bronchopulmonary dysplasia, tracheostomies, amputations, arthritis, or muscular dystrophy; (g) psychosocial disorders, such as reactive attachment disorder, infant autism, or childhood schizophrenia; or (h) atypical growth patterns consistent with a prognosis of developmental delay based upon parental and professional judgment, such as failure to thrive."

This statutory provision is aimed specifically at early intervention service eligibility which, admittedly, is different and usually more inclusive than eligibility for adult residential or case management services might be. However, the approach offers some innovative aspects which could possibly serve as a model for adult services eligibility. Here are the innovations in this document as we see them: (a) a more flexible approach to standard score ceilings, in that the cut-scores mentioned (minus 1.5 standard deviations equates to a standard score of 81.5, which is at the 11th percentile of the population; also mentioned is the 15th percentile, which is a standard score of 85.5) is considerably higher than the more usual 70–75 cut score, and likely to cover a substantial percentage of the population of children with FASD or other neurodevelopmental disorders, thereby considerably reducing the problem of false negatives; (b) the use of a broader list of deficit areas without limiting deficits to cognition only, and also defining cognition more broadly than just IQ; (c) the listing of many different types and categories of disorders strongly associated with ID, but including the expansive term "such as" rather than simply limiting the possibilities to a few named conditions; specifically mentioning FAS (and presumably other FASD conditions) under a category termed "other syndromes and conditions associated with delays in development"; and (e) mentioning the possibility that bases may include the possibility of observations by qualified assessors rather than limiting the bases to just measures or statistics.

Although our emphasis in this paper is on service eligibility formulas, it is worth mentioning that the issue of ID equivalency also comes up in the criminal arena, with FASD being a prime focus of such discussion. In the case of Brandy Aileen Holmes v State of Louisiana, a young woman with FAS in her early 20s was condemned to death in 2006 along with a male co-defendant after shooting an elderly man (who died) and his wife during a home-invasion robbery in 2003. In 2009, a petition to the Supreme Court of Louisiana asked for reconsideration of her conviction on the basis that Ms. Holmes' FAS diagnosis should have been taken into account in combination with her uncorrected IQ of 77 as a mitigating factor when determining her sentence. In particular, the defendant's apparent lack of empathy or remorse (symptoms of her FAS) were held against her as an aggravating factor. In an amicus brief from the National Organization for Fetal

Alcohol Syndrome (2009), it was argued that FASD causes adaptive deficits very similar to those found in people with ID and thus should have resulted in a finding of reduced moral culpability and execution exemption, as in the US Supreme Court's 2002 *Atkins v Virginia* decision, which exempted people with ID from facing execution. The Supreme Court of Louisiana rejected the *certiori* petition, but a dissenting opinion from the court's chief justice made a convincing case that Ms. Holmes likely qualified for a diagnosis of ID, given her FAS, low IQ, overt signs of brain damage, and poor adaptive behavior.

The State of Alaska, already a leader in recognizing FASD as a basis for ID Equivalence in determining human services eligibility, is also a leader in asserting the role of the disorder as a mitigating factor when determining criminal culpability and punishment. One assumes the advanced awareness of FASD in that state stems at least in part from the existence of a very large native Eskimo and Aleut Indian population, which is the US racial sub-group with the highest incidence of diagnosed FASD (Tenkku et al. 2009). In June 2012, both houses of the Alaska legislature unanimously passed SB 151: "An Act relating to mitigation at sentencing in a criminal case for a defendant found by the court to have been affected by a fetal alcohol spectrum disorder." According to the Alaska FASD Partnership Newsletter (2012), the bill was put forward by a large workgroup, including court personnel, concerned by "the large number of people affected by… FASD in the state's criminal justice system."

The essence of SB 151 is to allow judges flexibility in sentencing people with FASD if it is established that a defendant has the disorder and that it "substantially impaired the defendant's judgment, behavior, capacity to recognize reality, or ability to cope with the ordinary demands of life, and that the fetal alcohol spectrum disorder, though insufficient to constitute a complete defense, significantly affected the defendant's conduct." Three limiting aspects of the Act are that: (a) a judge is not required to use FASD as a mitigating factor, (b) the defendant would have to prove "by clear and convincing evidence" that he or she has FASD and it was a factor in the alleged criminal offense, and (c) the Act applies only to crimes "not against the person," and thus does not apply to sex acts, assault or homicide. Two other factors not in the Act but which limit its application are: (a) it takes resources to establish FASD convincingly, and these resources typically are not available for defendants accused of minor crimes, and (b) unless someone has previously been found to have an FASD, relatively few attorneys will recognize it as a possible diagnosis, especially for the large percentage of people in the spectrum who lack obvious physical signs (e.g., Alcohol Related Neurodevelopmental Disorder or ARND). In spite of the Act's limitations (which according to an Alaska lawyer informant, has resulted in its very infrequent application thus far), it represents a major step forward toward what its supporters term "smart justice," namely a more individualized and humane approach to punishment which recognizes that people with FASD (and organic disorders generally) should not automatically receive a lengthy, or any, mandated jail sentence for an act which reflects a brain-based condition for which they are not responsible.

Although the language in SB 151 does not explicitly make an argument for ID-Equivalence, it can be inferred from the comments of its supporters that such was the intent. SB 151 was intended to divert offenders with FASD from trial or jail, while another statute–Alaska 12:55.155 ("Factors in aggravation and mitigation")–was aimed at more serious offenses and the imposition of reduced sentences when certain conditions, including ID, were present. Under section (34) (d) (18), it was specified that a sentence less than the minimum might be imposed if a defendant committed an offense "while suffering from a mental disease or defect…that was insufficient to constitute a complete defense but that significantly affected the defendant's conduct." Mental disease or defect specifically included ID, as indicated in Alaska statute 12,47.130, and very similar language as found in SB 151 in describing "a disorder… that substantially impairs judgment, behavior, capacity to recognize reality, or ability to cope with the ordinary demands of life."

Conclusion: Seeking the ID Behavioral Phenotype

The move to devise ID-Equivalency pathways for service eligibility reflects a belief that direct real-world functioning (e.g., observed need for supports) should be given more weight than indirect performance on a measure (e.g., IQ score) that imperfectly predicts real-world functioning. In this concluding section, we suggest the parameters of what such a behavioral phenotype might involve and provide some suggestions for a DD services eligibility scheme that would better benefit people with FASD and other brain-based disorders. To some extent this involves a revisiting of some issues addressed in the Introduction, but with a consideration of lessons learned from a review of the diverse formulae that have been devised to get around the excessive reliance on IQ cut-offs.

A Broader Approach to Cognitive Impairment

A mistake made in some eligibility solutions is to assume that to eliminate excessive reliance on IQ and IQ ceilings requires one to get rid of an emphasis on cognition as a central feature of ID and ID-Equivalence. In fact, deficits in thinking, reasoning, learning and "intelligence" (broadly defined) are central to understanding ID, FASD and the pathway to ID Equivalence. What needs to be done, however, is to find a key to the cognitive essence of DD that is not grounded solely in IQ or an IQ ceiling. In fact, as discussed in the Introduction, full-scale IQ is an outmoded concept that is still taken seriously by very few leading cognitive psychology scholars, even if, because of simplicity and ease of decision-making, it is still central to judicial and administrative approaches to disability determination. Intelligence is just too broad-based a construct to be adequately summarized by a single score from a single test. Furthermore, critical aspects (such as social intelligence) are not covered at all by existing tests. As well, the idea that one can identify a specific ceiling score on a single test (or more than one test, for that matter) which can reliably discriminate between those who qualify or do not qualify for service eligibility is questionable to say the least. So, at a minimum, one should rely on a cross-battery approach, where information is integrated in a profile rather than in single-number manner, combining sub-scale scores from multiple measures of intelligence. Such an approach puts greater demands on an evaluator but certainly will provide a richer and fuller picture of the individual being evaluated.

A growing consensus is emerging in the field of ID, and also in the FASD field, that the quality of cognitive impairment that most contributes to everyday functioning difficulties in people with brain-based disorders, involves skills captured by models and measures of "executive functioning." This term refers to a set of skills that are controlled primarily by the prefrontal cortex, with numerous neuronal connections to other brain regions, and involves "a set of interrelated supervisory attention and control processes in the brain that are involved in the selection, planning, initiation, execution, monitoring, and troubleshooting of goal-directed behavior in non-routine situations" (Brown and Connor 2014). According to Shallice (1982), the primary role of the executive system is to respond effectively to novel situations that are poorly served by automatic and habitual responses. Such situations include those where planning and decision-making are required, those that are not well rehearsed, and those that contain sequences of actions that have not been performed previously. Executive functioning is particularly important in situations that require error detection and correction, such as dangerous or technically difficult situations (Brown and Connor 2014), and also plays an important role in resisting the temptation—often stemming from pressure or inducements from manipulative others—to engage in foolish, dangerous or illegal behaviors.

Executive functioning differs from what is typically thought of as intelligence, because it involves taking what one knows (intelligence) and translating it into action (adaptive behavior). Thus, the concept of executive functioning does a better job than intelligence in explaining why it is that people with ID and ID-Equivalence (i.e., most people with FASD) get into serious trouble in ambiguous or novel situations. People with FASD are by definition deficient in many areas of executive functioning, and the same is true of all individuals with ID. It is because people with FASD and other brain-based disorders inevitably have more serious limitations in executive functioning than might be inferred from their sometimes "borderline" IQ scores that knowledgeable scholars and professionals argue for a more inclusive approach to DD eligibility. Such an inclusive approach would provide a path to services for people with brain-based cognitive disorders who do not quite make the artificial and arbitrary IQ ceilings mentioned in too many eligibility statements.

Quality of Behavioral Incompetence

The behavioral essence of ID then, which flows directly from the above cognitive limitations especially in executive functioning, is a tendency to show very bad judgment in situations that are novel, ambiguous or anxiety-producing. As first stated by Spitz (1988), it is now understood that ID is less a disorder of learning (i.e. acquisition of rote schemas) and more a disorder of thinking (i.e. flexible adjustment of schemas to adapt to changing situations where habitual schemas no longer work). Such flexibility in solving novel problems, so central to "thinking," is in fact exactly what is tapped by measures of executive functioning. Because poor judgment in approaching novel problems will place a person at risk for some undesired outcome—ranging from task or role failure to a catastrophic physical or social event—it has been argued (Greenspan et al. 2011) that ID is a "common sense deficit disorder" in which common sense is defined as awareness of physical or social risk. In fact, it is to reduce, eliminate or provide a buffer against risk and its consequences that DD services are provided, whether to protect people from physical consequences (e.g., starvation, burning down the house, medical emergencies, getting run over) or social consequences (e.g., financial or sexual exploitation, imprisonment, loss of employment, severe harassment or abuse) stemming from a failure to appreciate danger and how to avoid it.

As mentioned by Tassé (2009) and by us earlier in this paper, adaptive behavior has been poorly defined, and measures that attempted to assess it often: (a) involve skills that are too low level (e.g., toileting, feeding) to apply to people with mild ID; (b) involve skills (e.g., self-direction, friendliness) that are not specific to ID or even to disability per se; (c)) fail to involve or give sufficient emphasis to skills such as social judgment that are particularly central to the ID behavioral phenotype, and (d) were developed more with service programming in mind and not for the purpose of diagnostic determination. Two problems with existing measures of adaptive behavior are: (a) there is almost no mention of risk-awareness or risk-vulnerability, and (b) the items or subscale scores do not translate directly into need for services, supports or funding levels Greenspan (2009).

In line with the fixation in the ID field on academic deficits and outcomes, social functioning is given very short shrift as it is only one out of 10 sub-scales on the ABAS-2. Further, it is conceptualized more in terms of "niceness" (which, however important for popularity and success, is not specific to ID) than in terms of "social judgment" (which also is important for success but is very specific to ID and ID-equivalence). Needless to say, poor social judgment is one of the main defining features of FASD. However, an aspect of poor social judgment that is particularly important diagnostically, and which is especially connected to poor executive functioning, is judgment about the likely bad outcomes flowing from particular courses of action. Measures of intelligence, and cognitive measures in general, ask subjects to come up the best solution to a problem, while the real challenge in terms of adaptive survival, is being able to avoid choosing the worst (i.e., most risky) solution to a problem. The construct of executive functioning, because it taps the ability to systematically foresee the future consequences of a course of action, has great explanatory and predictive value in identifying people likely to need supports to protect them from risk. Such a determination should not be based solely on scores on tests, whether cognitive or adaptive, but should involve qualitative information about episodes involving bad consequences reflecting social and practical judgment deficits. In the case of most people with FASD, such information is easy to obtain, as their lives have generally been a long litany of poor decisions and disastrous outcomes. The key to obtaining DD services through the ID-equivalence route is to help to convince a gate-keeping agency (such as the regional center that granted fifth category status for "Lisa" in spite of a relatively high IQ) that the person's history of behavioral mistakes reflects a cognitively-based unawareness of risk rather than a character-based lack of acceptance of social norms.

That is not always an easy thing to establish, however, as people with FASD often present with a complicated mixture of ability and inability, with maladaptive behaviors that give the misleading impression of being psychiatric rather than cognitive or neurodevelopmental in origin. Among other frequent misconceptions of FASD are that: (a) it is a temporary state that can be overcome, and (b) that persons with FASD (who because of modularity of brain lesions appear smarter than they are) are malingering their cognitive limitations.

Rediscovering the Neurological Basis of ID/DD

A curiosity of the definitional and diagnostic manuals on ID and DD is that the words "brain" or "neurological" are generally nowhere to be found, except to a limited extent in chapters devoted to possible biological causes. Thus, if one peruses the various AAMR/AAIDD classification manuals, the definitional criteria are completely functional i.e. emphasizing deficits in various skill domains. One will not see mention of the word "brain" anywhere. This is somewhat puzzling, as the skill deficits found in people with ID are clearly a sign of subnormal brain functioning. This absence of emphasis has been rectified to some extent in the ID section in DSM-5, which: (a) contains some mention of the role of the brain in ID, (b) locates ID within a new super-category termed "Neurodevelopmental Disorders" (previously it was located in a section dealing with disorders of childhood and adolescence), and (c) has added a parenthetical second name to the disorder, by replacing DSM-IV-TR's term "Mental Retardation" with the two-part name "Intellectual Disability (Intellectual Developmental Disorder)."

The addition of the parenthetical second name reflects a wish to eventually join ICD-11, which has signaled its intent to change the category name to "Intellectual Developmental Disorder" (Salvador-Carulla et al. 2011).This is much more than merely a semantic change, as it reflects a paradigm shift from a "disability" approach which emphasizes arbitrary psychometric (especially IQ) cut-offs, to a more clinical and qualitative "disorder" approach which emphasizes the medical, etiological and neurodevelopmental nature of the condition. This approach is reflected in a general disavowal of rigid IQ cutoffs and a suggestion that executive functioning is generally a more important indicator of ID than IQ. It is also reflected in a disavowal of a particular age-of-onset (such as 18), and replacement with a more general mention of slowed development including a failure to ever attain adult levels of functioning.

The failure in state DD statutes to mention the brain abnormality basis of ID is particularly notable in light of the fact that all of the developmental syndromes mentioned to illustrate the need for ID-equivalence—whether Prader-Willi, cerebral palsy, traumatic brain injury, or FASD—are diseases that affect brain processes and structures. A small number of statutes, however, after mentioning a few syndromes, do insert words along the lines of "[and] other neurological conditions" (State of Washington), or "...other neurologically disabling conditions closely related to mental retardation and requiring similar..." (Montana). In fact, such language is a holdover from the 1970 Federal DD legislation, which first introduced the term "Developmental Disabilities" (DD), and by operationally defining it as "mental retardation, epilepsy, cerebral palsy, and other neurological conditions originating before the age of 18," strongly implied a brain basis for ID. However, the 1975 revision of the Act added autism and dyslexia. This was curious, as dyslexia typically does not affect everyday adaptive functioning. Yet, the revision dropped "and other neurological conditions." Probably, the dropping of "other neurological conditions" was motivated by bureaucratic and economic concern about opening the floodgates to hordes of applicants, and possibly also by overly concrete equating of the term with "traumatic brain injury." However, this change (which has been modeled in virtually all state DD statutes) was unfortunate not only because it made it more difficult to take an expansive view of DD, but also because it removed an important (if obvious) statement, namely that ID is a disorder of brain development.

A Formula for DD Eligibility that Would Include Many with FASD

The ID section in DSM-5 represents a shift from a "disability" (numbers-based) to a "disorder" (clinical/medical) approach to the definition and diagnosis of ID (Greenspan and Woods 2014a, b). Central to the disorder approach is the notion that IQ and other test scores, while useful as a window into a person's level and profile of cognitive functioning, cannot be used rigidly as a make-or-break basis for ruling ID in or out. That is because ID is viewed as a reflection of underlying brain pathology, and such pathology usually manifests in a complex and variegated manner that cannot be captured by a single test score, especially one with limited content validity.

Here is our attempt at a definition of Intellectual Disability/Developmental Disability: "ID/DD is a developmental condition that involves incomplete development of brain functioning and which is reflected in significant deficiencies relative to peers in the acquisition of intellectual, cognitive and executive skills needed to succeed in various age-relevant community roles and to cope with everyday as well as unique challenges and dangers in physical, academic and social realms."

We believe that this definition is in the spirit of existing expanded DD eligibility formulae, but is superior in that it provides considerable flexibility by eschewing score ceilings. It also takes a broader approach to cognitive impairment than IQ, for example, by emphasizing executive functioning. It also links adaptive behavior (without using that poorly-defined term) to cognitive impairments, emphasizes the brain basis of the disorder (without limiting itself to a few causes such as Prader-Willi or even FASD), and indicates that what makes ID a disability are the dangers resulting from failure to recognize or avoid risk, including social risk. A much larger percentage of people with FASD would, we believe, be able to justly achieve DD eligibility using this formula. That assertion should be testable by applying the formulation to individuals discussed in this chapter or others who might be known to readers.

References

Administration for Community Living. 2013. *History of the DD Act*. Washington, DC: U.S. Department of Health and Human Services.

Alaska FASD Partnership. 2012. *SB 151 as a mitigating factor passes legislature*. Anchorage: Alaska FASD Partnership Newsletter.

American Psychiatric Association. 2013. *Diagnostic and statistical manual of mental disorders*, DSM-5, 5th ed. Alexandria: American Psychiatric Association.

Anastasi, A. 1983. What do intelligence tests measure? In *On educational testing: intelligence, performance standards, test anxiety, and latent traits*, eds. S.B. Anderson, and J.S. Hemlick, 5–28. San Francisco: Jossey-Bass, Inc.

Block, S.R. 2013. *What does it mean to revise Colorado's definition of intellectual and developmental disability?* Denver: Alliance for Developmental Disability.

Blumenfeld, S. 2011, Oct 10. Eugenics, or scientific racism, in American education. *New American*.

Brown, N.N., and P. Connor. 2014. *Impact of executive dysfunction on learning in children with fetal alcohol spectrum disorders (FASD)*. Unpublished manuscript, Seattle, Washington.

Ceci, S.J., and W.M. Williams. 1997. Schooling, intelligence and income. *American Psychologist* 52(10): 1051–1058.

Disability Law Center, Inc. (undated). Chart of state laws defining intellectual disability, Boston. http://www.dlc-ma.org/org/resources/intellectual_and_or_Developmental/chart_of_state_laws_defining_int.htm).

Disability Rights California. 2012. *5th category eligibility publication* (publication #5510.01). Sacramento: Disability Rights California.

Dykens, E. 1996, April 12. *Personal communication*. New Haven: Yale Child Study Center.

Edwards, W., and S. Greenspan. 2011. Adaptive behavior and fetal alcohol spectrum disorders. *Journal of Psychiatry and Law.* 39(4): 419–447.

Gettings, R.M. 2011. *Forging a federal-state partnership: A history of state and federal developmental disabilities policy*. Washington, DC: American Association on Intellectual and Developmental Disabilities.

Goddard, H.H. 1912. *The Kallikak family: A study in the heredity of feeble-mindedness*. New York: Macmillan.

Greenspan, S. 2009. Foolish action in adults with intellectual disabilities: The forgotten problem of risk-unawareness. In *International review of research in mental retardation*, ed. L.M. Glidden. New York: Elsevier.

Greenspan, S., and G. Olley. 2015. Variability of IQ scores. In *Intellectual disability and the death penalty*, ed. E. Polloway. Washington, DC: AAIDD.

Greenspan, S., and H.N. Switzky. 2006. Forty-four years of AAMR manuals. In *What is mental retardation?: Ideas for an evolving disability in the 21st century*, ed. H.N. Switzky, and S. Greenspan, 3–28. Washington: AAMR.

Greenspan, S., H.N. Switzky, and G.W. Woods. 2011. Intelligence involves risk-awareness and intellectual disability involves risk-unawareness: Implications of a theory of common sense. *Journal of Intellectual & Developmental Disability* 36(4): 1–33.

Greenspan, S., and G.W. Woods 2014a. Intellectual disability as a disorder of reasoning and judgment: The gradual move away from IQ ceilings. *Current Opinion in Psychiatry*.

Greenspan, S., and G. Woods. 2014b. Social incompetence of FASD offenders: Risk-awareness as a factor in criminal culpability. In *Brain disorders and the law,* eds. E. Jonsson, S. Clarren, and I. Binnie. Unpublished manuscript, Edmonton, Alberta.

Gresham, F.M., and D.J. Reschly. 2011. Standard of practice and Flynn effect testimony in death penalty cases. *Intellectual and Developmental Disability* 9(3): 131–140.

Grossman, H.J. 1983. *Classification in mental retardation*. Washington, DC: AAMD.

Luckasson, R., S. Borthwick-Duffy, W.H.E. Buntinx, D.L. Coulter, E.M. Craig, A. Reeve, et al. 1992. *Mental retardation: Definition, classification, and systems of supports*, 9th ed. Washington, DC: AAMR.

McGrew, K.S., and D.P. Flanagan. 1998. *The intelligence test desk reference: Gf-Gc cross-battery assessment*. Toronto: Longwoods Publishing.

National Council on Disability. 2012. *Exploring new paradigms for the developmental disabilities assistance and bill of rights act: Supplement to the 2011 NCD publication rising expectations: The developmental disabilities act revisited*. Washington, DC: National Council on Disability.

National Organization for Fetal Alcohol Syndrome. 2009. *Motion to the Louisiana Supreme Court for leave to file brief amicus curiae in support of condemned petitioner Brandy Aileen Holmes*. Washington, DC: National Organization for Fetal Alcohol Syndrome.

Salvador-Carulla, L., G.M. Reed, L.M. Vaez-Azizi, S.A. Cooper, R. Martinez-Leal, et al. 2011. Intellectual developmental disorders: Towards a new name, definition and framework for "mental retardation/ intellectual disability" in ICD-11. *World Psychiatry* 10: 175–180.

Schalock, R.L., S.A. Borthwick-Duffy, W.H.E. Buntix, D.L. Coulter, E.M. Craig, et al. 2002. *Intellectual disability: Definition, classification, and systems of supports*, 11th ed. Washington, DC: American Association on Intellectual Disability.

Scheerenberger, R.C. 1983. *A history of mental retardation*. Baltimore: P.H. Brookes.

Shallice, T. 1982. Specific impairments of planning. *Transactions of the Royal Society, Series B Biological Sciences* 298(1089): 199–209.

Smith, D. 1985. *Minds made feeble: The myth and legacy of the Kallikaks*. Rockville: Aspen.

Spitz, H.H. 1988. Mental retardation as a thinking disorder: The rationalist alternative to empiricism. In *International review of research in mental retardation*, vol. 15, ed. N.W. Bray, 1–32. San Diego: Academic Press.

Tassé, M.J. 2009. Adaptive behavior assessment and the diagnosis of mental retardation in capital cases. *Applied Neuropsychology* 16: 114–123.

Tenkku, L.E., D.S. Morris, J. Sala, and D.K. Xaverius. 2009. Racial disparities in pregnancy-related drinking reduction. *Maternal and Child Health Journal* 12(5): 604–613.

Zaharia, R., and C. Mosley. 2008. July. *State strategies for determining eligibility and level of care for ICF/ MR and waiver program participants.* New Brunswick: Rutgers Center for State Health Policy.

Zigler, E., D. Balla, and R. Hodapp. 1984. On the definition and classification of mental retardation. *American Journal of Mental Deficiency*, 89: 215–230.

Part V
Personal Perspectives and Stories

Minimizing Secondary Disabilities

Inger Thormann

Primary and Secondary Disabilities Stemming from FASD

While most people understand what a brain injury involves when it occurs after birth, prenatal brain injury is a different matter. There is no before-and-after behavior to compare, and many find it difficult to understand the severe impairments that may occur as a result of FASD. At birth, the symptoms may be vague and diffuse. They may even go unnoticed or be confused with the temporary aftereffects of a difficult delivery or the mother's medication prior to the delivery. When the child begins to display more obvious atypical behaviors, the explanation is often not sought in prenatal factors but in postnatal trauma.

Alcohol exposure during pregnancy may disturb normal cell division and the migration of brain cells. This may cause structural anomalies in brain development. Another possible consequence of fetal alcohol exposure is a disturbance of the brain's electrophysiological and neurochemical balance. As a consequence, messages are not conveyed as efficiently and as accurately as they should be and normally are. In some children with fetal alcohol effects, impaired connectivity in the brain causes brain receptors to receive erroneous messages.

It is, however, difficult to measure the structural and functional brain processes in large numbers of individuals with FASD. Instead, researchers and practitioners have relied on psychological testing to measure primary disabilities. This includes IQ tests, performance tests, adaptive behavior tests and behavioral observations as well as interviews with caretakers.

I. Thormann (✉)
Copenhagen, Denmark
e-mail: ingerthormann@yahoo.dk

M. Nelson and M. Trussler (eds.), *Fetal Alcohol Spectrum Disorders in Adults: Ethical and Legal Perspectives*, International Library of Ethics, Law, and the New Medicine 63, DOI 10.1007/978-3-319-20866-4_16

Studies of Primary and Secondary Disabilities

The Fetal Alcohol and Drug Unit at the University of Washington in Seattle has published three longitudinal studies of children with Fetal Alcohol Syndrome. Although these studies focus on primary disabilities, they also provided a glimpse of what was later termed *secondary disabilities.*

The first study—a ten-year follow-up study (Streissguth et al. 1985)—involved a group of eight children, who were among the first eleven children ever to be diagnosed with what was then termed fetal alcohol syndrome (FAS). Of the remaining three, two had passed away before the study, and one was unavailable for follow-up. The study found that the four persons who had an IQ below 70 were in better and more appropriate educational programs and led more stable lives than the four who had a higher IQ.

The four individuals who had an IQ above 70 were not officially categorized as having an intellectual disability, and they were not seen as having immediate problems or being at risk of developing future problems. However, they all had FAS.

One of the boys dropped out of school for an extended period; he resumed his education when he relocated to another state. One girl dropped out in mid-term and had a baby soon after. Another girl left after 9th grade and became a single mother. Only one of the four stayed in school.

These observations inspired a second study (Streissguth et al. 1991) focusing on adults and adolescents over 12 years of age, the age when children with FASD typically encounter problems, both at home and in school. The subjects were 61 adolescents and adults aged 12–40 years, all with FAS or fetal alcohol effects (FAE). Despite a mean chronological age around 17 years, the mean developmental age (Vineland Adaptive Behavior Scale) was 7 years. Academically, the group performed as expected: although some had reading and spelling skills at a 5th-grade level or slightly higher, the average level was 2nd–4th grade. The group's arithmetic skills lagged even further behind.

The 61 adolescents and adults were tested on three domains:

1. *Daily living skills*, which includes personal hygiene, cooking, cleaning, keeping doctor's and dentist's appointments, managing one's personal budget, personal safety.
2. *Socialization,* which includes appropriate interactions with others, friendships, visitors, contact with family, verbal communication skills.
3. *Communication skills,* which includes shopping, going to the post office, the library and the bank, dealing with public transport, securing positive contacts/neighborly relations.

The group performed best on daily living skills (mean at the nine-year level) and most poorly on socialization skills (mean around the six-year level). A few had age-appropriate daily living skills; none had age-appropriate socialization or communication skills.

As a group, the subset who did *not* have an intellectual disability, according to their IQ score, still had problems with considering the consequences of their actions, displaying appropriate initiative, responding appropriately to subtle social cues and establishing and maintaining reciprocal friendships.

Study Findings and the Development of Secondary Disabilities

Ann Streissguth has carried out numerous studies, and during the 1990s her findings caused her to take a growing interest in secondary disabilities.

She identified the following key challenges typically faced by persons with FASD:

1. *Poor judgment,* which puts them at increased risk of victimization
2. *Attention difficulties,* which causes them to lose focus often and be distracted
3. *Math problems,* which makes it hard for them to handle money
4. *Memory problems*, which makes it hard for them to learn from their experiences
5. *Difficulties with abstract thinking*, which makes it hard for them to consider consequences of their actions
6. *Problems with orientation in time and space*, which makes social interactions challenging
7. *Problems with control/impulsive behavior,* which leads to a low frustration threshold

These seven points describe the link between some of the cognitive impairments in persons with FASD and the resulting everyday challenges.

When these problem behaviors are exhibited by a person who clearly appears to have an intellectual disability, they hardly surprise us. In fact, we might feel sympathy and wish to offer some sort or care or support. However, these behaviors are much harder to accept from a person who otherwise speaks and acts normally, and who does not have appear to have an intellectual disability, either to a casual observer or in a formal IQ assessment. Some of the spontaneous reactions might be, "He's no Einstein!"—"Where does he get off?!"—"What horrible manners," or "Is this your idea of a joke?!"

Faced with a discrepancy between *how people behave* versus *how we expect them to behave*, we are left to draw our own conclusions. This is an example of the fundamental attribution error or correspondence bias, a term known from social psychology. If the person had recently been in a motor accident, we would probably associate the unexpected behavior with brain injury. But in the absence of a visible explanation, the average person is not likely to consider the presence of a brain injury. Knowledge of the diagnosis or about the manifestations of prenatal alcohol exposure, however, will probably enable the observer to arrive at a correct hypothesis.

It is this discrepancy between *the behavior we expect* of children, youths and adults with fetal alcohol effects and *the behavior we see from them* that shapes the climate for secondary disabilities. Other secondary disabilities may stem more directly from the children's experiences growing up and from their inability to look out for themselves. Ann Streissguth argues that some of the secondary effects could be mitigated if we had a better understanding of the primary effects.

In 1996, Ann Streissguth and her colleagues published a large-scale study of secondary disabilities (Streissguth et al. 1996). The study included 415 persons with FASD aged 6–51 years.

The study examined six secondary disabilities:

1. *Mental health problems,* defined as having received treatment for one or several disorders, for example attention deficit disorder/hyperactivity (ADHD), depression or contemplating or attempting suicide, panic reactions, psychoses (hallucinations), aggressive or obstinate behavior, sexual depravity etc.
2. *Disrupted school experience,* defined as being suspended or expelled or dropping out.
3. *Trouble with the law,* defined as involvement with the police or being charged or convicted of crime.
4. *Confinement*, defined as incarceration in the justice system or inpatient treatment for mental health problems or substance abuse.
5. *Inappropriate sexual behavior*, defined as repeated problematic sexual behaviors, for example sexual advances, sexual touching, promiscuity, exposure, compulsions, voyeurism, masturbation in public or incest, including convictions for sexual offences.
6. *Alcohol and drug problems*, defined as abuse of alcohol or drugs and inpatient treatment for substance abuse.

Mental health problems: More than 90 % of the children, adolescents and adults who took part in the study had mental health problems, and more than 80 % had received treatment. Among the children and adolescents, attention deficit disorder was the dominant issue, while depression was the most prevalent issue among the adults (more than 50 %).

Disrupted school experience: More than 60 % of the adolescents and adults had a disrupted school experience. Surprisingly, the same was true for 14 % of the children. Suspension was the most frequent occurrence across the age groups, but among adults, dropping out was a close second (almost 40 %). Individuals with a disrupted school experience were twice as likely to encounter learning and behavioral problems as the rest of the group. The most common problems across the age groups were failing to pay attention (70 %) and incomplete homework (55–60 %). The most common behavioral problem was being disruptive in class (55–60 %) and not getting along with peers (60 %).

Trouble with the law: 60 % of the young adults and as many as 14 % of the children had had trouble with the law, with shoplifting as the most common offence. Among adolescents and adults above 12 years of age, 60 % were referred to the juvenile justice system, and more than 40 % were given juvenile prison sentences; 46 % received a suspended sentence, and 39 % were sentenced to community service. The study found a clear correlation between a disrupted school experience and trouble with the law.

Confinement (incarceration or involuntary hospitalization or treatment): 50 % of the adolescents and adults but fewer than 10 % of the children had experienced some sort of confinement.

Inappropriate sexual behavior: 49 % of the adolescents and adults and 39 % of the children displayed inappropriate sexual behavior, which makes this category the second-most common secondary disability for children. The most common inappropriate sexual behaviors were sexual advances (18 %), sexual touching (16 %) and promiscuity (16 %).

Alcohol and drug problems: 35 % of the adolescents and adults (and none of the children) had problems in this area. Many in the surveyed group were completely abstinent. The most common reason stated for abstinence was lack of access to alcohol followed by personal conviction.

Ann Streissguth and her team were greatly surprised by the prevalence of secondary disabilities and the *detrimental effect on the individual person's quality of life.* Undoubtedly, better protection and support for this group of children, adolescents and adults requires increased understanding among teachers, caretakers, social workers and society at large.

As mentioned earlier, the permanent organic brain damage in individuals with FASD is often invisible, and the impairment does *not* automatically lead to special needs education or assistance. Unless the person in question has a severe intellectual disability or a very distinct physical characteristics, any educational or social support requires a presentation of some sort of evidence.

Risk Factors and Protective Factors in Relation to Secondary Disabilities

In order to understand the findings of the study fully, we need to examine the risk factors and the protective factors that affect the severity of the secondary disabilities; *risk factors* being anything that exacerbates the secondary disabilities, and *protective factors* being anything that mitigates the secondary disabilities.

In their analysis, Streissguth, Barr et al. identified five protective factors that had a universal preventive effect on the development of secondary disabilities:

1. a stable home with nurturing parents
2. a minimum of change in home and life environment
3. protection from violence
4. eligibility for special needs services
5. early diagnosis

A clear trend in the study was that those with the *lowest* level of primary impairments—those participants who did not have the full FAS diagnosis and who did not have an intellectual disability—faced the greatest difficulties and were most likely to encounter problems. The study also addressed the key issues of how children with FAE or FAS can develop resilience, and which factors can help reduce their secondary disabilities.

The evidence suggests that *resilience* develops in an environment that is simultaneously *protective* and *stimulating*, and which remains consistent over time. That sort of environment promotes the child's emotional and cognitive development.

A child with FAE/FAS generally has trouble dealing with change. The brain injury impairs the child's ability to understand and handle changes or replacements among key persons such as foster parents, caretakers and teachers.

Even if secondary disabilities cannot be avoided altogether, the risk can be minimized. Inadequate protective measures are the greatest risk factor.

Children and adolescents with less pronounced and obvious presentations of FASD symptoms are especially at risk of missing out on protective measures. Based on "normal" expectations, these children and youths are typically placed in mainstream classrooms rather than in individualized special needs programs. This means that the child/youth misses out on protective factors.

"Normal" expectations are also in play when the young person begins to go out, starts dating or moves away from home. To reduce the risk of secondary disabilities, the responsible caretakers need to use the knowledge that is available. Special needs education should always be a consideration, and specific programs should be explored. Similarly, when the young person moves away from home, both Streissguth's studies and my own study indicate that the demands and stress factors that the young person faces are greater than previously assumed. Therefore, some form of assisted living should be considered.

These aspects were considered for all the participants in the case study (Thormann 2006) reviewed below, including Martin.

Case: Martin

Martin comes to live at the Danish observation and treatment center *Skodsborg Observations- og Behandlingshjem* (SOBH) at the age of 4½ years. Until then he has been living with his alcoholic mother. He comes to the attention of Child Protective Services after he has spent three days alone in the apartment, when a neighbor eventually responds to his crying. At SOBH, Martin undergoes thorough observation and assessment, which concludes that his symptoms stem mainly from profound early neglect.

Martin is restless and hyperactive and has severe concentration problems. Occasionally, he retreats inside himself. He rocks his upper body back and forth, even in his sleep. Enveloped in his own sounds and movements, he sometimes retreats into his own world and is assessed as highly vulnerable. When he withdraws into his own world he is extremely sensitive to physical touching and easily startled.

In many areas, Martin is assessed to have age-appropriate skills when he receives adult support. He is extremely sensitive to sensory stimuli and unable to block out irrelevant stimuli; as a result, he is easily overstimulated.

Martin stays at SOBH from the age of 4½–6 years. He is found to have a very high anxiety level. Many things frighten him, especially new things and sounds. When he notices an unfamiliar sound he is unable to settle down until he has identified the source of the sound. Martin is wary of unfamiliar people. When he is scared he wraps both hands around his head for protection.

Martin develops an emotional bond with his primary caretaker, and his soft and loving side comes out more and more as he feels more secure. He needs support to stay focused on tasks and requires affirmation and ongoing adult support to complete them. Martin is a big boy with a strong appetite; he eats a lot and has a tendency to "suck" his food down, almost relying on sucking reflexes. He has to be reminded to use his teeth.

Martin receives remedial language training and is prompted to use verbal communication, as he is often quick to resort to kicking and hitting in conflict situations. Martin is profoundly affected by turmoil and noise; he says that it makes him nervous. He attends play therapy targeting his anxiety and his relationship with his mother who repeatedly fails to show up for pre-arranged visits. His rocking continues every night. The night watch soothes him. On one occasion his rocking is so pronounced that his bed collapses.

At the age of 6 years and 3 months, Martin is placed in a foster family. Due to his anxious and nervous character, the placement is preceded by a lengthy transition phase. The family already has one foster child and two children of their own. Martin settles in and thrives as the youngest of four children. From the outset, he is welcomed as a lovely and loving boy, and he develops rapidly. However, he remains an anxious boy, and he still rocks in his sleep, sometimes so much that he has to sleep on a mattress next to his foster parents' bed. A soothing hand can calm him down. Martin says that when he rocks he is often having violent dreams about being abandoned, as he was in real life as a young child. He improves his ability to express himself verbally when he is upset. But even the slightest change or disruption still triggers intense insecurity and anxiety in Martin. He is still a vulnerable boy.

At the age of 7 years, Martin attends kindergarten. Here, he exhibits concentration problems and insecurity and is referred for assessment. The assessment finds concentration problems, poor fine motor skills, poor body coordination and low muscle tone. Martin has a poor sense of space and direction and a reduced sense of his own body. He has a limited awareness of his own position in space and often bumps into others, which leads to conflicts. He is found to be intellectually age-appropriate with good general knowledge. At home, Martin thrives. He is closely attached to his foster parents and feels secure in the family. His biological mother is very happy with the placement and wants for him to stay with the foster family.

His mother still drinks when she is depressed; it is difficult to make appointments with her, and she often fails to show up for visits.

Martin receives remedial assistance in several areas. He is examined by a physiotherapist specializing in sensorimotor issues, and he sees a psychologist. Martin is not good at handling emotional pressure, and he has difficulty putting his feelings into words. Due to his own vague inner structure he needs a structured environment. He has a tendency toward perseveration: He has difficulty breaking a pattern and initiating a new one and finds it hard to vary his strategy in a given situation, even when his current strategy is not working. He sticks to the ineffective approach and generally needs adult support to change course.

Martin is in a mainstream class. The school describes him as follows:

> "Bright student, physically restless. He needs space around him, reacts strongly if others get too close. Hypersensitive to sensory stimuli and fragile. Needs adult support. Inner restlessness, hyperactive, loud. Has difficulty learning from his experiences, often repeats exactly the same sequence of events that he just apologized for. According to speech/hearing therapist Martin has language difficulties as a result of poor oral motor skills and poor coordination of cheeks, lips, tongue, pharynx and palate. In school, Martin benefits most from one-on-one sessions with an adult. (This is possible as he has been granted 7–10 weekly remedial lessons). He is motivated but unable to work on his own."

Martin attends an after-school program. He likes it there and is well liked by both the other children and the staff. In good periods, he is quite good at putting his actions into words. He is kind toward the other children and receptive to their moods. He is also very polite, sometimes excessively so, as he parrots adult language instead of using his own words. In bad periods, Martin has a low tolerance threshold and he might kick and hit the others. His behavior can be quite unpredictable. He requires adult support to avoid acting out and doing harm to himself and others.

In third through seventh grade his development continues along the same trajectory. His foster family is the secure base, where he thrives, and where there are few problems. He enjoys taking part in practical activities and chores.

Mealtime is a recurring source of conflict; his foster sister especially gives him a hard time. He eats nosily but fails to chew his food properly; instead he relies on something akin to sucking reflexes. Martin wants to eat properly but is physically incapable of doing so.

At the age of 11 years, Martin sees a relaxation therapist and for a while his sleep seems to improve, and he appears to be resting better at night. However, he soon reverts to his old pattern.

His extreme sensitivity to noise continues to inhibit him. For example, he is very good at playing the recorder, but he cannot attend music school because of the loud sounds from the other instruments. At the age of 11 years, Martin is evaluated for fetal alcohol effects. His mother's binge drinking in combination with his many neurological symptoms trigger the suspicion, which has in fact been present ever since his first arrival at SOBH at the age of 4½ years. In a week-long assessment procedure, Martin undergoes educational/psychological observation during the day and is also examined by a physician. He is found to be a likable boy who appears slightly immature and not quite age-appropriate. His face and appearance reflect his Greenlandic origin. A smooth philtrum is the only facial feature to suggest FAS. He is a big and handsome boy with no signs of growth deficiency.

He does however display many of the neurological symptoms typical of congenital alcohol syndrome. Martin is diagnosed with FAE—Fetal Alcohol Effects. The doctor fears that he will encounter increasing problems in his mainstream classroom as his classmates reach a limit for how much they are willing to tolerate, as both they and Martin get older.

Martin has developed a particularly close relationship with his foster father; they share many activities, including gardening, fishing and washing the car. He also has a couple of close friends.

At the age of 12 years, Martin is tested by a psychologist who concludes that Martin is going to be completely dependent on special needs education for the remainder of his school years.

At the age of 13 years, Martin is afraid of going to the bathroom on his own. He imagines rats coming out of the bathroom floor and up from the toilet. When his anxiety becomes overwhelming he sings a particular tune. Martin says that he sometimes hears voices that tell him to do bad things. He fights these voices, but when he has trouble gaining the upper hand with them he grows very anxious and unfocused and begins to sing his special tune. The foster parents are relieved that Martin shares this with them but are also concerned about his hallucinations.

Martin's social skills develop well. He continues to be kind, friendly and well-behaved; sometimes, however, his politeness seems superficial and studied. He seems to thrive in class and enjoys socializing. Despite his learning difficulties, his ability to pay attention and sustain focus in class improves. He continues to receive special needs education in language arts and math.

When Martin is 14 years old, the adults around him are uncertain about his future. Will he be able to continue in a mainstream setting? What sort of support or program might be appropriate?

He has a variety of problems, but his strength is that he is a very loving and kind boy with emotional resources. He is healthy and creative and still enjoys playing the recorder. He has few but close friends, and he has developed socially; in addition, he is handsome and charming. He has deep roots in his foster family and would prefer to stay there *forever*.

Martin lags far behind his classmates. He is still well-liked by his classmates and teachers, and therefore he stays in the same mainstream setting with ten hours of remedial teaching a week. However, Martin appears to be able to benefit most from a vocational path, and the options are investigated. Eventually, Martin begins to attend a socio-educational program with its own school. The program has a residential component, but Martin attends during the day and continues to live at home. Here he receives two school lessons a day; the rest of the time is occupied with workshop activities, including a mechanic's workshop and farming. According to the staff he thrives here and is well-liked. Martin's own assessment is more mixed:

He likes it when the staff sets a clear agenda, which also offers protection, but he is upset when the staff fails to grasp his particular needs.

At the age of 16 years, Martin visits Greenland together with his foster father. The purpose of the trip is to introduce him to his original culture, and the trip is very successful. It gives him a better understanding of his mother's problems in relation to the culture clash she experiences in Denmark.

When Martin is 16½ years old, the idea of continuation school is brought up, and in the summer when he turns 17 he begins at a continuation school for "youths with delayed development". He is motivated to receive help to learn more, to become more "normal", because that is his ultimate goal. He accepts his disability but feels both grief and anger at his reduced opportunities. He worries about the future.

Before Martin goes away to continuation school he asks for help with his rocking. He is embarrassed because his schoolmates can hear it when he rocks at night. He sees a psychiatrist and receives medication, which helps.

The reports from the continuation school are mainly positive, especially for the first six months. After New Year's Martin's difficulties increase. Martin has more and more conflicts with other students, uses bad language and provokes them. He repeatedly misreads social cues and situations, and combined with a hot temper (lack of impulse control), this creates several dangerous situations. In connection with the Easter break he is expelled from school.

Martin feels that he has been treated unfairly, and his foster parents share this view.

After two months at home, Martin finds a job in a salad factory where he chops and packages salads. He is praised for the quality of his work. The factory keeps him on after he has turned 18 and pays him the full adult rate.

Two weeks after Martin's 18th birthday his mother is found dead of alcohol-related causes. This marks the beginning of a difficult time for Martin and his foster family.

Martin thinks back to his childhood, with all its disappointments. He makes it through this difficult time, however, in part by focusing on positive memories about his mother and her original culture.

After a little more than a year at the factory, Martin experiences stress on the job. The machinery is noisy, the scene before his eyes is a blur, and he makes mistakes and is reprimanded. He works for hours without a break, because the salads have to be in the supermarkets by 11 o'clock. Martin starts work at 5 in the morning. His foster parents back his decision to quit, and his foster father helps him find another job, which he still has when the formal case study is concluded.

The new job offers circumstances that suit Martin well. He works alone but is able to request support and guidance when he needs it. The pace is moderate, and he has regular breaks. There is no noisy machinery, and Martin listens to music, which helps him filter out the normal noises that occur in any workplace. The job procedures are simple and concrete, and the result offers a concrete sense of satisfaction. Thus, he is able to hold a regular full-time job. His job is to cut Oasis floral foam into standardized blocks.

In connection with the completion of the case study Martin, now aged 19 years, undergoes a thorough neuropsychological evaluation. Here is a brief excerpt from the neuropsychologist's report:

> "In the present neuropsychological evaluation Martin appears an age-appropriate, handsome, well-dressed and charming young man. He appears polite and trusting but with relevant reservation toward strangers. He seems friendly and happy. He speaks in long, well-formed sentences. He has good insight into his own difficulties. He reports that he has trouble maintaining his concentration for sustained periods and that he has problems remembering things. He also reports that he tires easily when he has to concentrate on something. His working style is characterized by a willingness to cooperate with others. He appears to be good at concentrating over time. There are also no indications that he is easily distracted by irrelevant stimuli. Over time, however, there are slight signs that his mental stamina is less than would be expected.
>
> The psychological testing draws a picture of a young man whose general intellectual level is within the normal range, albeit at the lower end of the scale. However, Martin does have certain cognitive disorders. His attention span is short, which makes unable to handle more than a few elements at a time. Furthermore, his mental pace is slow, which slows down his information processing. These problem areas seem to be the cause of his learning and memory problems, as his learning and memory functions for visual and language content per se appear intact. These problem areas are also deemed to be the source of his math problems.
>
> In addition, Martin's thinking is generally characterized by a certain degree of perseveration and lack of flexibility, which limits the degree of complexity and abstraction he is able to handle. Martin's reading skills are fairly good, but spelling continues to be a challenge. The specific cognitive disorders that were identified are seen to be based on brain dysfunctions caused by FAE. Therefore, in a future job, Martin will require specially adapted conditions. These conditions will need to accommodate his tendency toward mental fatigue as well as his specific cognitive difficulties in the form of his lower mental pace, attention problems (low attention span and problems with shared attention, i.e. difficulty varying his attention focus) and his rigid thinking, which limits his capacity for handling more complex and abstract tasks."

Reflections on Martin's Life and Secondary Disabilities

Martin experienced severe neglect before an adequate intervention was put in place. He is noticed, and many people are concerned, but it is not until his mother physically disappears, and the 4-year-old boy is found alone in the apartment after a third consecutive day, when he is taken into care. Previously, there were concerns that Martin may have suffered the congenital effects of prenatal alcohol exposure, but the obvious signs of neglect during the first four years of his life blur the diagnostic image, and he is not diagnosed with FAE until he is 12 years old.

Martin spends almost 2 years at SOBH, and at the age of 6 years he is placed in the foster family where he still lives at age 19. Thus, the early neglect is replaced by convincing care and protection.

In many ways, Martin is a well-adapted boy and young man, but he has problems that are not visible to the outside world. This puts him at risk. Other people expect the same of him as they do of his peers. Good behavior and good looks do not exactly lower others' expectations of him—on the contrary. This leads to numerous experiences of failure for Martin, despite the high degree of care and protection he has received, especially from his foster parents. Throughout, a consultant from the association of foster homes has offered advice and support, and for several periods a psychological consultant has also been involved, providing individual therapy to Martin and supervision to the foster parents. Thus, many people have acted as Martin's advocates, and much has been done to minimize any secondary disabilities.

As mentioned earlier, Ann Streissguth argues that the basis for secondary disabilities is the gap between *what we expect* from a person with FASD and *what we see this person actually doing*.

At the age of 15 years, Martin repeatedly finds himself in situations where he is reminded of his shortcomings, and he experiences a series of setbacks. This makes him unhappy, and he sheds many tears about his situation. He realizes that he is unable to meet others' expectations of him, and that he is less capable than his peers. He is especially concerned that his memory is so poor, and that he keeps forgetting things. It is not only in the workshop he is reminded daily of his own shortcomings; in his foster family too, he is needled by his foster sister, who is two years his senior and who excels academically in high school.

A few examples may serve to illustrate how secondary disabilities can emerge:

> One day, Martin is assisting a mechanic who is working on a car. Without looking up, the mechanic asks Martin to fetch him a pair of pincers. Martin goes over to the wall-mounted tool board with its large array of different tools, but once he gets there he has forgotten what type of tool he is supposed to get. He therefore grabs three tongs at random and puts them in the pocket of his boiler suit. When he returns, the mechanic holds out his hand, and Martin reluctantly hands him one of the tools. In a less than gentle tone, the mechanic lets him know it is the wrong kind. Martin pulls out another tool and then another, but neither is the right kind. The mechanic gets angry and yells, "Is this your idea of a joke?!" This happens in the presence of a couple of Martin's peers. Martin leaves the workshop and takes the bus home to his foster family where he locks himself in his room, crying.

> A few weeks later, Martin is once again assisting a mechanic in the workshop. This time the mechanic needs some oil and hands Martin an oil can to be filled up at the oil pump. At the pump, Martin has to punch in a four-digit code, which has been told to him verbally. At the pump Martin finds that he cannot remember the code. He tries various combinations, initially trying to get it right but eventually he desperately tries random combinations. He feels an inner rage and anger toward the mechanic who asked him to do this. On his way back, he trips on a small set of steps; he has a bad fall and has to go to emergency room with chest pains. It turns out that he has broken two ribs.

For several days following the latter incident, Martin remains sullen and withdrawn. He does not want to go back to the workshop. His sick leave offers a natural break. Martin discusses the two incidents with his therapist. They analyze the incidents. What did Martin actually do, and could he have acted differently? At first, Martin is in low spirits, unhappy and despondent. But as the conversation unfolds, he comes up with ideas for alternative approaches: If the mechanic had had a set of cards with pictures of the various tools, Martin could have taken the card with him to the tool board and found the right one. And if the mechanic had written the code for the oil pump down he would have been able to handle that task too.

Both suggestions are good, but they require the mechanic to learn about the challenges Martin is facing—and why. Subsequently, a series of lessons are arranged for the program staff. Martin gives his consent for this to happen, albeit reluctantly. After all, he really wants to hide his disabilities. What he wants is to be normal.

Martin explains that he easily becomes stressed and nervous. As soon as he is given a task he is overwhelmed with apprehension that he will not be able to handle it; naturally, that makes it harder for him to carry out the task successfully.

When he works with his foster father, things are different. Over several weekends, Martin helped his foster father fix up an old Volvo. When he was shown how to do a particular task he was subsequently able to do it on his own. The hands-on demonstration works much better for him than the spoken word. When the two go hand in hand he does especially well.

As mentioned previously, Martin has difficulty forming visual imagery. When he hears the word *pincers*, as in the example from the workshop, he does not envision the pincers in his mind, so when he stands in front of the cluttered tool board he cannot compare it to his mental image because it was never formed.

If we consider the 15-year-old Martin and his frustration over being treated "wrong", compared to his abilities, the demands are too high. He is very vulnerable. After the negative incidents, he receives an extra dose of positive attention in his foster family, he sees a therapist, and his employers learn about FASD, and these protective factors help turn the situation around.

But new disappointments and failures await him.

Considerations about the balance between protective factors and demands are always essential in relation to vulnerable individuals. In Martin's case they are crucial. Martin needs the protection his foster family offers, but he is determined to be normal and wants independence from his over-protective foster parents. When he is a little over 19 years old he moves to Copenhagen to live in an apartment under "supervision" from a social worker. He distances himself from his foster parents. The specific trigger for his decision to move out is Martin's discovery that he is sexually attracted to other men. He chats online with guys, and this proves too much for his foster parents. Martin feels that he their control is excessive, and he moves out.

Martin pursues his goal of normalcy. He is in a stable and loving relationship and finds a full-time job in a supermarket, although he has been advised to limit his work to 24 hours a week. He wants to be able to buy nice clothes, have a trendy haircut and go out. He spends a lot of money and overdraws his credit card. A repayment plan is drawn up that stretches over several years, which is more than Martin can handle. He is put under administration by the bank, and around the same time he loses his job. He is tired and stressed and makes too many mistakes. As the final straw, his boyfriend breaks up with him. Martin gives up and takes his own life, when he is only 22 years old.

To briefly review the research discussed above, mainly by Ann Streissguth and colleagues, we can observe that

- Martin manages his everyday life well when he is in the familiar environment of his foster family
- Martin exhibits all the seven challenges identified by Streissguth and colleagues (poor judgment and difficulties with attention, math, memory, abstract thinking, orientation in time and space and impulse control)
- Of the six most common secondary disabilities Martin only has one: mental health problems
- Martin enjoys all the five protective factors mentioned on page 6 (stable home, stable life environment, protection from violence, special needs services and a relatively early diagnosis)

When a vulnerable young person like Martin with an invisible disability wants to be independent and experience *freedom*, there is no holding him back. He is no longer a minor, and he has the right to self-determination. The demands were too many and too high, and he lost the protection he had enjoyed when he broke with his foster family.

Closing Remarks

A study by Huggins et al. (2008) found that the suicide rate among adults with FASD aged 21–51 years is 23 %. This is five times the rate in the general U.S. population, which is 4.6 %. The high vulnerability in this group reminds us of the need to evaluate the balance between demands and protection constantly. One of our future challenges will be to create environments where people with disabilities can live independently in a context that offers a high degree of protection, for example in the form of a live-in professional.

Let us never waver in our efforts to overcome the secondary disabilities in vulnerable persons, including those with FASD.

References

Huggins, J.E., T. Grant, K. O'Malley, and A.P. Streissguth. 2008. Suicide attempts among adults with fetal alcohol spectrum disorders: clinical considerations. *Mental Health Aspects of Developmental Disabilities* 11(2): 33–41.

Rosett, H.L., and L. Weiner. 1984. *Alcohol and the fetus. A clinical perspective*. Oxford: Oxford University Press.

Streissguth, A.P. 1997. *Fetal alcohol syndrome. A guide for families and communities*. Baltimore: Paul H. Brookes Publishing Co.

Streissguth, A.P., J.M. Aase, S.K. Clarren, S.P. Randels, R.A. LaDue, and D.F. Smith. 1991. Fetal alcohol syndrome in adolescents and adults. *Journal of the American Medical Association* 265(15): 1961–1967.

Streissguth, A.P., S.K. Clarren, and K.L. Jones. 1985. Natural history of the fetal alcohol syndrome: A 10-year follow-up of eleven patients. *The Lancet* 13: 85–91.
Streissguth, A.P., Barr, H.M., Kogan, J., Bookstein, F.L. (1996). *Understanding the occurrence of secondary disabilities in clients with fetal alcohol syndrome (FAS) and fetal alcohol effects (FAE): Final report to the Centers for Disease Control on Grant No. R04/CCR008515 (Tech. Report No. 96-16).* Seattle: University of Washington, Fetal Alcohol and Drug Unit.
Streissguth, A.P., and J. Kanter. 1997. *The challenge of fetal alcohol syndrome. Overcoming secondary disabilities*. Seattle and London: University of Washington Press.
Thormann, I. 2006. *Medfødte Alkoholskader. Omsorg og behandling*. Copenhagen: Hans Reitzels Forlag.

The Challenges of Raising an FASD Child in Canada

Looking for immunity in the stars complex opportunities.
Jean-Paul Martino (1935–1996).

Abstract Parenting a fetal alcohol damaged adult is even more challenging than parenting the child. With little support services and no "road map", the task is time-consuming and nerve-wracking in ways that those who have not experienced it find difficult to understand. Jail looms like an inevitability, and who will take over when the aging parent can no longer provide a home and organizational support? This chapter was written by and provides an intimate perspective on one Canadian mother's struggles. The author has asked to remain anonymous.

A couple of weeks ago at about 10 p.m. there was a knock at the door that changed everything. It was the second knock that night, when I had been expecting a quiet evening.

The first knock was my son—my 20 year old "adult" who has the executive function of a 5 year old and still sucks his thumb. I had already gone to bed, exhausted from an exceptionally busy day. It was Friday and he had rushed out to meet a friend when I was about to start making dinner. He often does that. One phone call or text message and he is off, even if dinner's already on the table. No matter how many times I plead with him to eat before leaving, and to at least call once a day when he's gone, I may not catch hide nor hair of him for a week or more. Then he will show up hungry, wanting a shower and stoked about his latest misadventure.

This was a Friday night, the 4th of July. My boy had not been home on a Friday night since January, so I was relieved when he came in at a decent hour and said he wanted to rest and watch a movie. He even agreed to help clean the house on Saturday. I hoped we might finally be turning some kind of corner, until that second knock shattered my delusions.

The corner we turned was not the one I was hoping for. It was the police. They took him to jail. According to crumpled papers I had not noticed lying right there on a table by the front door, he had been charged with mischief a week earlier. He was supposed to go for finger-printing on July 2nd and to court at 9 a.m. on July

M. Nelson and M. Trussler (eds.), *Fetal Alcohol Spectrum Disorders in Adults: Ethical and Legal Perspectives*, International Library of Ethics, Law, and the New Medicine 63, DOI 10.1007/978-3-319-20866-4_17

4th. He said nothing about this to me. When I asked him about it later, he said he was afraid I would be angry. But that was not the only reason. He never brought me papers I needed to see, not even for my consent to school outings he was looking forward to. All the time he was growing up, I had to search his back-pack and ask his class mates to find out what was going on. He does not understand the link between papers and what might or might not happen.

Child-rearing specialists advocate tough love, but those who know more about FASD say it is better to hang on. Yes, the brain damage is permanent, but no, like everyone else, these kids can learn. They need some connection to properly functioning peers and adults yet the friends they find are often suffering from similar disabilities.

When I look back, I can see that the pattern of misadventures and disappearances that led to his arrest and that makes my life so uncertain developed gradually. He never could be counted on to come home reliably. He could not be sent on an errand or allowed to go to the park to be back in half an hour as promised. Nothing like that worked, ever. As the parent of another FASD child said, the closer to home the worse the sense of time. Nor could he explain where he was going or why. He could disappear out the door in the blink of an eye. I could not even plan dinnertime after they stopped driving him to and from special schools by bus.

One day when he was about 14 my twilight search for him hit a dead end. I had traced him to another neighbourhood and the kids who saw him last said he was running away with a class-mate. It was getting dark when I phoned the police. I lay in bed, reviewing horror stories about missing children. About 2 a.m. I finally got a call. Normal children do not understand my child's disability any more than adults do, and his lack of executive function gives them the courage to do what they might not otherwise try. This time, the other boy was afraid of being punished for missing his curfew. They had bus passes and 87 cents between them. So, following a garbled sense of geography, they decided to go to Winnipeg…to ride the roller-coaster at the West Edmonton Mall (a completely different city 1300 km away). Fortunately the classmate looked like a 10 year old and an alert bus driver in a distant suburb called the police. So that first bout of anguish lasted only a few painfully long hours. But that knot in my stomach that prevents sound sleep was to return time and time again as his absences stretched from hours to weeks.

In spite of all the misadventures that have happened since then, my child still does not understand why I cannot go on as usual when he does not call home. And I am so entrenched in my hope for normalcy that I forever fail to predict his next move. My words flow past him like water through a sieve. "Blah, blah, blah." One of the hardest things for me to come to terms with is that my looping connections full of causes and consequences do not make any sense at all to my child who is so good at living in the moment. If he has a phone, he might call me to say "I'm almost home". He is at Broadway and Alma, ten minutes away. "Don't start dinner", said my counsellor who has observed FASD for many years. And I laughed because the next call was 3 hours later from the other side of town. Of course the phone had not worked when I tried to call him. "What happened?" I ask when he finally comes home. "I saw a friend", he says, or something like that.

I have given up on asking names, addresses. He wants to be independent, so he resents my questions. He does not understand that just because *he* knows something does not mean that I know it too. So if he is okay, shouldn't that be obvious? Nor does he understand other people's priorities -whether for evil or for good. What did that job counsellor think when she came for him in the waiting room? He had just picked up the paper and he turned away from her to continue reading his horoscope. According to his diagnosis, his mental processing is extremely slow. Accurate enough when he has time to think things through, but transitions are difficult and he finds it hard to shift gears or prioritize. He really did want her to help him find a job. He wanted the help from all the other agencies whose appointments he has missed as well.

Now that he is 20, I am finally starting to get used to some parts of his idiosyncratic dynamic. Fetal alcohol damage varies from one person to the next, depending on what was developing at the time of exposure. Though he is not articulate, my child is cheerful, personable, physically fit and good looking. How many times have people told me, "He's 19, he'll grow out of it"? I suppose the quest for independence is hard on all parents, but most can at least hope their kids will eventually learn to fly. My parents could give me a sum of money to last for a term at university. Even if I was only getting B's instead of A's, I could get and keep a job or a place to live and they could find me if they needed to.

No matter how much money my child gets, be it one dollar or a thousand, it has never lasted more than about 6 hours. I set up a bank account for him when he was little and tried to teach him how to save for something he wanted. But $200 got drained on pizza and poutine within a day after his buddy showed him how to use his bank card. The institutional insistence that he became an adult at 19 took away the last remnants of my control. I tried in vain to get a couple of banks to give him an account that limited withdrawals to $20 per day. They will only do this for children, they said. He has no concept of what a credit rating is or why it could be useful. He has opened accounts at three different banks and, with the help of trusty bank machines and nameless "friends" who like to have "fun", he is over-drawn by a total of almost $2,000. He cashes his cheques at Money Mart and he's thinking of opening an account at Scotia Bank for the "free" movie passes. A phone company once gave him a "free" $200 phone on a 3-year $60/month plan when he had no job and no assets. It cost me a few hundred dollars to get him out of that one, but I have given up. He's out and about and I cannot stop him. It's a predatory world out there and I cannot protect him. So I now leave it up to big business to deal with the customers they deserve. At least when he loses his bank cards, there's no longer any worry that some stranger will ruin his credit. He's done that by himself.

The problem is not just with handling money. It's the sense of impending doom, the feeling of being in a car with no brakes. My child looks so normal that it is hard to get my mind around the profile of his disability. He was eighteen when he was assessed for FASD at the Asante centre and I fully expected a negative result. He always seems to be on the cusp of some kind of success that never quite materializes. I only sought testing because two friends who had FAS experience insisted. Alcohol Related Neuro-Developmental Disorder is what they found. ARND. I had never heard the term. The testers were very specific about the elements of his disability: very poor oral comprehension and short-term memory; extremely slow mental processing, in the bottom 3 % on some functions. But in areas related to visual/spatial relations he was well above average. They warned me that with this cognitive profile he was at high risk for drug and alcohol addiction. And for jail. At that point he was not into alcohol. He had seen too much chaos in his early childhood. He has memories of hiding behind a car while his dad's drunken girlfriend threw beer bottles at them. The sight of a six-pack of empties would send him into a panic. So marijuana was king. But then he discovered parties. He dropped his inhibitions and his idea of fun is certainly not mine.

It is July and I can no longer count the number of times since Christmas that he has drunk himself to oblivion. He has lost two cell phones, two pairs of glasses, two sets of identification, two wallets, not to mention various back-packs, articles of clothing and bus passes. He got two $173 fines for being on public transit without a ticket—a few days after I paid $124 for a two-zone bus pass…that he promptly sold to a friend. He was bear-maced at a bus stop by a random stranger and beaten up three or four times. Once or twice he was beaten up by police, though he has also managed to get police to drive him home a couple of times; once they even called me at 3 a.m., to tell me they had found him passed out at a bus stop (which turned out to be two blocks from home).

How to deal with all of this? We have certainly not been doing very well with the challenge of finding a place for people like him in our society with a high youth unemployment rate and a ready supply of in-the-moment hedonists. I could see jail coming long before it happened. I tried innumerable agencies. I even wrote to the minister responsible and they got back to me. They referred me to a social worker who found all the same agencies I found and got the same run-around. The few agencies that had programs that worked have had their funding cut. So, I am on my own with this. I have stopped replacing things. I cannot afford it. And to his credit my son paid for some replacements when he had a job. "Be there when he gets paid", said my savvy FASD/addictions family counsellor….and it worked, somewhat. If he re-paid me the consequences of all of his misadventures, there would be nothing left, no reward for the effort of learning a new job. Never mind what people say about getting your adult kids to pay their share of food and rent. I have never found a way to make that work. He qualifies for welfare, but giving him money just fuels the drunkenness and the wrong kinds of friends. It does not ensure his food or shelter or relieve the drain on my finances.

He is at a stage where he wants and needs some independence from me. He would probably do well in a home with the right kinds of supports. But he does not qualify for assisted living because his "adaptive functioning" is too high. His exceptionally low verbal and managerial skills balance against above average manual and performance abilities to give him a "normal" IQ when no skill measures in the normal range. He works well and can be very useful at the right kind of job if paired with a buddy to keep him on track. If only I had a stronger network of family and friends who live close by. If only I knew someone running the kind of business that could benefit from his talents. But my closest relatives are scattered across the country and my friends are academics reaching an age when they need help themselves.

When he was little it was not so difficult. I could swap off baby sitting with other parents. He is too old to have a "sitter" now. But he is not responsible enough to stay home alone, not even for an evening. He gets lonely so he might disappear and leave the door unlocked, or try a dangerous experiment or find a new "friend" who shows him how to sell my grandmother's silver spoons to get money so they can have "fun". If only I was not a single parent we could spell each other off for evenings, but FASD children have broken many a marriage. I am 66 years old and I worry about what will happen to him when I am gone.

Until he was diagnosed, I did not know why he was the way he is. When I took over raising him, he was eight and a half. I met him as a baby in foster care, during the brief time when I was a novice lawyer taking legal aid cases. His mother had died, intoxicated on a beach. His father was my client. As a 22 year-old, drifting without a job or a home, and charged with a crime, his father could not get custody. He was so proud of his baby, and we eventually got the baby sent back to his reserve. I passed the criminal case on to a more experienced lawyer, who passed the case to a junior who called me 2 days before trial to ask me what to do. His father ended up convicted of a crime he did not commit, based on a confession extracted by an ambitious cop whose deluge of verbiage overwhelmed my client's inarticulate denial. We did not know about fetal alcohol effects in the 1990s. My child's father has never been diagnosed, but in retrospect it is obvious and his conviction was a classic miscarriage of justice. (One can read all about the kinds of things that went wrong on Jonathan Rudin's website: Aboriginal Legal Services of Toronto, FASD and the Justice system, http://fasdjustice.ca/).

I felt badly about the wrongful conviction. I tried to give his father some support when he got out of jail and when I went to visit relatives on the prairies, I stopped by the reserve to see how his child was doing. He had a great foster family and I sent photos back to Vancouver for his dad. The reserve gave me the address of my former client's brother in Ontario and I got caught up in a family reunification, naively imagining that once people got to know each other they could just pick up and resume normal family life. The reserve was naive about family reunifications too and when my son's father moved back to Saskatoon and asked for his son, they gave him back. These removals from foster care (to be placed back with family) happened so many times to my son's foster mom that eventually she refused to take on a child unless he was signed over until adulthood. We understand more about attachment issues now.

At first things went well for this child and his dad. Welfare paid for their apartment and my son has fond memories of the sheet folded back when his dad put him to bed, of movies, the shampoo they used and fun things they did together. Dad got a job painting with his brother and I got panicked phone calls. "My kid stole the rent money and bought toys". When I was back in Saskatoon for a summer job, things were beginning to spiral out of control. Dad and uncle were drinking and doing drugs. The child was not getting to school on time, if at all. His friends' mothers complained that they had to feed him and keep him overnight when his dad could not be found. By this time I was "Aunty". "That boy needs a good woman to look after him", said his relatives and I did not get the hint. Maybe his dad's next girlfriend would be more stable.

The following March, the panicked phone call said the child had been apprehended. It took years before I got the full story. Dad said he was going out to get gas and left the child with his latest girlfriend. He did not come back for days. The girlfriend took her frustration out on the child, who was sexually molested and kept shut in a room. He thought of jumping from the window but was afraid of getting hurt. Eventually he snuck out wearing nothing but socks and a sheet in the cold March wind and found a neighbour to help. He was given back to his father, but it was only a matter of time before things got worse. I was living in a big house alone in Montreal. I thought all I needed to do was to keep him fed and clothed and get him to school on time. So, as a widow in my mid-fifties, I became a mom for the first time, by private arrangement with the family.

I loved school when I was young, but my child's hand was not dealt from the same deck of cards. When his mother drowned, he was a year and a half old and no family could be found to claim her body. His father's mother was raped by "a man with a bald head" while she was in residential school. The terror this induced ripples through the generations. We now realize that his father was probably also alcohol damaged before birth. Except for that brief time when he got care of his child, his father has couched surfed, living off one girlfriend after the next, each one failing in her hope to reform him. In his early forties now, he has become somber and strict with his children—my child and a younger half-sister. He has serious health problems and he is just out of his umpteenth addictions program. Now that I understand the profile of his disabilities, I know that he needed and still needs on-going support to lead a useful life. His trajectory looms in the background as I struggle to stop my child from being pulled into the same vortex. There is not much help out there, at least, not much of a kind that would be useful.

Last year I managed to get our child graduated from high school through a combination of brilliant alternate programs, some of which flopped when faced with his profile of disabilities. I loved school when I was young, but my child's primary school career included three of the worst years I have ever seen in any Canadian school. I will not get into the details, but I do think that he would have fared better with the structured practice I got in the olden days with spelling books and math texts that allowed us to learn and forget and relearn, teasing our brains into following models that could last for the rest of our lives. "You don't listen" his teachers would complain, but their punishments and "consequences" never managed to repair the pre-natal damage that he must learn to deal with.

The kinds of things that drew my child off track depended on every-day skills the rest of us take for granted. In grade 11, he was not ready for moving independently from class to class. When he could not find his way to the art room on his own, he just walked out the door and started skipping school. My child is an extrovert. He craves company and seems to compensate for his disability by finding a buddy to guide him. But it is hard finding friends as a teenager in a new city and when we moved back to Vancouver the pot smokers were the easiest to infiltrate.

What tricks he learned! The school gave out bus passes. He sold his on the internet and did not see what was wrong with that. "It was mine", he said. School punishments improved nothing but his rate of absenteeism. Never mind if I drove him. He would not stay.

Once after a prolonged disappearance when he was living "independently" by mooching off nameless acquaintances, he showed up with a higher-functioning friend who negotiated to rent my basement for $400 per month. "Is it OK if I just give you $200 to start?" Well, I did want to know where my child was, but as it happened, the $200 he "had" was a rent deposit that never materialized and they did not find work. I ended up supporting them both for four months, but did get them enrolled in adult education. This worked better than the classes full of students with disabilities. The predominantly immigrant students were there by choice. His new friends were less dysfunctional. They had ambitions. And though the buddy who came to live in my house did not, he did at least get my child to school without my driving him—until almost the end of term when he started skipping too and I kicked him out.

Despite all of these and various other misadventures, my child ended up with an honours diploma because I managed to get him into an automotive collision repair course where he was top in his class in welding. Without knowing it, the guys running that program probably have a lot of experience with people who have disabilities of the kind my child must cope with, and they were a god-send. My child got A's in all the practice classes, but did not get the first level of the trade because of the written test. He could apply the knowledge with speed, skill and grace but he could not represent it in words.

Surely it does not make sense to discard such talent. But from what I can see, it is very difficult for normal functioning youth to find a place in our society and with all the highs and lows I always feel that we are on the verge of drowning. Earlier this year, a god-sent Aboriginal counsellor found him a place and he was working steadily for the first time: manual labour, stripping wooden forms from concrete in high-rises. I would get up at 5 a.m and drive for an hour to get him there. The FASD/addictions councillor I go to said he only had the job because of me. Things never quite got routine. He never came home reliably, but I had hope. Then, 6 weeks in, I dropped him off as usual at 6:20 a.m. full of plans for the next pay-cheque in 2 days. He wanted to save up for an apartment with friends, some place closer to work. But neither his boss nor I heard anything from him for a number of days. I knew he was fine. There he was with a whiskey bottle on a friend's Face Book page, but it was a month of sparse contact. He managed to get his last cheque from his boss, but it evaporated into the party with his friends of the moment where he was staying until they were evicted.

These friends may also have fetal alcohol effects or comparable disabilities. He was with them on the night when they got charged with mischief. They were observed running down the street kicking cars. Two lost their side-mirrors. My child was too drunk to remember anything that happened. But even if he had not been drunk, he would not have understood the significance of the papers he was handed on his release. It was the eviction of his friends that ended the party and sent him home again. The 20 hours he spent in jail probably did him no harm.

Alone, in an uncomfortable cell without his glasses, he could not even read the graffiti on the walls. In 20 hours he had nothing to eat but two meals consisting only of two granola bars and a juice box and, if I had not been there to get him, he would have been out on the street in a dysfunctional part of town with no money or bus fare to get home. "How did I end up here?", I wondered as I waited for him.

These low functioning people are set up to fail, but there have been some improvements. My child's case was sent to a community court where the judge, prosecutor and duty counsel were all women who did not imagine they could solve social problems with the gruff pound of a gavel. They modeled the kind of behaviour we would like those charged to exhibit, explaining why someone could not be released from jail or why sentencing of another could not be waived. My child had no record and, armed with his diagnosis, the prosecutor was quick to suggest diversion. We met with an Aboriginal counsellor who understood my child's priorities and knew how to talk to him. They set up a plan. He was to get three kinds of counselling to help him find housing, a job and to learn how to deal with identity issues. I thought we were on track.

My child has found himself a new buddy, a 19-year-old who has just "aged out" of the foster care system so he is alone in the world after out-growing a series of group homes. He may well be the child of a fetal alcohol mom, but he is the first buddy my child has had who does not seem to drink or do drugs. I said they could both stay at my place until they found somewhere together and at first things looked good. With the buddy in tow, my child started to come home at a reasonable hour. But three weeks down the road, everything seems to have fallen in my lap again. None of the agencies have done much to follow through. My son and his friend have updated their resumés and supposedly spend their days looking for work and for apartments they do not have rent or damage deposits for. The job search seems to consist of sending out applications from youth drop-in centres and nobody calls. Last night they crawled in through a window well past midnight and at 10 a.m. they are still asleep. They need a program that will help them get established but I have not been able to get them into one. I can certainly understand why employers prefer to hire responsible, high functioning immigrants.

I used to be judgmental about parents who moved and did not give their dysfunctional kids a forwarding address, but now I understand.

I need my sleep.

I feel like I am drowning and yet I am trying to relax, as we drift towards the next statistically proven probability in a society whose failure to see the link between behaviours and outcomes seems to mirror that of my alcohol damaged 20 year old child.

I Did not Take on This Job of Parent in Order to Raise Criminals

Dianne O'Connor

I did not take on this job of parent in order to raise criminals. Looking at my adult children now, you would not be able to tell me which one became a drug addict, or a thief, or an abusive person, or someone who would never be able to care for their own children. You would see handsome, beautiful adults with broad smiles and life dancing in their eyes.

Little did I know when I became a foster and adoptive parent to children born impaired by prenatal exposure to drugs and alcohol that my family would be under tremendous stress: that we would be living paycheque to paycheque, surrounded by case workers, protective workers, therapists, behaviorists, police officers, crisis teams, and psychiatrists in order to get my children to where they are today.

When I started married life with my husband Mickey over 39 years ago, I believed that we could do some good for the abused and unwanted children in our community. We both thought just a little love; a hot meal and all would be good. When, in 1981 we began fostering a five-year old boy, we believed that if we could get the children before they reached 5 years of age, we could make a difference. Now we know we need to get to the moms before they get pregnant to tell them not to drink alcohol, smoke cigarettes, or use illegal drugs before conception since all six of my adoptive children have varying degrees of Fetal Alcohol Spectrum Disorders or FASD. Within my family of nine children, six have FASD. All have been involved with the law for various reasons. One daughter, who joined the Marine Corps, was thrown in the brig for using and selling a control substance. She is now homeless, an alcoholic, in and out of drug courts and addiction centers, and is currently awaiting sentencing for criminal mischief. She just finished her probation for car theft. One son, at the age of 13, violently and defiantly confronted a police officer when he was accused of stealing cell phones. Although scared, at age 19, he continues to have swagger when it comes to talking to the police. Another daughter was arrested when her involvement with her birth family

D. O'Connor (✉)
Scotia, New York, USA
e-mail: doconno2@nycap.rr.com

M. Nelson and M. Trussler (eds.), *Fetal Alcohol Spectrum Disorders in Adults: Ethical and Legal Perspectives*, International Library of Ethics, Law, and the New Medicine 63, DOI 10.1007/978-3-319-20866-4_18

turned violent. Her disabilities prevent her from parenting her two children. The children are living with us and we deal with family court regularly. Another son started stealing from us at age seven and turned to stealing the medicine out of a locked box in our cupboard to sell to friends. Today, he breaks into houses to steal because he does not want to get a job or fears he will fail at a job. He is currently lives out of a car with his girlfriend and other family members. And one daughter was involved with Child Protective Services over her poor choices in parenting, living situation and domestic abuse. We raised her son for nearly 2 years.

This road has been frustrating and exhausting. I have spent over $21,500 in lawyer fees just in the last few years with issues of custody for our two grandchildren. While our friends and family talk about retirement, we talk about 15-month well care checkups and the latest fiasco with the police. There is no end in sight for us. My husband, now 64 years old, believes he will have to work until he is 75 just to make ends meet for us and our growing family and to save enough for retirement.

We are parenting children who did not respond to "normal" parenting approaches. We needed to provide a level of intervention that required a one-on-one intervention, 24 hours a day, seven days a week. One mom describes her son as, "a good kid with a good heart, but it takes so much energy for our children to make the right decisions and some days they just don't have it. He wants to do the right thing, but makes mistakes." Our children tend to make the kind of mistakes that can be life threatening or can get them in great trouble. Exhaustion, isolation, and constantly running interference to prevent our children from risky or inappropriate behaviors became the world we lived in. School became a respite for us from the consistent level of supervision- at least until the telephone or cellphone call communicating that one of the children is suspended or expelled from school, is being held at the police station or jail, or is on their way to the psychiatric hospital. We, the caregivers, then receive the brunt of their emotionally charged behaviors once they come home and are able to "let go". I felt deeply that each call reminded me of my failure and I would spend what little time I had asking myself what I could have done differently or better.

To add to my feelings of guilt, outsiders typically consider me a bad parent when my children are in trouble. A birth mother from Australia articulates the shame and blame she feels whenever her son gets in trouble with the law. She states, "I have never been able to deny it. I feel I am responsible for my son's disability and for his daily struggles. The fact is he is now in prison because he developed secondary disabilities caused by FASD. Every time he looks in the mirror, he sees a poor tortured soul." The embarrassment and shame that comes along with being a parent of a child who has involvement with the various legal systems can be unbearable and financial draining.

When my children were in elementary school, someone suggested I read the book "The Broken Cord" by Michael Dorris (1989) and my life changed. I was reading about Dorris's son, Adam, and thinking, this is **my** daughter or son. Why didn't someone tell me about alcohol-exposed pregnancies before now? Dorris describes his adoptive son as being in a fog most of the time; as living slightly out of focus. This explanation was very insightful for me when dealing with my own alcohol-affected children.

Now I recommend this book to anyone thinking of adopting a child with FASD. Although the book is very negative, since Dorris's son is struck by a car and killed when walking home from work one day after many years of struggle, I think prospective parents need to start with the worst-case scenario. Parents and other caregivers can then move on to what the research and years of experience has to offer new parents. We have been there and done that. We have made all the mistakes in parenting these children. One of the most enduring qualities of an individual with FASD is that they probably will not remember the bad things, so every day we get a "do over" to get it right.

The Center for Disease Control longitudinal study of secondary disabilities associated with those prenatally exposed to alcohol conducted by Ann Streissguth et al. (1996) should be required reading for parents and caregivers of individuals with FASD. In her book, *Fetal Alcohol Syndrome, A Guide for Families and Communities,* Streissguth described the challenges faced by 415 persons with FASD who were studied by her and her colleagues at the University of Washington. She reported that over 60 % had some involvement with the law, and 40 % of those were incarcerated at some point in their lives. Other statistics included about 50 % of those studied over the age of 12 showed inappropriate sexual behaviors, over 60 % of those over the age of 12 had been expelled from school, or dropped out and almost a third of the 415 individuals abused alcohol and/or drugs. In all, 94 % of those studied showed some form of mental health issues.

Families and caregivers have come to realize that their loved ones with FASD are emotionally and cognitively about half of their chronological age. One mom from Kentucky says about her son, "He lost most of his elementary school friends and at age 19, he hangs out with children who are 15 years old. I worry about what will happen to him when he is 30, 40, or 50 years old. I worry about who will advocate for him when I am gone. I do not hold out hope the judicial system will be much help to him". Ann Streissguth (1997) writes that families who maintain clear and consistent expectations and behavioral consequences are the most successful when raising a child with FASD. Structure is the key. Not only does it take a village, it takes an army. Doctor, therapy, psychiatrist, evaluations, home visits, and blood work appointments took what little time I had during a typical day after school let out. I have never been able to work outside the home for more than a few hours a day, usually dashing to work after the last school bus left in the morning and returning home from work before the first bus arrived in the afternoon, -a lifestyle I called "bus to bus". Raising enough money to keep this family financially afloat fell to my husband.

As a society, we need to change our expectation for these children, as I have had to do. I had to think of what mattered to my children before thinking of what I thought they needed. It is a chore for some of my children to go out to dinner with the family. Rules for being in public may be stricter than the rules at home. Rules like sitting up straight, using a napkin, and cutting up your food takes concentration.

My children would rather stay home than to have to think, using precious mental energy, of what they need to do next so that they do not displease anyone. If I could not manipulate the environment to fit, then my children would not be successful. This takes a tremendous amount of energy. Many professionals think it is poor parenting and does not allow the child to learn from consequences. For those who say these things, they have not raised a child with FASD. My parenting decisions have nothing to do with my children's motivation or willingness to do the right thing. It has to do with difficulty they have processing the information from the world around them and trying to come up with an appropriate plan of action. For my children, the plan is the last thing they remember, no matter how the outcome turned out. So they tend to do the same thing wrong over and over again.

Many of their "best intention" plans still results in meeting up with the law, regardless of our best intentions to keep them safe from harm. Many times, hungry for attention and eager to fit in with their peers, the prenatal affected person follows the peers who accept him, is led into poor choices and becomes a statistic. Today, we can add cyber crimes to the long list of criminal activities that our children and adults with FASD can potentially get involved with.

Family members and caregivers may appear physically exhausted and emotionally distant when loved ones who have FASD are involved with criminal justice systems. It often becomes a constant battle to protect and educate those around the individual with FASD to ensure they are safe and able to meet their full potential. The police and the court systems have little understanding of their issues and often even less patience. One mother was told by police officers that her son needed a beating so he would learn not to break the law again. A father in the US speaks to how the judge was disrespectful to the family by allowing statements from the victim of his son's criminal behavior to be read into the court records, yet his parents were unable to speak on their son's behalf about his impairments. Eventually, the family of the child with FASD was sued in civil court for damages.

It is important that the courts and law enforcement agencies understand how parents and caregivers of individuals with FASD feel like when the law is involved. After years of learning about the disabilities associated with FASD, parents and caregivers develop educational and remedial plans to protect their loved one at home, in school, and in the community.

It is also important to hear from the individuals with FASD. What I have heard from families is their children have no idea what is happening. Some behave as if being arrested is like having a new notch in their belt, something they can use to brag about in order to get friends. Others are so scared they will say anything to the police just so they can go home. One 24-year-old man was arrested for flirting with an underage girl. When police questioned him about grooming the girl for sexual activity, he said yes he was. He did not understanding the question and believing he was innocent, willing to cooperate with the officer. Once he understood the consequences of his actions resulted in a lengthy prison sentence, he made a plan for suicide while in his jail cell. Police questioned another 13-year-old boy with FASD for 4 ½ h, without his parents or an attorney, because the boy did not think to ask for them. When the parents tried to get into see him, they were told they were interfering. The parents tried to present documentation to the lawyer that could help their son, and the lawyer rejected the paperwork. The lawyer told them he felt the boy "looked normal". It was during the assessment process that the boy was found to have major learning deficits; then adaptations were made. Regardless, this child was traumatized.

Some families told me about positive results from court involvement. Judges and lawyers behaved respectfully; and when a disability lawyer was hired, the outcome was better for the individual with FASD. Lawyers from NAMI were specific examples of appropriate legal representation for their children. One family noted that it was the court who ordered a full psychological workup on their son and with that testing, they found out their son had FASD. This finding led to a more positive outcome.

Overwhelmingly, families stressed that the legal systems needs to learn about FASD and what the core behaviors that are presented by individuals affected by prenatal alcohol exposure are and how they affect that child's decision-making process. No family insisted that their children should not be accountable for their actions; they urge that appropriate action should be taken.

Jail is not the solution for a person prenatally exposed to alcohol who is involved with the criminal justice system. Although our loved ones with FASD may be able to recite the law, they do not understand the law or what is right or wrong. They have inadequate coping strategies, poor memories, and are easy victims to those who would take advantage of them. Lack of appropriate treatment when these children are younger, leads to the secondary disabilities that the legal system sees later on. A Kentucky mother states, "By the time our children hit the court system, I can guarantee that parents have tried over and over again to implement meaningful interventions and are shuffled from one agency to another, none of which communicate with each other, none which look at our children and our families as a whole entity. They look at our children much like our children see the world…in parts. Then our children hit the legal system, and the judge and lawyers are limited by 'the law', which has in itself been completely bastardized from what is right, to who can win a legal game. Our children get caught in the crossfire. They don't understand the rules of the 'game' and get condemned for it.

So who is to blame for the continued involvement with the law among youth and young adults with FASD? Many would say it is our parenting style. We are either too strict or too lenient with our children. I have been told that only way for my children to learn is to let them fail. I have been told that a few days in jail would do them some good. The most hurtful of comments are the ones that tell me to cut them loose since they were not really our children anyway. In the end, our loved ones with FASD have permanent brain damage that will never be repaired. Parents, caregivers, and other caring persons can only provide appropriate interventions and treatment to help them be the best they can be. They need protection and supervision, not harsh punishment. Finally, we as parents are looking for accountability with compassion for our loved ones with FASD.

In conclusion, I urge the legal systems to understand that my children can learn rote skills, but only after much repetition and consistency. Higher levels of problem solving, such as learning from mistakes, takes longer and does not follow the concept that wanting to please others leads them to learn to do the right thing.

My children know they are doing wrong by our expressions and our words, not because they a sense of social consciousness in their brain. This does not mean they are without a conscience. They have a conscience. They feel remorse very deeply. But they have difficulty relating their behavior to it consequences. So no amount of "after the fact" remorse leads them to stop breaking the rules.

Over the years, my husband Mickey and I have changed not only our parenting style, but our attitudes towards parenting and raising a family as well. Our house was always lively. There was always someone to play with, and there was always laundry to do. There were many wonderful celebrations, and several very low points. There was a balance to our lives and I would say today, there was more good than bad.

Raising a large family with family members who have disabilities has many challenges. I became a sensory decorating expert, ensuring that bedrooms and living areas of our home were functional, durable and met everyone's sensory needs for sleep and play. We bought a pop-up trailer and made off to the Adirondacks for family camping trips. It was a freeing experience for most of the children. My youngest found he didn't need to wear shoes and ran bare-footed through the woods and along the beaches. Swimming all day was part of the experience and the children would stay in the water for hours. Of course, then there were the summertime pickup baseball games, where my daughters would rule the field and astonish all those playing with their abilities. We did whatever we could to give our children a variety of experiences.

For Mickey and I, life was stressful. One or both of us always needed to think about how to manipulate the environment so the children could be successful. If we went out for dinner, we had to go to a restaurant that served us fairly quickly, was away from the bar and any smoking, and away from the kitchen so the noises of preparing the meals would not startle anyone. If we visited friends, we went in two vehicles, made sure snacks and sensory items were available, and one of us became the designated "Jiminy Cricket" or external brain that they need.

The financial burden of caring for children with disabilities can be overwhelming. I could never work full time with the number of appointments, school meetings, and lack of knowledgeable babysitters. Repairs for damage or lost property along with replacing stolen merchandise drained any savings accounts we had. As grandchildren came along, we would pay for day care, clothing, and many time medical appointments since the parents were unable to pay for the basic needs themselves. One grandchild lived with us for nearly 2 years with no financial agreement with any agency or with the parents. We met all his needs through our own family budget.

In the last few years, two of my grandchildren have come to live with us as kinship foster children. Their mother, my daughter, struggles to maintain her day-to-day existence and although she truly loves her children, will never really be able to raise them alone. To protect our rights as the grandparents, I have had to secure a lawyer to represent us in family court. First, I secured a lawyer in the county where my daughter and grandson were while went through family court proceedings to get him back into our county. It took nearly four months until we brought our grandson home and the lawyer bills for that service were nearly $5,000.

Once in our home and in the county where we live, we secured another lawyer and have been paying lawyer fees for the last two plus years while the father of the children is afforded a court appointed attorney to file for custody every 90 days. Our bill with the lawyer is over $17,000. I am the only person in that family courtroom who pays for lawyer fees, and through my taxes, I am also paying for everyone else's court appointed attorneys. I should also mention, we are in appeal on one aspect of the family court decisions.

One suggestion I would make is to learn everything you can about FASD. Although known and studied for years, this disability is not as well-known as other disabilities like autism or cerebral palsy, yet it affects so many people, from infants to the elderly. FASD may mimic other disabilities, but it is unique in many ways. The lack of understanding gets our loved ones into trouble. Over stimulation, poor information processing and lack of self-control can be addressed when the helping person knows about the brain damage they are dealing with, but unaddressed, the person with FASD can develop a low self-esteem and depression. Many times, people affected by prenatal alcohol exposure are thought to be uncooperative, unwilling and aggressive when really they are scared, misunderstood, and don't understand what you want them to do. FASD is an invisible disability, affecting the brain, which causes brain-based behaviors.

Looking back, I am not sure Mickey and I would do anything differently. We had three birth children before we fostered and then adopted our other children. We could gauge what was normal behavior and development and what was not. Since no child comes with a manual on how to raise them, we have learned along with our children what works best for them.

Our parenting may be different for each child, but that is what each child needed. One good thing about our children and their disabilities is that you get a "do over" as parents every day because our children do not hold grudges for long and most time will forget what has happened.

I would however, recommend to align yourself with an agency early on. Mickey was one of those parents who thought families should care for their own and not let outsiders know what was happening. He changed his mind after we had our first respite weekend with our daughter. We were able to reconnect with each child left at home in a positive way. After that weekend, Mickey saw the value in allowing outside help into our lives. For many families, their first attempt at engaging outside help is when their loved one is in crisis. Families do not know where to turn.

Another time when support is needed is during the transition to adult services. By applying for and planning for the service needed for your loved one with FASD and your family to be successful, when crisis does happen, you are better able to handle what may happen next. We have raised our family with help from several people and agencies. We have met wonderful people who love our family and have been very good to us.

Others looking in on our family may see enabling, ignoring, favoritism, or perceived neglect, but all my children have made it through high school. We have held each child to a high standard of conduct and moral thinking. At their employment, they are diligent and conscientious of doing a good job. They are good to others and are genuinely caring, compassionate adults.

That is all we have ever asked of them, and they have responded.

References

Dorris, M. 1989. *The broken cord*. NY: Harper Perennial.

Streissguth, A. 1997. *Fetal Alcohol Syndrome*. A guide for families and communities. In ed. Paul H. Brookes Publishing Co. MD.

Streissguth, A.P., H.M. Barr, J. Kogan, and F.L. Bookstein. 1996. Understanding the occurrence of secondary disabilities in clients with fetal alcohol syndrome (FAS) and fetal alcohol effects (FAE). In Final Report to the Centers for Disease Control and Prevention (CDC). Seattle, Washington: University of Washington, Fetal Alcohol & Drug Unit.

MIX
Papier aus verantwortungsvollen Quellen
Paper from responsible sources
FSC® C105338

If you have any concerns about our products,
you can contact us on
ProductSafety@springernature.com

In case Publisher is established outside the EU,
the EU authorized representative is:
Springer Nature Customer Service Center GmbH
Europaplatz 3, 69115 Heidelberg, Germany

Printed by Libri Plureos GmbH
in Hamburg, Germany